Digital Control of Electrical Drives

Power Electronics and Power Systems

Series Editors: M. A. Pai, University of Illinois at Urbana-Champaign, Urbana, Illinois; Alex Stankovic, Northeastern University, Boston, Massachusetts

---

Digital Control of Electrical Drives
Slobodan N. Vukosavić
ISBN 978-0-387-48598-0

Three-Phase Diode Rectifiers with Low Harmonics
Predrag Pejović
ISBN 978-0-387-29310-3

Computational Techniques for Voltage Stability Assessment and Control
Venkataramana Ajjarapu
ISBN 978-0-387-26080-8

Real-Time Stability in Power Systems: Techniques for Early Detection of the Risk of Blackout
Savu C. Savulesco, ed.
ISBN 978-0-387-25626-9

Robust Control in Power Systems
Bikash Pal and Balarko Chaudhuri
ISBN 978-0-387-25949-9

Applied Mathematics for Restructured Electric Power Systems: Optimization, Control, and Computational Intelligence
Joe H. Chow, Felix F. Wu, and James A. Momoh, eds.
ISBN 978-0-387-23470-0

HVDC and FACTS Controllers: Applications of Static Converters in Power Systems
Vijay K. Sood
ISBN 978-1-4020-7890-3

Power Quality Enhancement Using Custom Power Devices
Arindam Ghosh and Gerard Ledwich
ISBN 978-1-4020-7180-5

Computational Methods for Large Sparse Power Systems Analysis: An Object Oriented Approach
S.A. Soman, S.A. Khaparde, and Shubha Pandit
ISBN 978-0-7923-7591-3

Operation of Restructured Power Systems
Kankar Bhattacharya, Math H.J. Bollen, Jaap E. Daalder
ISBN 978-0-7923-7397-1

Transient Stability of Power Systems: A Unified Approach to Assessment and Control
Mania Pavella, Damien Ernst, and Daniel Ruiz-Vega
ISBN 978-0-7923-7963-8

Maintenance Scheduling in Restructured Power Systems
M. Shahidehpour and M. Marwali
ISBN 978-0-7923-7872-3

Continued after index

Digital Control of Electrical Drives

Slobodan N. Vukosavić
The University of Belgrade

Slobodan N. Vukosavić
University of Belgrade
Faculty of Electrical Engineering
Bulevar Kralja Aleksandra 73
11120 Belgrade
Serbia

Series Editors:
M. A. Pai, Professor Emeritus
Dept. of Electrical and Computer Engineering
University of Illinois at Urbana-Champaign
Urbana, IL 61801

Alex M. Stankovic, Professor
Dept. of Electrical & Computer Engineering, 440DA
Northeastern University
360 Huntington Ave.
Boston, MA 02115

ISBN 978-1-4419-3854-1 e-ISBN 978-0-387-48598-0

Printed on acid-free paper.

Softcover reprint of the hardcover 1st edition 2007

9 8 7 6 5 4 3 2 1

springer.com

Contents

Preface

This book is intended for engineering students in the final years of undergraduate studies. It is also recommended for graduate students and engineers aspiring to work in intelligent motion control and digital control of electrical drives. By providing a bridge between control theory and practical hardware aspects, programming issues, and application-specific problems, the book is intended to help the reader acquire practical skills and become updated regarding concrete problems in the field.

Basic engineering principles are used to derive the controller structure in an intuitive manner, so designs are easy to recall, repeat and extend. The book prepares the reader to understand the key elements of motion control systems; to analyze and design the structure of discrete-time speed and position controllers; to set adjustable feedback parameters according to design criteria; to identify, evaluate, and compare closed-loop performances; to design and implement nonlinear control actions; to devise and apply antiresonant compensators; and to generate speed reference profiles and position trajectories for use within motion-control systems. The Matlab tools are used extensively through various chapters to help the reader master the phases of design, tuning, simulation, and evaluation of speed and position controllers.

Key motion-control topics, such as nonlinear position control, control of mechanical structures with flexible couplings, compliance and mechanical resonance problems, and antiresonant solutions, are introduced in a systematic manner. A set of exercises, problems, design tasks, and computer simulations follows each chapter, enabling the reader to foresee the effects of various control solutions and actions on the overall behavior of motion-controlled systems. In addition to control issues, the book contains an extended introduction to the field of trajectory generation and profiling. In the closing chapters, the reader is given an overview of coding the control algorithms on a DSP platform. The algorithm coding examples are included, given in both assembly language and C, designed for fixed point DSP platforms. They offer a closer look into the characteristics and peformance of contemporary DSP cores and give the reader an overview of the present performance limits of digital motion controllers. Most of the control solutions presented in the book are supported by experimental evidence obtained on test rigs equipped with typical brushless DC and AC servo motors, contemporary servoamplifiers, and suitable mechanical subsystems.

Readership

This book is primarily suited for engineering courses in the third and fourth year of undergraduate studies. It is also aimed at graduate students who want to deepen their understanding of electrical drives and drive control; and at practicing engineers designing and using motion-control systems and digital controlled electrical drives. The book provides a bridge between control theory, practical hardware aspects, programming issues, and application specific problems. The subject, related problems, and solutions require interdisciplinary understanding among control, electrical, and mechanical engineers.

Prerequisites

Required background includes fundamental engineering subjects typically covered during the first and second years of undergraduate engineering curricula. Prerequisites include basics and common principles of control engineering, power conversion, and electrical machines, as taught in undergraduate introductory courses. A distinctive feature of the book is that it does not require that the reader be proficient in control theory, electrical drives, and power electronics. The theoretical fundamentals are reviewed and included in the book to the extent necessary for understanding analysis and design flow. Most chapters include a brief theoretical introduction. Wherever possible, the theory is reviewed with reference to practical examples. Limited reader preparation in the use of the Laplace transform and *z*-transform can be partially compensated for by adequate skills in the use of relevant computer tools.

Objectives

- *Understanding of basic elements and of key control objectives in motion-control systems.* Analysis, design, and evaluation of discrete-time speed and position controllers. Parameter-setting procedures driven by design criteria. Reader's ability to design, evaluate, and compare closed-loop performances, to synthesize and implement nonlinear control actions, to devise and apply antiresonant compensators, and to generate speed reference profiles and position trajectories for use within motion-control systems.

- *Understanding feedback signal acquisition and sampling process.* Distinguishing between the high-frequency range of unmodeled dynamics and the bandwidth of interest. Recognizing noise and quantization problems. Designing sampling circuits and filters. Selecting the sampling frequency.

- *Using the Laplace and z-transforms to convert differential and difference equations into their algebraic form.* Dealing with the complex representation of signals and transfer functions in the analysis, design, and evaluation phases. Relating the response character and bandwidth to the placement of the closed-loop poles and zeros.
- *Designing the control structures to suppress relevant load disturbances and to eliminate the tracking error for given reference profiles.* Formulating performance criteria and deriving optimized feedback. Understanding the nonlinearities of the system and designing control countermeasures, aimed at preserving the stability and improving response. Analyzing and evaluating the mechanical resonance and torsional oscillations. Designing and implementing the antiresonant compensators. Specifying speed and position trajectories. Generating and interpolating reference profiles.

- *Mastering the phases of design, tuning, simulation, and evaluation of speed and position controllers, assisted by Matlab and Simulink tools.* Gaining insight into coding the control solutions on contemporary fixed point DSP platforms in assembly language and in C. Appreciation of the performance limitations of DSP cores and of digital motion controllers.

Field of application

This book offers a comprehensive summary of discrete-time speed and position controllers. Control of the speed of a moving part or tool and driving its position along predefined trajectories are the fundamental elements of motion-controlled systems and an integral part of many manufacturing processes. The skills acquired in this book will prepare the reader to design the structure of motion controllers, set adjustable feedback parameters according to design criteria, design nonlinear control actions and antiresonant compensators, generate reference profiles and trajectories, and evaluate performances of motion-control systems. Such skills are required to increase the speed of motion, reduce the cycle time, and enhance the accuracy of production machines. Improvements and advances in control technology and electrical drives are continuously sought in a number of industries. High-performance position-controlled feed drives and automated spindles with tool exchange are required in the metal processing industry. An increase in precision and a reduction in cycle time is required in packaging machines; plastics injection molding; the glass, wood, and ceramics industries; welding, manipulating, and assembly robots in the automotive industry; metal-forming machines; and a number of other tasks in processing machines.

Lighter and more flexible mechanical constructions introduce new challenges. Conflicting motion-control requirements for decreased cycle time, increased operating speed, and increased accuracy are made more challenging by mechanical resonance, finite resolution, sensor imperfection, and noise. Motion-control solutions and systems are in continual development, requiring the sustained efforts of control, electrical, and mechanical engineers.

Acknowledgment

The author is indebted to Prof. Milić R. Stojić, Prof. Aleksandar Stanković, Prof. Emil Levi, Dr. Ing. Vojislav Aranđelović, Ing. Ljiljana Perić, Ing. Mario Salano, Ing. Michael Morgante, and Dr. Ing. Martin Jones who read through the first edition of the book and made suggestions for improvements.

1 Speed Control

This chapter explains the role of speed-controlled drives in general automation and industrial robots, identifies the basic elements of the speed-controlled system, defines the control objective, and devises control strategies. Fundamental terms related to continuous- and discrete-time implementation are defined. An insight is given into the role and characteristics of the torque actuator, comprising the servo motor and the power converter. Separately excited DC motor coupled with an inertial load is analyzed as a sample speed controlled system.

1.1 Basic structure of the speed-controlled system

In the realm of motion control, the task of controlling the speed of a moving object or tool is frequently encountered. The actual speed of rotation or translation should be made equal to the set speed. The difference between the actual and set speed is known as the *speed error*. It is the task of the speed controller to keep the speed error as small as possible, preferably equal to zero. To achieve this result, the controller generates the torque/force reference. To begin with, let us consider the system where the rotational speed ω is controlled, with the inertia of the moving parts J, the friction coefficient B, and the load torque T_L. The rate of change of the actual speed ω is given in Eq. 1.1, where T_{em} represents the driving torque. The necessary elements of a speed-controlled system are given in Fig. 1.1.

The desired speed (ω^* in Fig. 1.1) is referred to as the *speed reference* or the *set point*. When the desired speed changes in time, the speed-reference change is called the *reference profile* or *trajectory* $\omega^*(t)$. The speed error $\Delta\omega$ is found to be the difference between the set speed and the speed feedback ω_{fb}. The error discriminator is shown as the leftmost summation junction in Fig. 1.1. The speed controller, represented by the transfer function $W_{SC}(s)$, processes the error signal and generates the torque reference T_{ref}, the latter producing the driving torque T_{em}.

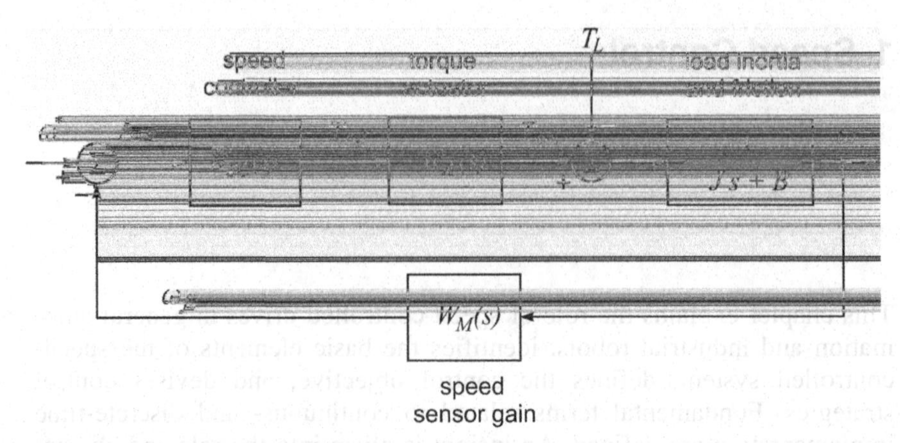

Fig. 1.1. Basic elements of the speed-controlled system.

$$J\frac{d\omega}{dt} = T_{em} - T_L - B\omega \tag{1.1}$$

The torque T_{em} is the system's driving force, and its role is to make the actual speed ω track the reference ω^* in the presence of disturbances and the load torque T_L variations. As inferred from Eq. 1.1, the driving torque should compensate for the load changes T_L, suppress the effects of friction $B\omega$ and other secondary phenomena, and provide the inertial component $Jd\omega/dt$ in the phases of acceleration and braking.

In practical implementations, T_{ref} is a digital signal brought to the input of the torque actuator, represented by block $W_A(s)$ in Fig. 1.1. In order to facilitate the speed control task, it is desirable to use actuators where the actual torque T_{em} tracks the reference T_{ref} accurately and without delays. Hence, the ideal torque actuator's transfer function is $W_A(s) = 1$ or $W_A(s) = K_M$ = const. Most actuators make use of power amplifiers with sufficiently large bandwidth and electric motors. The power amplifier supplies the motor windings with appropriate voltages and currents, thus enabling the motor to generate the desired driving torque T_{em} at its output shaft. The motor shaft is coupled to the load either directly or through a mechanical transducer that may convert the rotation into translation, thus providing the driving force instead of the driving torque.

A power amplifier makes use of semiconductor power switches (such as transistors and thyristors), inductances, and capacitors and performs the power conversion. It changes the voltages and currents of the primary power source into the voltages and currents required for the motor to generate the desired torque T_{em}. In most cases, the primary power is obtained either from a utility connection (AC) or from a battery (DC). Given the potential use of both AC and DC motors, power amplifiers may be requested to perform DC/DC, DC/AC, AC/DC, or AC/AC power conversion. The power amplifier is connected to the electric motor, and the combination of the two is referred to as an electric drive.

Most available electric drives provide the torque T_{em}, which responds to the command T_{ref} with a time lag ranging from several tens to several hundreds of microseconds. The motor torque is determined by the current circulating in its windings. Consequently, the torque response time depends upon the current control-loop bandwidth, and is, therefore, limited. Hence, the desired transfer function ($W_A(s) = K_M =$ const.) can hardly be achieved. On the other hand, the desired speed-loop response is measured in tens of milliseconds. In most cases, delays introduced by practicable torque actuators are negligible compared with the dynamics of the mechanical subsystem and the desired response time of the speed loop. In such cases, the speed loop analysis and tuning can be performed under the assumption that the torque actuator has a static gain K_M and no associated dynamics or delays.

The speed feedback ω_{fb} (Fig. 1.1) is obtained at the output of the block $W_M(s)$. The feedback signal ω_{fb} is not an exact copy of the actual speed ω, due to a limited resolution of some shaft sensors, owing to the need to filter out the noise and high-frequency content, and due to specific techniques of speed signal acquisition and/or reconstruction. The transfer function $W_M(s)$ describes the signal processing within the shaft sensor and the associated circuits. If we consider a brushed tachogenerator with an RC low-pass network, said transfer function becomes $W_M(s) = 1/(1+sRC) = 1/(1+s\tau)$. In cases when electromagnetic resolvers are used [2], the function $W_M(s)$ is more complex. In the design and tuning of speed controllers, the transfer function $W_M(s)$ must be taken into consideration. In cases when time constants involved in feedback filtering and processing are found to be considerably smaller compared with the desired speed response times, the function $W_M(s)$ can be neglected and considered equal to one ($\omega_{fb} = \omega$). Specifically, the tacho-filtering RC network with $\tau = 100$ μs can be ignored in designing a speed controller with a desired rise time of $\tau_R = 10$ ms.

The speed control system given in Fig. 1.2 is used in the preliminary analysis of speed controllers. It has an idealized speed measurement system ($\omega_{fb} = \omega$) and a torque actuator that provides a driving torque T_{em} equal to the reference T_{ref}. The system makes use of a separately excited DC motor that drives an inertial load J. The excitation current i_p and motor field Φ_p are assumed to be constant. Therefore, the torque is in direct proportion to the armature current i_a. For the given driving torque T_{em}, the armature current i_a must be equal to $T_{em}/(k_m\Phi_p)$, where k_m is the motor torque constant.

For this reason, the torque reference T_{ref}, derived from the speed controller $W_{SC}(s)$, becomes the armature current reference $I_a^* = T_{ref}/(k_m\Phi_p)$.

For the sake of simplicity of introductory considerations, the power amplifier supplying the armature current in Fig. 1.2, is reduced to an idealized, controllable current source. In practice, the DC drive power amplifiers operate on the basis of commutating the switching power transistors or thyristors, and they are associated with an analog or digital current controller. The amplifier supplies the armature voltage u_{AB} to the motor. The armature current changes according to the equation $L_a \mathrm{d}i_a/\mathrm{d}t + R_a i_a = u_{AB} - e_a$, where L_a and R_a denote the armature inductance and resistance, while $e_a = k_e\Phi_p\omega$ represents the back electromotive force induced in the armature winding. The current controller actuates the power switches in order to obtain the voltage u_{AB} that compensates e_a and suppresses the error $\Delta i = I_a^* - i_a$. The current controller produces the voltage reference u^*_{AB} by multiplying the error by the proportional and integral gains. With sufficiently high loop gains, the error Δi has negligible values. In such cases, the impact of the electromotive force e_a on the armature current can be neglected. Further considerations assume an ideal current controller where $I_a^* = i_a$.

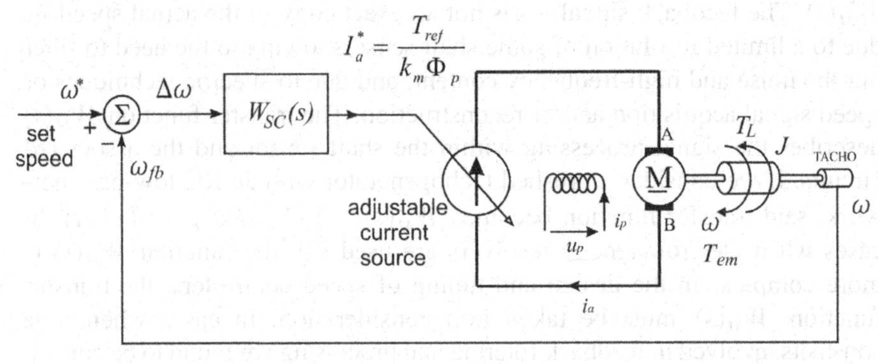

Fig. 1.2. Separately excited DC motor supplied from a controlled current source, used as the torque actuator in a simple speed-controlled system.

Most contemporary speed controllers are implemented in a digital manner; that is, they reside within the program memory of microcontrollers and digital signal processors (DSP) dedicated to motion-control tasks. In other words, their control actions take place at discrete, equally spaced time instants, paced by the interrupt events of a microcontroller/DSP. The analysis, synthesis, and tuning of such discrete-time (or *digital*) speed controllers involves the *z*-domain representation of relevant signals and transfer functions (i.e., *z*-transform). Prior to digital speed controllers, the speed control functions $W_{SC}(s)$ were historically implemented in a continuous domain, mostly by means of analog electronic circuitry comprising operational amplifiers, resistors, and capacitors. Ancestors to digital controllers, the continuous-domain speed controllers are frequently referred to as *analog*. The analysis of analog speed controllers involves *s*-domain representation of signals and functions (i.e., Laplace transform).

To facilitate understanding of the basic concepts, analog speed controller analysis, synthesis, and parameter setting are discussed in the next two chapters. Digital, discrete-time implementation is considered from Chapter 4 onward. Chapter 2 explains the basic concepts of the speed controller design. Necessary control actions are inferred from the speed-controlled system in Fig. 1.2. Hence, it is assumed that the speed measurement system introduces no delays ($\omega_{fb} = \omega$). In the same way, an ideal torque actuator is assumed with $I_a^* = i_a$ and $T_{em} = T_{ref}$. The load is assumed as inertial, having no friction ($B = 0$). The analysis given in Chapter 2 considers the basic proportional and integral control actions, derives key transfer functions, formulates design goals, gives insight into the closed-loop bandwidth and parameter setting, and discusses the impact of various disturbances on the speed controller structure.

In Chapter 3, the impact of the dynamics and transfer functions related to the speed feedback acquisition and torque actuation on the design of analog speed controllers is examined. For simplicity, the traditional DC drive with analog speed control is taken as a design example. The delays in torque actuation are discussed and derived for the most common power amplifiers used in conjunction with the speed controlled DC drives. The parameter-setting procedures frequently used for tuning conventional PI analog controllers are reviewed and discussed, including the *double ratios*, *symmetrical*, and *absolute value* optimum. The bandwidth and performance limits are attributed to intrinsic drawbacks of the analog implementation. At the end of Chapter 3, the place and role of the analog speed controller within the conventional cascaded structure of motion-control systems is outlined, along with the description and the need for feedforward control actions. Chapter 4 and the succeeding chapters discuss digital, discrete-time speed controllers.

Problems

P1.1

Consider the speed-controlled system in Fig. 1.2, comprising the speed controller $W_{SC}(s)$, the separately excited DC motor with $k_m \Phi_p = 1$ Nm/A, the inertial load with parameters $J = 0.1$ kgm^2 and $B = 0$, and a power amplifier that can be modeled as a controllable current source. The power amplifier provides the armature current $i_a(t) = T_{ref}(t)/k_m/\Phi_p$. The torque reference is obtained from the proportional speed controller as $T_{ref}(t) = K_P \Delta\omega$. The gain K_P is set to 10 Nm/(rad/s). The load torque is constant and equal to $T_L(t) = T_{LOAD} = 10$ Nm. Determine the difference between the speed reference ω^* and the actual speed ω in the steady state. Note that the torque developed by the separately excited DC motor equals $T_{em} = k_m \Phi_p i_a$.

P1.2
For the system described in problem P1.1, determine the closed-loop transfer function $W_{SS}(s) = \omega(s)/\omega^*(s)$. Calculate the bandwidth frequency f_{BW} from the condition $|W_{SS}(j2\pi f_{BW})| = 1/\text{sqrt}(2)$.

P1.3
For the system described in P1.1 and P1.2, determine the output-speed transient response to a step change in the reference speed by using the Matlab command *step* (). Estimate the rise time τ_R (i.e., the time interval required for the output speed to change from 10% to 90% of its steady-state value) from the figure. Compare the value $f_X = 0.3/\tau_R$ to the bandwidth frequency f_{BW} obtained in P1.2.

P1.4
Assuming that the previous system has friction $B = 1$ Nm/(rad/s), determine the closed-loop transfer function $W_{SS}(s) = \omega(s)/\omega^*(s)$. Given the speed reference of $\omega^*(t) = \Omega^* = 100$ rad/s and with $T_L = 0$, calculate the steady-state values of the output speed and speed error.

P1.5
For the system described in P1.4, calculate the bandwidth frequency f_{BW} from the condition $|W_{SS}(j2\pi f_{BW})| = 1/\text{sqrt}(2)$.

2 Basic Structure of the Speed Controller

In this chapter, the basic speed-controller design concepts are analyzed. Considering proportional and integral control actions, the key transfer functions are derived, and design goals formulated. An insight is given into the closed-loop bandwidth and the parameter setting. Discussing the impact of various disturbances on the speed controller structure, the feedforward control and internal model principle are explained. Laplace transform basics and familiarity with computer simulation tools are required to understand the developments and examples in chapters 2 and 3.

2.1 Proportional control action

The role of the speed controller is to generate the torque reference signal (Fig. 1.2) in a way that makes the actual speed ω track the reference ω^* in the presence of disturbances and load torque T_L variations. If we consider the load with negligible friction coefficient B, the rate of change of the controlled speed ω is given in Eq. 2.1:

$$J\frac{\mathrm{d}\omega}{\mathrm{d}t} = T_{em} - T_L \,. \tag{2.1}$$

Whenever the driving torque T_{em} overwhelms the load T_L, the speed ω is bound to increase. Given a constant or slowly varying load T_L, the rate of change $\mathrm{d}\omega/\mathrm{d}t$ is proportional to the driving torque T_{em}. The speed controller $W_{SC}(s)$ design must ensure that the actual speed tracks the reference ω^*. Therefore, the simplest design decision is to generate the torque reference in proportion to the detected speed error, $T_{ref} = K_P(\omega^* - \omega) = K_P\Delta\omega$. It is expected that positive errors $\omega^* - \omega$ produce positive $T_{em} = T_{ref}$ and, therefore, a positive rate of change $\mathrm{d}\omega/\mathrm{d}t$ which, in turn, drives the actual speed ω towards the reference ω^* and reduces the error $\Delta\omega$. A similar line of thought applies to the cases when the error $\Delta\omega$ is negative. The design decision $T_{ref} = K_P\Delta\omega$ is referred to as *proportional control law*. Without being aware of that law, a car driver applies proportional control while pressing the accelerator/brake pedal in order to bring the speed close to the desired level. With the actual speed ω being the sole state variable within the control object $1/Js$ in Fig. 2.1, the proportional control action represents the state feedback controller. Hereafter, the performance of the speed controller with proportional action is investigated. The analysis is based on a simplified speed-controlled system, as represented in Fig. 1.2, having an idealized speed measurement system ($\omega_{fb} = \omega$), an instant response torque actuator ($T_{em} = T_{ref}$), and a load with no friction ($B = 0$). A block diagram of the proportional speed controller is given in Fig. 2.1.

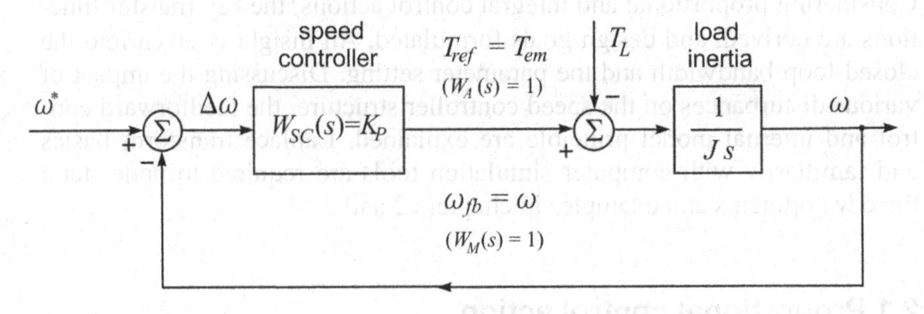

Fig. 2.1. Proportional speed controller applied to the system comprising a load inertia with no friction ($B = 0$) and a separately excited DC motor with instant torque response ($T_{em} = T_{ref}$) and idealized speed measurement.

2.1.1 Open-loop and closed-loop transfer functions

For the system in Fig. 2.1, the driving torque T_{em} is proportional to the speed error (Eq. 2.2). The transfer function of the speed controller $W_{SC}(s)$ is defined by Eq. 2.3. In Eq. 2.4, the transfer function $W_P(s)$ stands for the control object (plant), which is a plain integrator, since the load has an inertia J and no friction.

$$T_{em} = T_{ref} = K_P\left(\omega^* - \omega_{FB}\right) = K_P\left(\omega^* - \omega\right) = K_P \Delta\omega \tag{2.2}$$

$$W_{SC}(s) = \frac{T_{ref}(s)}{\Delta\omega(s)} = K_P \tag{2.3}$$

$$W_P(s) = \frac{\omega(s)}{T_{em}(s) - T_L(s)} = \frac{1}{Js} \tag{2.4}$$

For the case when the load torque equals zero and the feedback is broken ($\omega = 0$), the open-loop transfer function $W_S(s)$ (Eq. 2.5) relates the system output $\omega(s)$ to the reference input $\omega^*(s)$. In other words, the transfer function $W_S(s)$ relates the system output $\omega(s)$ to the speed error $\Delta\omega(s)$ in conditions when $T_L = 0$. For the system shown in Fig. 2.1, the open-loop signal transfer is defined by the series connection of the speed controller and the control object, namely, $W_S(s) = W_{SC}(s)\,W_P(s)$. In the developments that follow, some properties of the open-loop system transfer function $W_S(s)$ and the closed-loop system transfer function $W_{SS}(s) = W_S(s)/(1 + W_S(s))$ will be exploited in accessing the speed error and the closed-loop dynamics. The closed-loop transfer function for the system in Fig. 2.1 is given by Eq. 2.6, which defines the signal flow from the speed reference input ω^* to the system output ω in conditions when the load torque is zero and the feedback loop is closed.

$$W_S(s) = \left.\frac{\omega(s)}{\omega^*(s)}\right|_{(\omega_{fb}=0,\ T_L=0)} = \left.\frac{\omega(s)}{\Delta\omega(s)}\right|_{T_L=0} = W_{SC}(s)W_P(s) = \frac{K_P}{Js} \tag{2.5}$$

$$W_{SS}(s) = \left.\frac{\omega(s)}{\omega^*(s)}\right|_{T_L=0} = \frac{1}{1+\frac{J}{K_P}s} = \frac{1}{1+s\tau}, \quad \left(\tau = \frac{J}{K_P}\right) \tag{2.6}$$

A frequently encountered speed reference waveform is the Heaviside step, $\omega^*(t) = \Omega^*$ for $t > 0$. Being a causal function, $\omega^*(t) = 0$ for $t < 0$. The Laplace transform (i.e., complex image) of such a reference is $\omega^*(s) = \Omega^*/s$. Under the assumption $T_L = 0$, the complex image of the system output $\omega(s)$ is given by Eq. 2.7. The time response $\omega(t)$ is obtained as the inverse Laplace transform of $\omega(s)$ and is given in Eq. 2.8, where the constant $\omega(0)$ stands for the initial condition. The initial condition $\omega(0)$ is the value of the speed found at the instant $t = 0$, inherited from the system events prior to the step $\omega^*(s) = \Omega^*/s$. In the response given in Eq. 2.8, the initial value $\omega(0)$ decays exponentially to zero, while the output $\omega(t)$ converges towards the reference Ω^*. The steady-state value of the output is $\omega(\infty) = \Omega^*$. Hence, we conclude that the proportional speed controller with no load torque and with a Heaviside reference input exhibits no error in the steady state ($\Delta\omega = 0$). Note that the steady-state value $\omega(\infty)$ of the time function $\omega(t)$ can be found directly from the complex image $\omega(s)$ (see Eq. 2.9).

$$\omega(s)\Big|_{T_L=0} = W_{SS}(s)\omega^*(s) = W_{SS}(s)\frac{\Omega^*}{s} \tag{2.7}$$

$$\omega(t) = \mathcal{L}^{-1}\left[\frac{\Omega^*}{s}\frac{1}{1+s\tau}\right] = \omega(0)\,e^{\frac{-t}{\tau}} + \left[\Omega^* - \omega(0)\right]\left(1-e^{\frac{-t}{\tau}}\right) \tag{2.8}$$

$$\omega(\infty) = \lim_{s\to 0}\left(s\,\omega(s)\right) = \lim_{s\to 0}\left(s\frac{\Omega^*}{s}\frac{1}{1+s\tau}\right) = \Omega^* \tag{2.9}$$

The time constant $\tau = J/K_P$ defines the speed of response or the bandwidth of the proportional speed controller. With a step response having an exponential form, such as the one in Eq. 2.8, the actual speed reaches the value of $(1 - e^{-1})\Omega^*$ (i.e., 63.21%) with a time delay of τ. Higher values of proportional gain K_P would result in a faster response. The response speed is frequently described by means of the closed-loop bandwidth, briefly explained below. The bandwidth is understood as the range of frequencies the input signal ω^* may assume while passing from the speed reference input to the output ω without excessive attenuation. Most systems have a low-pass nature. Hence, the bandwidth is normally defined as the interval starting from zero frequency (DC) and ending at the maximum frequency ω_{BW}, referred to as the *cutoff* frequency or the bandwidth frequency.

Several definitions of the frequency ω_{BW} exist in practice. Outlined in short, the most common interpretation defines the bandwidth ω_{BW} as the frequency of sinusoidal reference input $\omega^*(t)$ that results in the system output $\omega(t)$ attenuated (reduced in amplitude) by a factor of sqrt(2) (corresponding to –3db). The proper delimiting of the bandwidth frequency for a practical speed-controlled system requires the following steps and considerations:

- It is assumed that the speed reference is a sinusoidal signal $\omega^* = A \sin(\omega_e t)$.
- The amplitude A is to assume a moderate value that does not saturate the torque actuator. Namely, the values of T_{em} required for the proper tracking of $\omega^* = A \sin(\omega_e t)$ should not go beyond the intrinsic limits of the torque actuator.

- With an increase of the excitation frequency ω_e, the amplitude of the system output ω would vary. Given the low-pass nature of the system, a decrease in amplitude is to be expected. Depending on the poles and zeros in the transfer function $W_{SS}(s)$, the output amplitude may exhibit a resonant peak (increase) next to the bandwidth frequency, followed by eventual decline.
- The bandwidth ω_{BW} is defined as the excitation frequency ω_e at which the output amplitude drops down to 70.71% of the input A.

For the closed-loop transfer function of the proportional speed-controlled system having a single, real pole (Eq. 2.6), the bandwidth frequency ω_{BW} is given in Eq. 2.10. By considering sinusoidal excitation $\omega^*(t)$ and sinusoidal output $\omega(t)$, we can derive their Fourier transforms $\omega^*(j\omega)$ and $\omega(j\omega)$. If we replace the argument s with $j\omega$, the transfer function $W_{SS}(j\omega)$ provides the ratio $\omega(j\omega)/\omega^*(j\omega)$. Therefore, the closed-loop bandwidth frequency ω_{BW} can be determined as the one driving the $W_{SS}(j\omega)$ magnitude down to 0.7071. Finally, the bandwidth of the system given in Fig. 2.1 equals $\omega_{BW}=K_P/J$.

$$W_{SS}(j\omega)=\frac{1}{1+j\omega\tau} \Rightarrow \left|W_{SS}(j\omega)\right|^2=\frac{1}{1+(\omega\tau)^2} \Rightarrow \omega_{BW}=\frac{1}{\tau} \tag{2.10}$$

The proportional gain of the speed controller has a direct impact on the bandwidth ($\omega_{BW}=K_P/J$). Apparently, the bandwidth of the system in Fig. 2.1 is not limited, as the gain K_P can be arbitrarily chosen. In practice, the range of applicable gains is always limited. The effects that limit the gain include, but are not confined to the noise problems. Specifically, a certain level of parasitic components exists in the feedback signal ω_{fb} and hence in the speed error $\Delta\omega$ as well. Referred to as noise, the parasitic signals usually have an amplitude smaller than ω_{fb} by several orders of magnitude. Even when multiplied by a moderate K_P gain, the noise contribution to the torque reference T_{ref} is insignificant in most cases. However, extremely high values of the gain would result in significant noise content in T_{ref} and T_{em} signals, impairing the system's capability to track the reference input and placing into question the integrity of the electric motor and associated power amplifier. For these and other reasons, the feedback gains, such as the gain K_P and the close-loop bandwidth ω_{BW}, are subjected to limits.

2.1.2 Load rejection of the proportional speed controller

The closed-loop transfer function $W_{SS}(s)$ defines the complex image of the system output $\omega(s)$ for the given speed reference $\omega^*(s)$, under the assumption that the load torque T_L is equal to zero. The load torque, itself, affects the output speed ω and may be the origin of speed errors $\Delta\omega$. Therefore, it is of interest to determine, for any speed-controlled system, the transfer function $W_{LS}(s) = \omega(s)/T_L(s)$ relating the complex images of the system output $\omega(s)$ and the load torque $T_L(s)$. Ideally, the $W_{LS}(s)$ should be equal to zero. In this case, the load torque T_L variations would not produce any variation of the speed. Systems with the $W_{LS}(s)$ close to zero are referred to as *rigid* or *stiff*, reflecting the fact that their loads $T_L(t)$ have little or no influence on controlled speed. In terms of control theory, the load torque $T_L(t)$ is an external disturbance, unrelated to the speed reference and the internal variables. It can be predictable (deterministic) or stochastic in nature. The transfer function $W_{LS}(s)$ reflects the output speed sensitivity to load disturbances. For the system in Fig. 2.1, the transfer function $W_{LS}(s)$ is given in Eq. 2.11. The function $W_{LS}(s)$ is inversely proportional to the gain K_P. Hence, the stiffness (rigidity) of the system is directly proportional to an applicable gain.

$$W_{LS}(s) = \left.\frac{\omega(s)}{T_L(s)}\right|_{(\omega^*=0)} = -\frac{1}{K_P}\,\frac{1}{1+s\tau} \tag{2.11}$$

In Eq. 2.12, the complex image of the output speed $\omega(s)$ is given for the case when both the speed reference $\omega^*(s)$ and the load torque $T_L(s)$ are present. It is of interest to determine the steady-state value of the output speed in the case when both the speed reference and load torque are the Heaviside step functions with amplitudes Ω^* and T_{LOAD}, respectively. The value $\omega(\infty)$ is given in Eq. 2.13.

$$\omega(s) = W_{SS}(s)\omega^*(s) + W_{LS}(s)T_L(s) = \frac{1}{1+s\tau}\left[\omega^*(s) - \frac{T_L(s)}{K_P}\right] \tag{2.12}$$

$$\omega(\infty) = \lim_{s\to 0}\left(s\,\omega(s)\right) = \lim_{s\to 0}\left[\frac{s}{1+s\tau}\left(\frac{\Omega^*}{s} - \frac{T_{LOAD}}{sK_P}\right)\right] = \Omega^* - \frac{T_{LOAD}}{K_P} \tag{2.13}$$

From Eq. 2.13, we conclude that the proportional speed controller cannot ensure zero error in the steady state. In the presence of the load torque, the actual speed will deviate from the speed reference. The speed deviation is inversely proportional to the feedback gain (T_{LOAD}/K_P). The gain K_P is subjected to limits and cannot be increased such as to make the speed error $\Delta\omega = T_{LOAD}/K_P$ negligible.

The presence of the steady-state speed error can be predicted from the block diagram in Fig. 2.1. If we consider the steady-state operation in the case when a constant load torque T_{LOAD} is applied, it is obvious that the driving torque $T_{em} = T_{LOAD}$ must be present in order to balance the load torque and keep the speed constant. Given the speed controller transfer function $W_{SC}(s) = K_P$, it is clear that the speed error $\Delta\omega = T_{LOAD}/K_P$ must be fed to the speed controller input in order to obtain the desired output $T_{ref} = T_{em}$. At this point, it is of interest to suggest the necessary modification of $W_{SC}(s)$, suited to eliminating the steady-state speed error in the presence of a constant load-torque disturbance. The block $W_{SC}(s)$ in Fig. 2.1 should be capable of providing the torque reference $T_{ref} = T_{LOAD}$ at the output, while having the speed error $\Delta\omega = 0$ at the input. Therefore, $W_{SC}(s)$ should be enhanced by adding an integral control action. An integrator supplied with $\Delta\omega = 0$ at input keeps the output T_{ref} constant. The effects of adding the integral control action to the speed controller will be discussed further on.

2.1.3 Proportional speed controller with variable reference

The task of tracking a variable reference is frequently referred to as the *servo problem*. When the desired speed changes in time, the speed reference change is called the *reference profile* or *trajectory* $\omega^*(t)$. The ability of the speed controller to track the desired profile is measured by the speed error $\Delta\omega$, which should be as small as possible.

The speed reference increasing at a constant rate, also known as the *ramp*, is given in Eq. 2.14, along with its complex image (i.e., Laplace transform). If we consider the speed controller with the proportional action, given in Fig. 2.1, and assume that the load torque is absent, the Laplace transform of the output speed is derived in Eq. 2.15. The complex image of the speed error $\Delta\omega$ is given in Eq. 2.16.

$$\omega^*(t) = \mathrm{A}^* t, \quad \omega^*(s) = \mathscr{L}\left[\mathrm{A}^* t\right] = \int_0^{+\infty} \mathrm{A}^* t e^{-st}\, \mathrm{d}t = \frac{\mathrm{A}^*}{s^2} \tag{2.14}$$

$$\omega(s)\Big|_{T_L=0} = W_{SS}(s)\,\omega^*(s) = \frac{1}{1+s\tau}\frac{\mathrm{A}^*}{s^2} \tag{2.15}$$

$$\Delta\omega(s)\big|_{T_L=0} = \omega^*(s) - \omega(s) = [1 - W_{SS}(s)]\omega^*(s) = \frac{s\tau}{1+s\tau}\frac{A^*}{s^2} \tag{2.16}$$

Provided that sufficient time has passed ($t >> \tau$) from the instant $t = 0$, when the reference ramp $\omega^*(t) = A^* t$ is applied and the system is put into motion, and given the fact that the system in Fig. 2.1 is stable, the steady-state condition is reached. The output speed will track the reference. Any eventual speed error $\Delta\omega$, as well as the driving torque T_{em}, will assume constant values. The steady-state speed error is derived in Eq. 2.17:

$$\Delta\omega(\infty) = \lim_{s\to 0}(s\Delta\omega(s)) = \lim_{s\to 0}\left[\frac{s^2\tau}{1+s\tau}\frac{A^*}{s^2}\right] = A^*\tau = A^*\frac{J}{K_P}. \tag{2.17}$$

The error $\Delta\omega$ in tracking the ramp with slope A^* is proportional to the load inertia and inversely proportional to the feedback gain K_P. An increase in K_P would reduce the tracking error. However, since the range of applicable gains is limited due to the stability condition, the tracking error cannot be completely removed. Therefore, in order to provide error-free slope tracking, the structure of the speed controller $W_{SC}(s)$ needs to be changed.

The limited ability of the speed controller to track the reference can result in a sustained tracking error, as shown in the previous expression. There is a class of references that may result not only in a limited steady-state tracking error (Eq. 2.17), but, also, in a complete lack of tracking capability and a progressive increase of the error $\Delta\omega$. As an example, consider the proportional speed controller (Fig. 2.1) with a parabolic speed reference $\omega(t) = Y^* t^2$. The complex image of the speed error $\Delta\omega(s)$ and the value $\Delta\omega(\infty)$ are given in Eq. 2.18.

$$\begin{aligned}
&\omega^*(t) = Y^* t^2,\ \omega^*(s) = \frac{2Y^*}{s^3} \\
&\Rightarrow \omega(s)\big|_{T_L=0} = W_{SS}(s)\omega^*(s) = \frac{1}{1+s\tau}\frac{2Y^*}{s^3} \\
&\Delta\omega(s)\big|_{T_L=0} = \frac{s\tau}{1+s\tau}\frac{2Y^*}{s^3} \\
&\Rightarrow \Delta\omega(\infty) = \lim_{s\to 0}\left[\frac{s^2\tau}{1+s\tau}\frac{2Y^*}{s^3}\right] = \infty
\end{aligned} \tag{2.18}$$

The steady-state tracking error $\Delta\omega(\infty)$ in Eq. 2.18 is infinite, meaning that the difference $\Delta\omega(t) = \omega^*(t) - \omega(t)$ would progressively increase, even in the case of a stable system. The developments that follow in this chapter will illustrate the speed controller enhancements that improve the speed profile tracking capability.

2.1.4 Proportional speed controller with frictional load

In the analysis given in Sections 2.1.1–2.1.3, it is assumed that the motor is coupled to an inertial load exhibiting no friction. The load torque $T_L(t)$ is considered to be an external disturbance, unrelated to the speed reference or internal system variables. It is of interest to investigate the effects that the frictional load may have on the performance of the proportional speed controller. In the following considerations, it is assumed that the friction coefficient B cannot be neglected. The transfer function $W_P(s)$ of the inertialfrictional load is given in Eq. 2.19.

$$W_P(s) = \frac{\omega(s)}{T_{em}(s)}\bigg|_{(T_L=0)} = \frac{1}{Js+B}, \quad W_{SC}(s) = K_P \tag{2.19}$$

In the case where the load has no friction (Eq. 2.5), the open-loop transfer function $W_S(s)$ becomes an integrator $1/s$. When we take into account the friction B, the $W_S(s)$ obtains a real pole $s = -B/J$:

$$W_S(s) = \frac{\omega(s)}{\omega^*(s)}\bigg|_{(\omega_{fb}=0,\, T_L=0)} = \frac{\omega(s)}{\Delta\omega(s)}\bigg|_{T_L=0} = W_{SC}(s)W_P(s) = \frac{K_P}{Js+B} \tag{2.20}$$

The closed-loop transfer function, given in Eq. 2.21, is similar to the one given in Eq. 2.6. The time constant τ_1 in the former equation is smaller than τ in the latter. The coefficient $K_1 < 1$ multiplies the expression in Eq. 2.21. This means that the steady-state value of the output speed will not reach the reference.

$$W_{SS}(s) = \frac{\omega(s)}{\omega^*(s)}\bigg|_{T_L=0} = \frac{W_S(s)}{1+W_S(s)} = \frac{K_P}{K_P+B}\,\frac{1}{1+\dfrac{Js}{K_P+B}}$$

$$W_{SS}(s) = K_1\frac{1}{1+s\tau_1}, \quad \left(K_1 = \frac{K_P}{K_P+B} < 1, \quad \tau_1 = \frac{J}{K_P+B}\right) \tag{2.21}$$

For the speed reference, assuming the Heaviside step form $\omega^*(s) = \Omega^*/s$, the complex image of the speed error is given in Eq. 2.22. The steady-state speed error $\Delta\omega(\infty)$ is found in Eq. 2.22. Due to the presence of the friction B, the steady-state error assumes a nonzero value, directly proportional to the speed reference and the friction coefficient, and inversely proportional to feedback gain K_P.

$$\Delta\omega(s)\Big|_{T_L=0} = \omega^*(s) - \omega(s) = \left[1 - W_{SS}(s)\right]\omega^*(s) = \frac{B+Js}{K_P+B+Js}\,\frac{\Omega^*}{s} \quad (2.22)$$

$$\Delta\omega(\infty) = \lim_{s\to 0}\left(s\Delta\omega(s)\right) = \frac{B}{K_P+B}\,\Omega^* \quad (2.23)$$

2.2 The speed controller with proportional and integral action

In the previous section, it was demonstrated that the speed controller with proportional action alone cannot suppress the speed error $\Delta\omega = \omega^* - \omega$ in cases when either the load torque T_L is present or the friction $B\omega$ assumes a value that cannot be neglected. A nonzero steady-state speed error remained, even in the case when both $T_L = 0$ and $B = 0$, and the speed reference input has the form of a ramp function ($\omega^*(t) = A^* t$). The previous analysis concluded that the elimination of such an error requires the speed controller $W_{SC}(s)$ to be enhanced by adding an integral control action. In this section, the speed control system with proportional and integral (PI) control actions is investigated. The open-loop and the closed-loop system transfer functions are derived and the load rejection is investigated. The ability of the PI controller to track the speed reference profile is analyzed for ramp-shaped and parabolic references. This analysis is based on the diagram in Fig. 2.2, which represents the speed controlled system comprising the PI speed controller, the idealized speed measurement system ($\omega_{fb} = \omega$), the torque actuator with an instant response ($T_{em} = T_{ref}$), and the load with inertia J and friction $B > 0$.

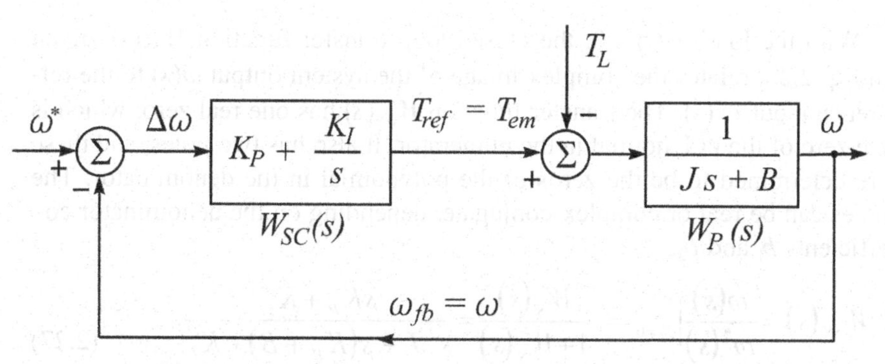

Fig. 2.2. The speed controller with proportional and integral action. The speed-controlled system comprises the load inertia J and friction B, and a separately excited DC motor with the instant torque response ($T_{em} = T_{ref}$) and idealized speed measurement.

2.2.1 Transfer functions of the system with a PI controller

The speed control system with proportional and integral (PI) control actions is represented in Fig. 2.2. The transfer function of the load (plant) is given in Eq. 2.24. The speed controller transfer function is given in Eq. 2.25. The open-loop transfer function $W_S(s)$, given in Eq. 2.26, describes the input-output signal flow under conditions when the feedback loop is opened and the load torque is absent. The open-loop transfer function has one negative, real zero ($-K_I/K_P$), one pole in the origin ($s = 0$), and one real, negative pole ($-B/J$).

$$W_P(s) = \left.\frac{\omega(s)}{T_{em}(s)}\right|_{(T_L=0)} = \frac{1}{Js+B} \tag{2.24}$$

$$W_{SC}(s) = K_P + \frac{K_I}{s} = \frac{sK_P + K_I}{s} \tag{2.25}$$

$$\begin{aligned} W_S(s) &= \left.\frac{\omega(s)}{\omega^*(s)}\right|_{(\omega_{fb}=0,\, T_L=0)} = \left.\frac{\omega(s)}{\Delta\omega(s)}\right|_{T_L=0} \\ &= W_{SC}(s)W_P(s) = \frac{sK_P + K_I}{s(Js+B)} \end{aligned} \tag{2.26}$$

With the load $T_L(t) = 0$, the closed-loop transfer function $W_{SS}(s)$, given in Eq. 2.27, relates the complex image of the system output $\omega(s)$ to the reference input $\omega^*(s)$. The transfer function $W_{SS}(s)$ has one real zero, which is the zero of the polynomial in the numerator. It also has two poles, and these are determined to be the zeros of the polynomial in the denominator. The poles can be real or complex-conjugate, depending on the denominator coefficients b_1 and b_2.

$$W_{SS}(s) = \left.\frac{\omega(s)}{\omega^*(s)}\right|_{T_L=0} = \frac{W_S(s)}{1+W_S(s)} = \frac{sK_P + K_I}{s^2 J + s(K_P + B) + K_I}$$
$$= \frac{1+s\frac{K_P}{K_I}}{1+s\frac{K_P+B}{K_I}+s^2\frac{J}{K_I}} = \frac{1+sa_1}{1+sb_1+s^2 b_2} \tag{2.27}$$

2.2.2 Load rejection with the PI speed controller

It is of interest to investigate the effects of introducing the integral action on the speed controller's capability to keep the steady-state speed insensitive to the constant, or slowly varying, load disturbances. The transfer function $W_{LS}(s) = \omega(s)/T_L(s)$ relates the complex images of the system output $\omega(s)$ and the load torque $T_L(s)$ in the condition when $\omega^*(t) = 0$. It is desirable to keep $W_{LS}(s) = 0$, since, in this case, load torque T_L variations would not produce any variation of the controlled speed $\omega(t)$. Such a system would be infinitely stiff or rigid, and its sensitivity to load disturbances would be nil. For the system in Fig. 2.2, the transfer function $W_{LS}(s)$ is given in Eq. 2.28.

$$W_{LS}(s) = \left.\frac{\omega(s)}{T_L(s)}\right|_{(\omega^*=0)} = -\frac{s}{K_I + s(K_P + B) + s^2 J}$$
$$= -\frac{1}{K_I}\frac{s}{1+sb_1+s^2 b_2} \tag{2.28}$$

When both the speed reference and load torque are present in the form of Heaviside step functions with amplitudes Ω^* and T_{LOAD}, respectively, the complex image of the output speed is given in Eq. 2.29.

$$\begin{aligned}\omega(s) &= W_{SS}(s)\omega^*(s) + W_{LS}(s)T_L(s)\\ &= \frac{1+sa_1}{1+sb_1+s^2b_2}\omega^*(s) - \frac{s}{1+sb_1+s^2b_2}\frac{T_L(s)}{K_I} \\ &= \frac{1+sa_1}{1+sb_1+s^2b_2}\frac{\Omega^*}{s} - \frac{1}{1+sb_1+s^2b_2}\frac{T_{LOAD}}{K_I}\end{aligned} \tag{2.29}$$

Equation 2.30 gives the steady-state value of the output speed $\omega(t)$. While both the speed reference input Ω^* and load torque are present, the steady-state output equals Ω^*. Hence, the load torque $T_L(t) = T_{LOAD}$ does not affect the controlled speed in the steady state. A comparison of expressions 2.13 and 2.30 shows that the load rejection capability of the speed controller is improved, due to the introduction of the integral action.

$$\omega(\infty) = \lim_{s\to 0}\left(s\,\omega(s)\right) = \lim_{s\to 0}\left[\frac{(1+sa_1)\Omega^* - \dfrac{sT_{LOAD}}{K_I}}{1+sb_1+s^2b_2}\right] = \Omega^* \tag{2.30}$$

Notwithstanding the presence of the integral control action, load torque disturbances other than $T_L(s) = T_{LOAD}/s$ may produce an error in the output speed. With the ramp-shaped load torque, where $T_L(t) = T_{RAMP}\,t$ and $T_L(s) = T_{RAMP}/s^2$, the steady-state speed error is proportional to the slope T_{RAMP} and inversely proportional to the gain K_I (Eq. 2.31).

$$\begin{aligned}\omega(\infty) &= \lim_{s\to 0}\left(s\,\omega(s)\right) \\ &= \lim_{s\to 0}\left[\frac{(1+sa_1)\Omega^* - \dfrac{T_{RAMP}}{K_I}}{1+sb_1+s^2b_2}\right] = \Omega^* - \frac{T_{RAMP}}{K_I}\end{aligned} \tag{2.31}$$

2.2.3 Step response with the PI speed controller

In this section, the step response of the output speed is analyzed. The relation between the feedback gains K_I, K_P, closed-loop poles, and character of the step response is investigated. Suppression of the overshoot in the step response by means of dislocating the proportional action is reviewed.

Practicable closed-loop bandwidths and damping coefficients are discussed and explained, and a rule of thumb for parameter setting is suggested.

With the speed reference $\omega^*(t) = \Omega^*\, h(t)$ and load torque $T_L(t) = 0$, the complex image $\omega(s)$ of the output speed is given in Eq. 2.32. The step response of the output speed $\omega(t)$ is obtained as the inverse Laplace transform of $\omega(s)$. The inverse Laplace transform of the complex image $\omega(s)$ is given in Eq. 2.33. The constant γ is a real positive number securing the convergence of the integral, known as the axis of the absolute convergence. The desired time response $\omega(t)$ can be obtained by performing the integration in Eq. 2.33.

$$\omega(s) = W_{SS}(s)\omega^*(s) = \frac{1+sa_1}{1+sb_1+s^2b_2}\frac{\Omega^*}{s} \tag{2.32}$$

$$\omega(t) = \mathscr{L}^{-1}\left[\omega(s)\right] = \frac{1}{2\pi j}\int_{\gamma-j\infty}^{\gamma+j\infty}\omega(s)e^{st}\,\mathrm{d}s \tag{2.33}$$

The integration can be avoided in cases when the complex image $\omega(s)$ can be split into several parts, their inverse Laplace transforms [1] being available in Appendix 2. The inverse Laplace transforms $f(t)$ of several complex functions $F(s)$, relevant for solving Eq. 2.33, are given in Eq. 2.34.

$$\begin{aligned}
&\mathscr{L}^{-1}\left[\frac{s+\alpha}{(s+\alpha)^2+\omega_0^2}\right] = e^{-\alpha t}\cos(\omega_0 t),\\
&\mathscr{L}^{-1}\left[\frac{\omega_0}{(s+\alpha)^2+\omega_0^2}\right] = e^{-\alpha t}\sin(\omega_0 t),\\
&\mathscr{L}^{-1}\left[\frac{s\sigma+\omega_n^2}{s\left(s+2\sigma s_n+\omega^2\right)}\right] = 1-e^{-\sigma t}\cos(\omega_0 t),\ \omega_0=\sqrt{\omega_n^2-\sigma^2}.
\end{aligned} \tag{2.34}$$

It is evident in Eq. 2.34 that the presence of oscillations within the time response $f(t)$, their frequency, and their decay rate depend on the zeros of the $F(s)$ denominator. It can be demonstrated that the complex image $F(s)$, having denominator zeros $s_{1/2} = -\sigma \pm \mathrm{j}\omega_0$, corresponds to response $f(t)$ in the time domain, comprising the oscillations at the frequency ω_0. The amplitude of said oscillations exhibits an exponential decay ($e^{-\sigma t} = e^{-t/\tau}$) with time constant $\tau = 1/\sigma$.

The transfer function $W_{SS}(s)$ in Eq. 2.32 has the second-order denominator $d(s) = 1 + b_1 s + b_2 s^2$, having the zeros s_1 and s_2, given in Eq. 2.35. When $b_1^2 - 4b_2 > 0$, and with both $b_1 > 0$ and $b_2 > 0$, the zeros are real negative numbers. With $b_1^2 - 4b_2 < 0$, the zeros are conjugate complex numbers having the form $s_{1/2} = -\sigma \pm j\omega_0$. In this case, the real component equals $-b_1/2/b_2$. When the roots are complex, their absolute value $\omega_n = |s_1| = |s_2| = (\sigma^2 + \omega_0^2)^{0.5}$ is called the *natural frequency*. The oscillations within the time response are expected at a reduced frequency $\omega_0 = (\omega_n^2 - \sigma^2)^{0.5}$. The denominator of $d(s)$ of the closed-loop transfer function $W_{SS}(s)$ is, at the same time, the characteristic polynomial of the closed-loop system. The zeros s_1 and s_2 are referred to as the *closed-loop poles*. In the matrix representation of the system, the poles s_1 and s_2 can be found as the eigenvalues of the system matrix. Besides poles, the $W_{SS}(s)$ may have zeros, and these are the zeros of the polynomial in the nominator of $W_{SS}(s)$.

$$s_1 = -\frac{b_1}{2b_2} + \frac{1}{2b_2}\sqrt{b_1^2 - 4b_2} = -\frac{B+K_P}{2J} + \frac{1}{2}\sqrt{\left(\frac{B+K_P}{J}\right)^2 - \frac{4K_I}{J}}$$
$$s_2 = -\frac{b_1}{2b_2} - \frac{1}{2b_2}\sqrt{b_1^2 - 4b_2} = -\frac{B+K_P}{2J} - \frac{1}{2}\sqrt{\left(\frac{B+K_P}{J}\right)^2 - \frac{4K_I}{J}} \qquad (2.35)$$

The closed-loop poles s_1 and s_2 depend on the plant parameters (B, J) and the control parameters K_P, K_I, also called the *feedback gains*. In this way, the step response of the output speed $\omega(t)$ can be controlled by means of tuning the feedback gains to the desired values. If we consider $\omega(s)$, given in Eq. 2.32, the step response $\omega(t)$ can be expressed in the form shown in Eq. 2.36, clearly indicating the impact of the zeros s_1 and s_2 on the time-domain response.

$$f(t) = \omega(t) = C_0 h(t) + C_1 e^{s_1 t} + C_2 e^{s_2 t}$$
$$e^{(\sigma + j\omega_0)t} = e^{-\sigma t}\cos(\omega_0 t) + j e^{-\sigma t}\sin(\omega_0 t) \qquad (2.36)$$

Whenever the closed-loop poles are real, the step response will be exponential and without oscillations. With complex poles $s_{1/2} = -\sigma \pm j\omega_0$, oscillations at the frequency ω_0 will take place, exponentially disappearing in time. A larger real component of the poles $\mathrm{Re}(s_{1/2}) = -\sigma$ will result in a faster oscillation decay. The ratio $|\mathrm{Re}(s_{1/2})| / |s_{1/2}| = \sigma/\omega_n$ is known as the *damping factor* and is denoted by $\xi = \sigma/\omega_n$. For damping factors $\xi > 1$, the closed-loop poles are real and the step response is strictly exponential, involving no oscillations. When $\xi < 1$, the poles are conjugate complex. With a smaller damping factor ξ, the oscillations are more emphasized. The influence of the damping factor on the step response is illustrated in Figs. 2.3–2.5.

If we consider a second-order polynomial $d(s)$ in the denominator, having a pair of conjugate complex zeros $s_{1/2} = -\sigma \pm j\omega_0$, and introduce the natural frequency $\omega_n = (\sigma^2 + \omega_0^2)^{0.5}$ and damping factor $\xi = \sigma/\omega_n$, then $d(s)$ can be rewritten as shown in Eq. 2.37. Frequently encountered in textbooks and articles, this presentation of $d(s)$ relates the polynomial coefficients to the undamped natural frequency ω_n and the damping coefficient ξ. By means of the denominator of the transfer function $W_{SS}(s)$ in Eq. 2.27, it is possible to observe the impact of the feedback gains on ξ and ω_n.

$$\begin{aligned} d(s) &= (s - s_1)(s - s_2) = s^2 + (s_1 + s_2)s + s_1 s_2 \\ &= s^2 + 2\sigma s + \omega_n^2 = s^2 + 2\xi\omega_n s + \omega_n^2 \end{aligned} \tag{2.37}$$

The closed-loop transfer function 2.27 can be rewritten by multiplying both the numerator and denominator by K_I/J. The denominator assumes the form shown in Eq. 2.38. Within the same expression, it is evident that the integral gain K_I determines the natural frequency ω_n of the closed-loop poles, while the proportional gain K_P can be used for tuning the damping factor ξ.

$$\begin{aligned} d(s) &= s^2 + \frac{K_P + B}{J} s + \frac{K_I}{J} = s^2 + 2\xi\omega_n s + \omega_n^2 \\ \Rightarrow \omega_n &= \sqrt{\frac{K_I}{J}}, \quad \xi = \frac{K_P + B}{2J\omega_n} \end{aligned} \tag{2.38}$$

The effects of the damping factor ξ on the step response are investigated in Figs. 2.3–2.5. On the left, the closed-loop poles are shown in the s-plane, while the step-response waveform is given on the right side of each figure. The step response is obtained from Eq. 2.39. In this expression, it is assumed that the closed-loop transfer function $W_{SS}(s)$ has two poles, while the step $\omega^*(t) = \Omega^* h(t)$ is fed into the reference input.

$$\omega(s) = \frac{\Omega^*}{s} W_{SS}(s) = \frac{1}{s} \frac{\omega_n^2}{d(s)} = \frac{1}{s} \frac{\omega_n^2}{s^2 + 2\xi\omega_n s + \omega_n^2} \tag{2.39}$$

In Fig. 2.3, the closed-loop poles are complex conjugate, located in the left half of the s-plane, while the damping factor $\xi < 1$ is equal to $\cos(\alpha)$, where α is the angle between the radius ω_n and the negative side of the real axis. A significant overshoot is observed in the step response (on the right in Fig. 2.3) with subsequent oscillations decaying exponentially. In Fig. 2.4, disposition of the closed-loop poles and the step response are given for the case when both poles are real and equal $s_1 = s_2$ and where $\alpha = 0$ and $\xi = 1$. Finally, Fig. 2.5 displays the pole placement and the step response for the case when s_1 and s_2 are real and different, and where $\xi > 1$. In the latter cases, the step response is delayed and strictly exponential, with no oscillations present.

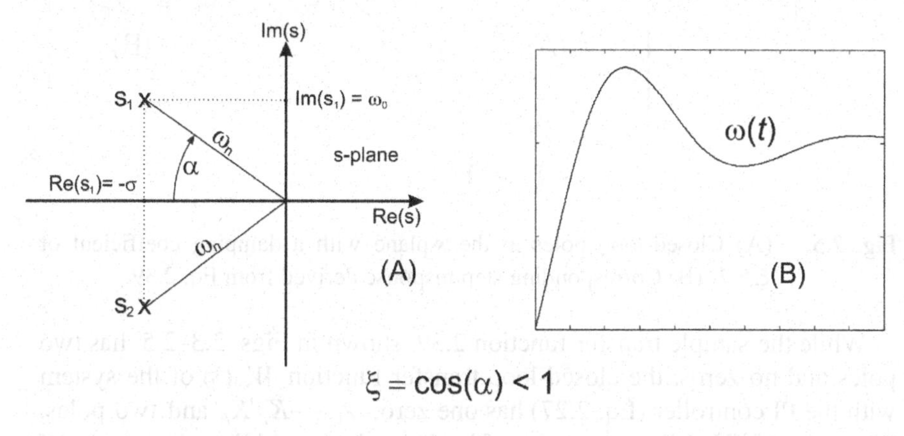

Fig. 2.3. (A) Closed-loop poles in the s-plane with a damping coefficient of $\xi < 1$. (B) Corresponding step response derived from Eq. 2.39.

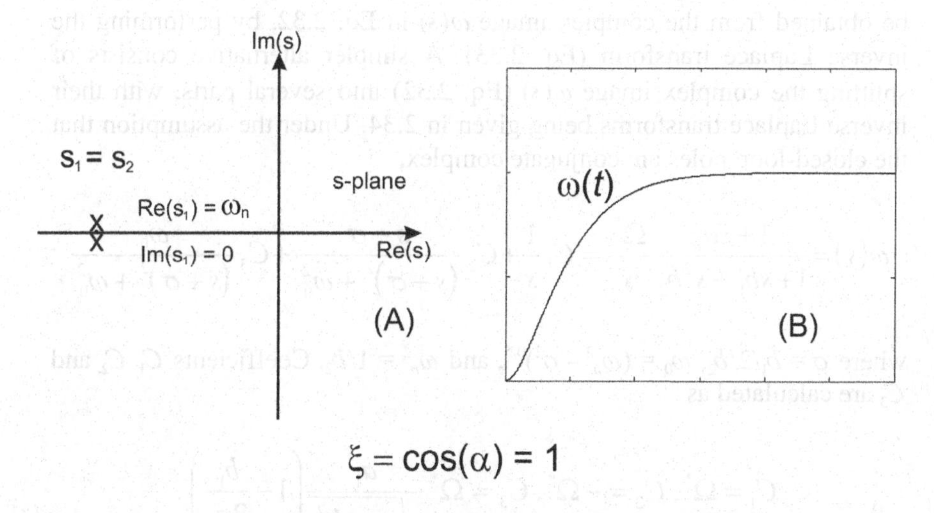

Fig. 2.4. (A) Closed loop poles in the s-plane with a damping coefficient of $\xi = 1$.(B) Corresponding step response derived from Eq. 2.39.

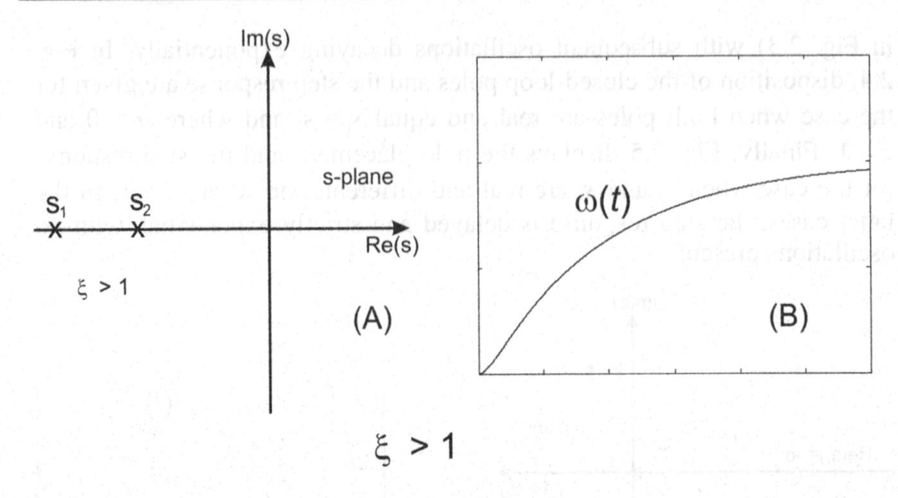

Fig. 2.5. (A) Closed-loop poles in the *s*-plane with a damping coefficient of $\xi > 1$. (B) Corresponding step response derived from Eq. 2.39.

While the sample transfer function 2.39, shown in Figs. 2.3–2.5, has two poles and no zeros, the closed-loop transfer function $W_{SS}(s)$ of the system with the PI controller (Eq. 2.27) has one zero, $z_1 = -K_I/K_P$ and two poles. The poles of $W_{SS}(s)$ are the zeros of its denominator, while z_1 is the zero of the polynomial in the numerator. The presence of z_1 will make the step response shown in Eq. 2.33 somewhat different. The step response $\omega(t)$ can be obtained from the complex image $\omega(s)$ in Eq. 2.32, by performing the inverse Laplace transform (Eq. 2.33). A simpler alternative consists of splitting the complex image $\omega(s)$ (Eq. 2.32) into several parts, with their inverse Laplace transforms being given in 2.34. Under the assumption that the closed-loop poles are conjugate complex,

$$\omega(s) = \frac{1+sa_1}{1+sb_1+s^2b_2}\frac{\Omega^*}{s} = C_1\frac{1}{s} + C_2\frac{s+\sigma}{(s+\sigma)^2+\omega_0^2} + C_3\frac{\omega_0}{(s+\sigma)^2+\omega_0^2}$$

where $\sigma = b_1/2/b_2$, $\omega_0 = (\omega_n^2 - \sigma^2)^{0.5}$, and $\omega_n^2 = 1/b_2$. Coefficients C_1, C_2 and C_3 are calculated as

$$C_1 = \Omega^*,\; C_2 = -\Omega^*,\; C_3 = \Omega^* \frac{a_1}{\sqrt{b_2 - \frac{b_1^2}{4}}}\left(1 - \frac{b_1}{2a_1}\right).$$

Applying Eq. 2.34, the step response of the system in Fig. 2.2 is found as

$$\omega(t) = \Omega^* h(t) - \Omega^* e^{-\sigma t}\cos(\omega_0 t) + \Omega^* \frac{a_1}{\sqrt{b_2 - \frac{b_1^2}{4}}}\left(1 - \frac{b_1}{2a_1}\right) e^{-\sigma t}\sin(\omega_0 t). \tag{2.40}$$

When the closed-loop poles are negative real numbers ($\xi > 1$), the step response $\omega(t)$ can be obtained in a similar manner, by splitting the closed-loop transfer function into parts having the form $C_n/(1+s\tau_n)$, and using the inverse transform $\mathcal{L}^{-1}[1/(1+s\tau_n)] = \exp(-t/\tau_n)$.

The step-response waveform $\omega(t)$ is tested with four different settings of the feedback gains. The resulting traces are shown in Fig. 2.6. The traces are obtained assuming that the friction B is negligible compared with K_P. The integral gain K_I is kept constant, maintaining the natural frequency $\omega_n = (K_I/J)^{0.5}$ at a constant value. The proportional gain K_P is varied, affecting the damping factor ξ (Eq. 2.38) and changing the closed-loop zero $z_1 = -K_I/K_P$.

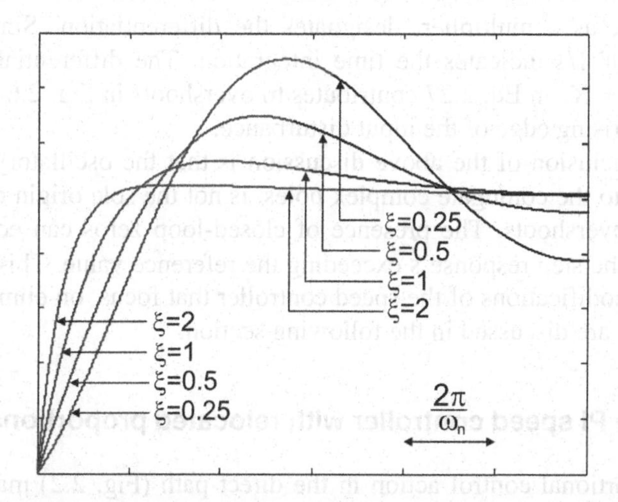

Fig. 2.6. Step response of the closed-loop system in Fig. 2.2. The natural frequency ω_n and the gain K_I are kept constant (Eq. 2.38), while the gain K_P assumes four different values, affecting the damping factor ξ and changing the value of the transfer function zero $z_1 = -K_I/K_P$.

For the damping of 0.25 and 0.5, the responses in Fig. 2.6 contain a large overshoot. With $\xi = 1$, the closed loop poles s_1 and s_2 are equal negative real numbers and the response is exponential, having no oscillations. However, a slight overshoot in the step response is persistent. When $\xi = 2$, the closed-loop poles are real and different, and a very small overshoot is still present. The presence of an overshoot in the step response is usually associated with complex poles, leading to $\xi < 1$ and involving oscillatory components.

If we probe further into the issue of the overshoot being present in the exponential responses $\xi = 1$ and $\xi = 2$ in Fig. 2.6, it is interesting to compare them with the step responses shown in Figs. 2.3–2.5. The exponential step responses in Figs. 2.4 and 2.5 do not have an overshoot. The latter are obtained from the transfer function in Eq. 2.39, having two poles and no zeros. Hence, it is the closed loop zero $z_1 = -K_I/K_P$ of the closed-loop transfer function $W_{SS}(s)$ in Eq. 2.27 that makes the difference and contributes to the existence of an overshoot, even in cases when the closed-loop poles are real.

The numerator $n(s) = sK_P + K_I$ of $W_{SS}(s)$ in Eq. 2.27 contributes to the derivative action of the transfer function. Such a statement is inferred from properties of the Laplace transform: namely, when the complex image $F(s)$ of the time function $f(t)$ is known, the Laplace transform of the derivative $\mathrm{d}f(t)/\mathrm{d}t$ can be found to be $\mathscr{L}(\mathrm{d}f(t)/\mathrm{d}t) = sF(s) - f(0_-)$. Hence, the Laplace operator s, as a multiplier, designates the differentiation. Similarly, the presence of $1/s$ indicates the time integration. The differential nature of $n(s) = sK_P + K_I$ in Eq. 2.27 contributes to overshoots in Fig. 2.6 by emphasizing the rising edge of the input disturbance.

The conclusion of the above discussion is that the oscillatory response, attributed to the conjugate complex poles, is not the sole origin of the step-response overshoots. The presence of closed-loop zeros can equally contribute to the step response's exceeding the reference value. This issue, and possible modifications of the speed controller that focus on eliminating the overshoot, are discussed in the following section.

2.2.4 The PI speed controller with relocated proportional action

The proportional control action in the direct path (Fig. 2.2) makes use of the speed error $\Delta\omega$. When placed in the feedback path, the control action is proportional to the feedback signal ω_{fb}. Replacing the action from the direct path into the feedback path alters the numerator of the closed-loop transfer function and removes the closed-loop zero. At the same time, the denominator and closed-loop poles remain unchanged. The closed-loop transfer function $W_{SS}(s)$ for the system in Fig. 2.7 is given in Eq. 2.41, revealing the same denominator as in Eq. 2.27 and having no closed-loop zeros.

The effects of removing the closed-loop zero are investigated by checking the response of the output speed to the step change in the speed reference. The step responses are shown in Fig. 2.8. The integral gain determining the natural frequency ω_n is kept constant, while the proportional gain K_P is varied, so as to obtain responses for four different values of the damping factor. Note that the time scaling is slightly changed with respect to Fig. 2.6. The exponential responses, obtained with $\xi = 1$ and $\xi = 2$, have no overshoot, while the responses obtained with the conjugate complex poles ($\xi = 0.5$ and $\xi = 0.25$) do retain an overshoot, inversely proportional to the damping factor ξ.

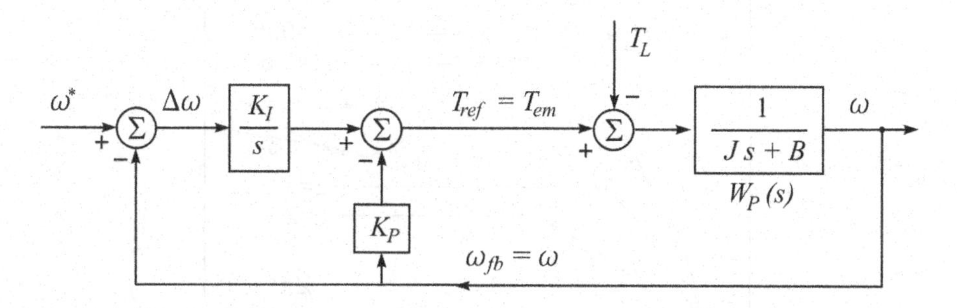

Fig. 2.7. The PI speed controller with the proportional action relocated into the feedback path. The poles of the closed-loop transfer function $W(s)$ are not changed, while the closed-loop zero $z_1 = -K_I / K_P$ is removed.

$$W_{SS}(s) = \frac{\omega(s)}{\omega^*(s)}\bigg|_{T_L=0} = \frac{K_I}{s^2 J + s(K_P + B) + K_I}$$

$$= \frac{1}{1 + s\dfrac{K_P + B}{K_I} + s^2\dfrac{J}{K_I}} = \frac{1}{1 + sb_1 + s^2 b_2} \tag{2.41}$$

If we compare the step responses in Figs. 2.6 and 2.8, it is evident that the relocation of the gain K_P removes at the same time the closed loop zero z_1 and the overshoot in the step responses obtained with $\xi \geq 1$. However, the overshoot suppression is obtained at a cost: the responses in Fig. 2.8 are slower, due to the absence of the zero z_1 in the closed-loop transfer function $W_{SS}(s)$. The speed controller given in Fig. 2.2 and the one presented in Fig. 2.7 are both in use, depending on the application needs.

With the proportional gain relocated from the direct path into the feedback loop, the pulsations of the driving torque T_{em} are reduced. In Fig. 2.7, the speed reference signal $\omega^*(t)$ is not multiplied by K_P. Therefore, the fluctuations of the reference do not have a direct contribution to T_{em} pulsations. On the other hand, in motion-control systems where the speed controller is one of the inner loops, the absence of the closed-loop zero z_1 makes the task of tuning the outer loops more difficult to achieve.

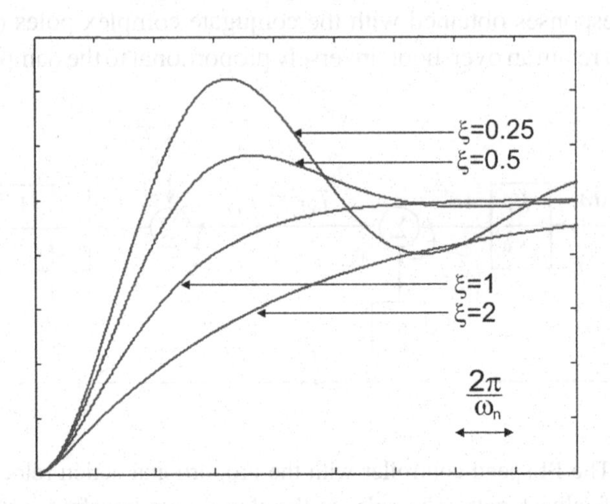

Fig. 2.8. The step response of the closed loop system in Fig. 2.7. The proportional gain is relocated in the feedback path. The natural frequency ω_n and the gain K_I are kept constant (Eq. 2.38), while the gain K_P assumes four different values, affecting the damping factor ξ.

2.2.5 Parameter setting and the closed-loop bandwidth

The previous section has shown that the character and speed of the step response depend on the closed-loop poles s_1 and s_1, the latter being the roots of the denominator $d(s)$ of the transfer function $W_{SS}(s)$. The poles can be real or conjugate complex. The real poles result in a relatively slow step response, with no oscillations and without the overshoot. Conjugate complex poles provide a faster response, with an overshoot inversely proportional to the damping factor. The natural frequency ω_n and the damping factor ξ (Eq. 2.38) are determined by the plant parameters J and B, and by the feedback gains K_P and K_I:

$$\omega_n = \sqrt{\frac{K_I}{J}}, \quad \xi = \frac{K_P + B}{2J\omega_n}.$$

The closed-loop bandwidth ω_{BW} is related to the natural frequency ω_n. The bandwidth ω_{BW} can be defined as the frequency ω_e of the sinusoidal reference input $\omega^*(t) = A^* \sin(\omega_e t)$, which results in the system output $\omega(t) = A\sin(\omega_e t - \varphi)$ attenuated by $A/A^* = 1/\text{sqrt}(2)$, corresponding to –3db. The transfer function $W_{SS}(j\omega)$ is the ratio between the complex images $\omega(j\omega)$ and $\omega^*(j\omega)$. Hence, the bandwidth frequency can be found by solving $|W_{SS}(j\omega_{BW})| = 0.707$. In general, the ratio ω_{BW}/ω_n is not constant. Variations in the damping factor ξ can produce resonant peaks (increase) in the amplitude characteristics $|W_{SS}(j\omega)|$ and, consequently, can shift the decline towards higher frequencies. At the same time, the presence of zeros in $W_{SS}(s)$ leads to higher values of ω_{BW}. For reference, the ratio ω_{BW}/ω_n is determined below for the closed-loop transfer function in Eq. 2.41, having two real, equal closed-loop poles $s_1 = s_2 = \omega_n$ ($\xi = 1$). In Eq. 2.42, the closed-loop bandwidth is found to be $\omega_{BW} = 0.6436\ \omega_n$.

$$W_{SS}(s) = \left.\frac{\omega(s)}{\omega^*(s)}\right|_{T_L=0} = \frac{1}{\left(1+\frac{s}{s_1}\right)\left(1+\frac{s}{s_2}\right)} = \frac{1}{\left(1+\frac{s}{\omega_n}\right)^2}$$
$$|W_{SS}(j\omega)| = \left|\frac{\omega(j\omega)}{\omega^*(j\omega)}\right| = \frac{1}{\left(1+\left(\frac{\omega}{\omega_n}\right)^2\right)} \Rightarrow \omega_{BW} = \sqrt{\sqrt{2}-1}\ \omega_n \tag{2.42}$$

In a number of applications, satisfactory results are obtained with conjugate complex poles, keeping the damping factor $\xi \in [0.5\ ..\ 1]$. The response speed and the closed-loop bandwidth are determined by the natural frequency ω_n. According to Eq. 2.38, the natural frequency of the closed-loop poles is proportional to $(K_I/J)^{0.5}$. Hence, the integral gain should be set to obtain the desired loop bandwidth. At the same time, the gain K_P should be changed to obtain the desired damping factor $\xi \approx 1$. From Eq. 2.38, the damping coefficient is directly proportional to K_P and inversely proportional to $K_I^{0.5}$ ($\xi^2 \sim (K_P/\omega_n)^2 \sim K_P^2/K_I$). Therefore, in order to keep the damping factor in the prescribed zone, the ratio K_P^2/K_I of the feedback gains should be preserved as well.

2.2.6 Variable reference tracking

It is worthwhile to investigate the capability of the system with the PI controller to suppress the speed errors while operating with a variable reference. The problem of tracking the speed reference profile $\omega^*(t)$ is often called the *servo problem.* For a given speed reference $\omega^*(s)$, the speed error is found to be $\Delta\omega(s) = W_E(s)\omega^*(s)$, with the error function $W_E(s)$ being equal to $1-W_{SS}(s)$. With proportional action in the direct path (Fig. 2.2), the speed controller's error function is given in Eq. 2.43 as $W_{E1}(s)$. The gain K_P in the feedback path results in the error function $W_{E2}(s)$:

$$W_{E1}(s) = \frac{\Delta\omega(s)}{\omega^*(s)}\bigg|_{T_L=0} = \frac{s\dfrac{B}{K_I} + s^2\dfrac{J}{K_I}}{1 + s\dfrac{K_P + B}{K_I} + s^2\dfrac{J}{K_I}}$$

$$W_{E2}(s) = \frac{\Delta\omega(s)}{\omega^*(s)}\bigg|_{T_L=0} = \frac{s\dfrac{K_P + B}{K_I} + s^2\dfrac{J}{K_I}}{1 + s\dfrac{K_P + B}{K_I} + s^2\dfrac{J}{K_I}}. \tag{2.43}$$

In the subsequent analysis, the reference profiles considered are the step reference ($\omega^*(t) = \Omega^* h(t)$), the ramp reference ($\omega^*(t) = A^* t\, h(t)$), and the parabolic reference ($\omega^*(t) = Y^* t^2 h(t)$). The complex images of these references are $\omega^*(s) = \Omega^*/s$, $\omega^*(s) = A^*/s^2$, and $\omega^*(s) = 2Y^*/s^3$, respectively. Provided that the system is stable and sufficient time has passed ($t >> \tau$) from the origin $t = 0$, the steady-state condition is reached, and any eventual speed error reaches the value of $\Delta\omega(\infty)$.

The speed error $\Delta\omega(\infty)$ is found by calculating the limit value in Eq 2.44. For the step reference $\omega^*(s) = \Omega^*/s$, the tracking error $\Delta\omega(\infty)$ is given in Eq. 2.45 for both presentations of the speed controller. Regardless of the location of the proportional action and friction coefficient B, the steady-state tracking error is equal to zero.

$$\Delta\omega(\infty) = \lim_{s\to 0}\left(s\,\Delta\omega(s)\right) = \lim_{s\to 0}\left(s\,W_E(s)\omega^*(s)\right) \tag{2.44}$$

$$\lim_{s\to 0}\left(sW_{E1}(s)\frac{\Omega^*}{s}\right) = 0, \quad \lim_{s\to 0}\left(sW_{E2}(s)\frac{\Omega^*}{s}\right) = 0 \tag{2.45}$$

When the speed reference profile assumes the form of a ramp $\omega^*(t) = A^* t\, h(t)$, the steady-state tracking error $\Delta\omega(\infty)$ is given in Eq. 2.46 for both cases. With the proportional action in the direct path, the error is proportional to the friction B. Therefore, in cases when the friction is negligible, the PI controller in Fig. 2.2 tracks the ramp $\omega^*(t) = A^* t\, h(t)$ without an error. On the other hand, the PI controller with the proportional action relocated in the feedback path (Fig. 2.7) sustains an error $\Delta\omega(\infty) = A^*(K_P + B)/K_I$, a nonzero value even in the case of a load with no friction.

The error in tracking the parabolic reference $\omega^*(t) = Y^* t^2 h(t)$ is given in Eq. 2.47. Notwithstanding the implementation of the proportional action, the PI speed controller cannot track the parabolic reference and suffers an ever increasing tracking error.

$$\lim_{s\to 0}\left(sW_{E1}(s)\frac{A^*}{s^2}\right) = A^*\frac{B}{K_I}, \quad \lim_{s\to 0}\left(sW_{E2}(s)\frac{A^*}{s^2}\right) = A^*\frac{K_P + B}{K_I} \tag{2.46}$$

$$\lim_{s\to 0}\left(sW_{E1}(s)\frac{2Y^*}{s^3}\right) = \infty, \quad \lim_{s\to 0}\left(sW_{E2}(s)\frac{2Y^*}{s^3}\right) = \infty \tag{2.47}$$

2.3 Suppression of load disturbances and tracking errors

The task of a speed controller is to generate the torque reference that drives the output speed towards the reference ω^* and keeps the speed error $\Delta\omega$ as small as possible, preferably zero. The speed error is caused either by the input disturbance (i.e., the speed reference) or by changes in the load torque T_L. The proportional and integral control actions of the PI controller, discussed in the previous sections, are capable of suppressing a range of input and load disturbances. In the presence of step disturbances $\omega(t) = \Omega^* h(t)$ and $T_L(t) = T_{LOAD} h(t)$, the PI speed controller provides the steady-state output without an error ($\omega(t) = \Omega^*$ in Eq. 2.30). On the other hand, there are disturbances that cannot be suppressed by the PI controller. Some of them result in a finite speed error (Eq. 2.46), while others may prevent the system from reaching the steady state (Eq. 2.47).

Suppression of the speed error can be achieved by selecting the proper structure of the speed controller. Given the reference trajectory $\omega^*(t)$, the structure can be designed to ensure $\Delta\omega(\infty) = 0$. In the same way, for a known load disturbance $T_L(t)$, the controller structure can be devised to suppress the steady-state error. The resulting structures of the speed controller may be different from the ones given in Fig. 2.2 and Fig. 2.7. In Section 2.3.1, the discussion focuses on designing a controller structure compatible with a given speed reference profile. In Section 2.3.2, the structure is devised to suppress the effects of known load disturbances on the steady-state output.

2.3.1 The proper controller structure for the given reference profile

The ability of the PI controller to suppress the speed error while tracking the speed reference profile $\omega^*(t)$ has been investigated in the previous section. The stepwise reference profile is tracked without the steady-state error. With the ramp profile (Eq. 2.46), the speed error is absent only in the case when the friction B is negligible and the proportional action is in the direct path. The PI speed controller is not capable of tracking the parabolic profile, since the tracking error would progressively increase (Eq. 2.47). Given the speed reference profile $\omega^*(t) = Ct^n$, it is evident that the difficulty in tracking the reference increases with n. The Laplace transform of t^n being $n!/s^{n+1}$, a range of speed reference profiles can be expressed in the following form:

$$\omega^*(t) = \sum_{k=0}^{n} C_k t^k h(t), \quad \omega^*(s) = \mathcal{L}^{-1}\left[\omega^*(t)\right] = \sum_{k=0}^{n} C_k \frac{k!}{s^{k+1}}.$$

The speed error in tracking such profiles would depend on n, the maximum value of the exponent k in the expression t^k. Addition of the integral action to the proportional speed controller enabled zero-free reference tracking for $k = 0$ and also for $k = 1$, provided that $B = 0$. It is expected that the speed references comprising t^k with $k > 1$ may require additional controller enhancements. At this point, it is of interest to determine the structure of the speed controller that would lead to a zero tracking error, with the reference profile having the factor t^k. Considerations given in Sections 2.2.6 and 2.1.3 are generalized below for an arbitrary value of k. Without lack of generality, it is assumed that all speed controller actions are placed into the direct path, as indicated in Fig. 2.9, which presents the block diagram that is the basis for further considerations. In this diagram, it is assumed that the load torque disturbance $T_L(t)$ is absent. In the next step, it is demonstrated that the reference tracking capability of speed-controlled systems can be judged from the properties of the open-loop transfer function $W_S(s)$.

The open-loop transfer function $W_S(s)$ is rearranged as $K{\cdot}num(s)/den(s)/s^R$, where K represents the gain, R is the exponent of the factor s^R in the denominator, *num*(*s*) is the polynomial in the numerator and *den*(*s*) is the polynomial in the denominator. It is assumed that the latter two can be expressed as

$$num(s) = 1 + \sum_{k=1}^{m} a_k s^k, \quad den(s) = 1 + \sum_{k=1}^{n} b_k s^k,$$

namely, that their coefficients with s^0 are equal to one. The exponent R is known as the *astatism* [1, 3]. As an example, the open-loop transfer function of the system with PI controller, as given in Eq. 2.26, can be rearranged as

$$W_S(s) = W_{SC}(s)W_P(s) = \frac{sK_P + K_I}{s(Js+B)} = \frac{(K_I/B)}{s}\frac{1+s(K_P/K_I)}{1+s(J/B)} \tag{2.48}$$

and it has the astatism $R = 1$. With an astatism of 1, the system is capable of tracking the reference $\omega^*(s) = 1/s^R = 1/s$, namely, the step reference, as already proven in Eq. 2.45. Below, we will show that zero-free tracking of the reference $\omega^*(s) = 1/s^R$ requires a transfer function with astatism R.

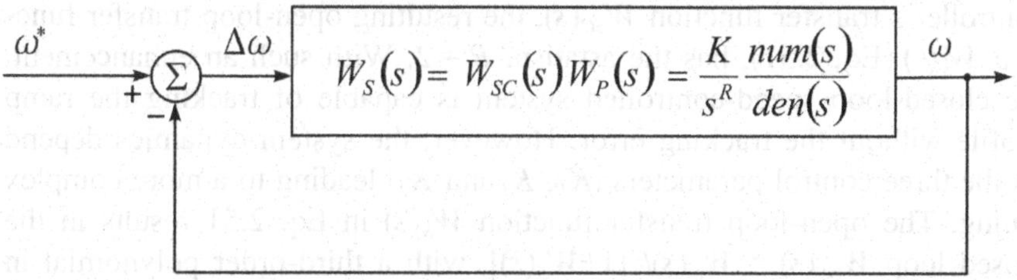

Fig. 2.9. The open-loop transfer function $W_S(s)$ embodies the speed controller $W_{SC}(s)$ and the transfer function $W_P(s)$ of the mechanical subsystem (load).

For the system in Fig. 2.9, the complex image of the speed error is given in Eq. 2.49. If we consider a reference profile $\omega^*(s) = K^*/s^{RR}$ applied to the system with the astatism *RS*, the complex image of the steady-state tracking error is derived and given in Eq. 2.50. With $s \to 0$, both polynomials become $num(s) = 1$ and $den(s) = 1$. Therefore, $\Delta\omega(\infty)$ is obtained only with $RS \geq RR$. Hence, the astatism of the system must be equal to, or must exceed, the exponent R in $\omega^*(s) = 1/s^R$.

$$\Delta\omega(s) = \omega^*(s)\frac{1}{1+W_S(s)} = \omega^*(s)\frac{s^R den(s)}{s^R den(s) + K\, num(s)} \tag{2.49}$$

$$\Delta\omega(\infty) = \lim_{s\to 0}\left(s\,\Delta\omega(s)\right) = \lim_{s\to 0}\left(s\,\frac{K^*}{s^{RR}}\,\frac{s^{RS}\,den(s)}{s^{RS}\,den(s) + K\,num(s)}\right) \tag{2.50}$$

The required astatism for the step, ramp, and parabolic reference profile is summarized in Table 2.1. It is of interest to confirm that $R_{min} = 2$ is required to track the ramp reference. By examining the leftmost expression in Eq. 2.46, it is found that the speed controller, given in Fig. 2.2, tracks the reference profile $\omega^*(s) = A^*/s^2$ with the tracking error $\Delta\omega(\infty) = A^*B/K_I$. If we check the open-loop transfer function $W_S(s)$ (Eq. 2.48) for $B > 0$, we find that it has the astatism $RS = 1 < R_{min} = 2$. On the other hand, without the mechanical friction ($B = 0$), $W_S(s)$ assumes the astatism $RS = 2$, and the tracking error $\Delta\omega(\infty) = A^*B/K_I$ becomes zero.

When required, the necessary astatism can be achieved by enhancing the speed controller's structure with the necessary control actions. Additional control actions are of the form of K_R/s^R, where R is the desired astatism. To demonstrate this, the open-loop transfer function $W_S(s)$ is considered (Eq. 2.48), having initially $R = 1$. If we add the control action K_2/s^2 to the speed controller's transfer function $W_{SC}(s)$, the resulting open-loop transfer function $W_S(s)$ (Eq. 2.51), has the astatism $R = 2$. With such an enhancement, the closed-loop speed-controlled system is capable of tracking the ramp profile without the tracking error. However, the system dynamics depend on the three control parameters (K_P, K_I and K_2) leading to a more complex tuning. The open-loop transfer function $W_S(s)$ in Eq. 2.51 results in the closed loop $W_{SS}(s) = W_S(s)/\,[1+W_S(s)]$, with a third-order polynomial in the denominator (i.e., a third-order characteristic polynomial). Hence, there are three closed-loop poles determining the step response character.

$$W_S(s) = W_{SC}(s)W_P(s) = \left(K_P + \frac{K_I}{s} + \frac{K_2}{s^2}\right)\left(\frac{1}{Js + B}\right)$$
$$= \frac{(K_2/B)}{s^2}\,\frac{1 + s(K_I/K_2) + s^2(K_P/K_2)}{1 + s(J/B)} \tag{2.51}$$

The results obtained in Eqs. 2.49–2.50, and those summarized in Table 2.1, can be used to determine the structure of a speed controller capable of tracking specific speed reference profiles. With the profile $\omega^*(t)$ comprising the element t^n, the complex image $\omega^*(s)$ has the factor $n!/\,s^{n+1}$. According to Eq. 2.50, for the tracking error $\Delta\omega(\infty) = 0$ to be achieved, the speed controller $W_{SC}(s)$ needs to be extended with the control action K/s^{n+1}. Hence, the controller structure must comprise the complex image of the input disturbance $\omega^*(s)$, and then the corresponding profile $\omega^*(t)$ will be tracked with no error. However, the tracking capability comes at a cost. The introduction of the elements K/s^{n+1} into the transfer function $W_{SC}(s)$ makes stability assurance and parameter tuning troublesome. The problem complexity increases with exponent n. Therefore, it is preferable to use the profiles $\omega^*(t)$ which demand the smallest possible value of n.

Table 2.1. The minimum astatism R_{min} of the speed controller required to achieve the speed error of $\Delta\omega = 0$ in tracking the step, ramp and parabolic references.

The speed reference profile	Complex image of the speed error $\Delta\omega$	R_{min}
$\omega^*(t)=\Omega^* h(t)$ $\quad \omega^*(s)=\frac{\Omega^*}{s}$	$\Delta\omega(s)=\frac{\Omega^*}{s}\frac{s^R}{s^R+K\frac{num(s)}{den(s)}}$	1
$\omega^*(t)=\mathrm{A}^* t\, h(t)$ $\quad \omega^*(s)=\frac{\mathrm{A}^*}{s^2}$	$\Delta\omega(s)=\frac{\mathrm{A}^*}{s^2}\frac{s^R}{s^R+K\frac{num(s)}{den(s)}}$	2
$\omega^*(t)=\Upsilon^* t^2 h(t)$ $\quad \omega^*(s)=\frac{\Upsilon^*}{s^3}$	$\Delta\omega(s)=\frac{\Upsilon^*}{s^3}\frac{s^R}{s^R+K\frac{num(s)}{den(s)}}$	3

A range of speed reference profiles can be expressed in the form of a weighted sum of elements t^k, with k ranging from 0 to n:

$$\omega^*(t) = \sum_{k=0}^{n} C_k t^k \tag{2.52}$$

It can be shown that the value of n corresponds to the order of the highest nonzero time derivative of $\omega^*(t)$. The time derivative of the function in Eq. 2.52 is given in Eq. 2.53 for an arbitrary order p:

$$\frac{d^p \omega^*}{dt^p} = \sum_{k=p}^{n} C_k \frac{n!}{(n-p)!} t^{k-p}. \tag{2.53}$$

From this expression, it is found that the nth-order time derivative of the same function is a constant (Eq. 2.54), and the derivative $n + 1$ is equal to zero:

$$\frac{d^n \omega^*}{dt^n} = C_n n!, \quad \frac{d^{n+1} \omega^*}{dt^{n+1}} = 0. \tag{2.54}$$

Given Eq. 2.54, the speed reference profile $\omega^*(t)$ with $d^k\omega^*/dt^k \equiv 0$ for each $k > n$ can be successfully tracked with a speed controller comprising a weighted sum of control actions $1/s^p$, with p ranging from 0 to $n + 1$ (Eq. 2.55). Hence, the speed reference profiles with an abundance of nonzero higher-order time derivatives are more difficult to track, as they call for a more complex $W_{SC}(s)$:

$$W_{SC}(s) = \sum_{k=0}^{n+1} \frac{K_k}{s^k} = K_0 + \frac{K_1}{s} + \frac{K_2}{s^2} + \ldots + \frac{K_{n+1}}{s^{n+1}}. \tag{2.55}$$

Moreover, not all reference profiles have a limited number of nonzero time derivatives. A sinusoidal profile $\omega^*(t) = \sin(2\pi f t)$ has an infinite number of higher-order time derivatives. A speed controller with an infinite number of control actions of the form $1/s^p$ cannot be implemented in practice due to stability problems.

With larger values of n, the speed controller $W_{SC}(s)$ in Eq. 2.55 results in *conditional* stability: in conditions where the initial gain setting (K_0 .. K_{n+1}) results in a stable response, the reduction of certain gains K_k brings the system into instability. On the other hand, an *absolutely* stable system preserves stability in cases when the gains are decreased. The proportional speed controller (Fig. 2.1) and PI controller (Fig. 2.2) are *absolutely* stable. The relevant closed-loop transfer functions $W_{SS}(s)$ in Eq. 2.6 and Eq. 2.27 retain stable poles, even in cases when the gains K_P and K_I are reduced with respect to the initial setting.

In most motion-control applications, the astatism is $R \leq 2$. Therefore, while tracking the profile $\omega^*(t) = \sin(2\pi f t)$, the speed-controlled system will exhibit the speed error $\Delta\omega \neq 0$. This problem is addressed in the following section.

2.3.2 Internal Model Principle (IMP)

A simplified block diagram of the speed-controlled system is given in Fig. 2.10. The torque actuator is assumed as ideal ($T_{em} \equiv T_{ref}$), the delay in feedback acquisition is neglected ($\omega_{fb} = \omega$), and nonlinear aspects such as the torque limit are also neglected. It is assumed that both the speed controller and control object are linear, expressed in terms of their transfer functions $W_{SC}(s) = N_{SC}(s)/D_{SC}(s)$ and $W_P(s) = N_P(s)/D_P(s)$, respectively. In cases where the control object is an inertial load with friction $B\omega$, $N_P(s) = 1$ and $D_P(s) = Js + B$. It is of interest to design the speed controller polynomials $N_{SC}(s)$ and $D_{SC}(s)$ in such a way that the load disturbance $T_L(s)$ does not produce the steady-state error $\Delta\omega(\infty)$.

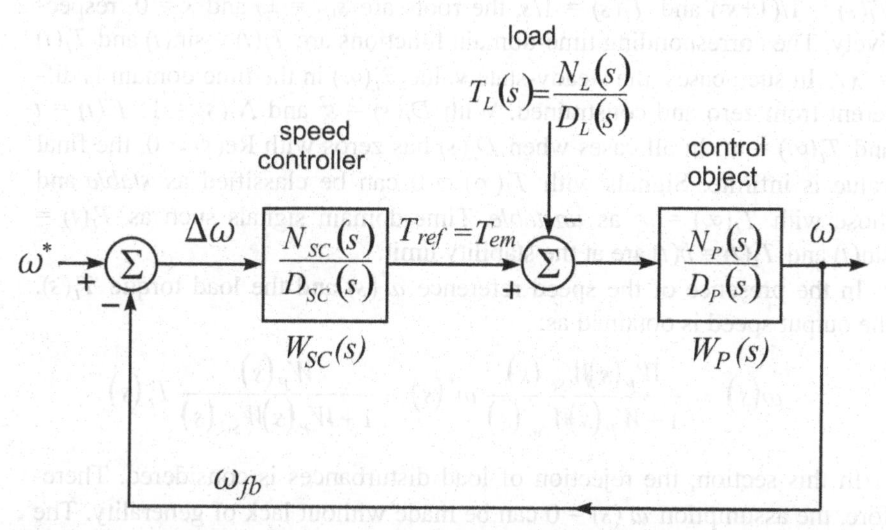

Fig. 2.10. A simplified block diagram of the speed-controlled system, with the load disturbance $T_L(s)$ expressed in terms of polynomials $N_L(s)$ and $D_L(s)$.

The complex image of the given load disturbance $T_L(t)$ can be obtained by the Laplace transform. It is assumed that $T_L(t) = 0$ for $t < 0$. The complex image $T_L(s)$ can be expressed in terms of the numerator $N_L(s)$ and denominator $D_L(s)$:

$$T_L(s) = \mathscr{L}\left[T_L(t)\right] = \frac{N_L(s)}{D_L(s)}.$$

With $T_L(t > 0) = T_L(\infty) = 1$, one obtains $T_L(s) = 1/s$, with $N_L(s) = 1$ and $D_L(s) = s + 0$. In case of a ramp signal, $T_L(t) = t$, $N_L(s) = 1$ and $D_L(s) = s^2 + 0 \cdot s + 0$.

The speed controller $W_{SC}(s)$ has to remove the impact of $T_L(t)$ on the steady-state value of the output $\omega(\infty)$. Hence, the design of the polynomials $N_{SC}(s)$ and $D_{SC}(s)$ has to consider the complex image $T_L(s)$ and the polynomials $N_L(s)$ and $D_L(s)$. The disturbances $T_L(t)$ can be evaluated based upon the roots of $D_L(s) = 0$.

It is interesting to note that the zeros of denominator $D_L(s)$ determine the final value $T_L(\infty) = 0$ of the signal in the time domain. The complex image $T_L(s)$, with all denominator zeros in the left half of the s-plane (Re(s) < 0), corresponds to $T_L(t)$, which decays in time and reaches $T_L(\infty) = 0$. An example of this is $T_L(s) = 1/(1+s)$, resulting in $T_L(t) = \exp(-t)$.

The equation $D_L(s) = 0$ may have its roots on the imaginary axis. With $T_L(s) = 1/(1+s^2)$ and $T_L(s) = 1/s$, the roots are $s_{1/2} = \pm j$ and $s = 0$, respectively. The corresponding time-domain functions are $T_L(t) = \sin(t)$ and $T_L(t) = h(t)$. In such cases, the steady-state value $T_L(\infty)$ in the time domain is different from zero and constrained. With $D_L(s) = s^2$ and $N_L(s) = 1$, $T_L(t) = t$ and $T_L(\infty) = \infty$. In all cases when $D_L(s)$ has zeros with Re(s) > 0, the final value is infinite. Signals with $T_L(\infty) = 0$ can be classified as *stable* and those with $T_L(\infty) = \infty$ as *unstable*. Time-domain signals such as $T_L(t) = \sin(t)$ and $T_L(t) = h(t)$ are at the stability limit.

In the presence of the speed reference $\omega^*(s)$ and the load torque $T_L(s)$, the output speed is obtained as:

$$\omega(s) = \frac{W_P(s)W_{SC}(s)}{1+W_P(s)W_{SC}(s)}\,\omega^*(s) - \frac{W_P(s)}{1+W_P(s)W_{SC}(s)}\,T_L(s).$$

In this section, the rejection of load disturbances is considered. Therefore, the assumption $\omega^*(s) = 0$ can be made without lack of generality. The output speed is expressed in terms of the polynomials $D_L(s)$, $D_{SC}(s)$, $D_P(s)$, $N_L(s)$, $N_{SC}(s)$, and $N_P(s)$:

$$\omega(s) = -\frac{\dfrac{N_P(s)}{D_P(s)}}{1+\dfrac{N_P(s)\,N_{SC}(s)}{D_P(s)\,D_{SC}(s)}}\frac{N_L(s)}{D_L(s)}$$

$$= -\frac{D_{SC}(s)N_P(s)}{D_{SC}(s)D_P(s)+N_P(s)N_{SC}(s)}\frac{N_L(s)}{D_L(s)}.$$

In order to achieve $\omega(\infty) = \omega^*(\infty) = 0$, the design of the polynomials D_{SC} and N_{SC} must ensure that all the zeros of denominator $(D_{SC}D_P+ N_{SC}N_P)D_L$ are stable ($\mathrm{Re}(s) < 0$). In cases when the load disturbance signal is unstable, this condition is not guaranteed. Namely, with $T_L(t) = \sin(t)$, the zeros $s_{1/2} = \pm j$ of the polynomial $D_L(s) = 1+s^2$ become poles of the complex image $\omega(s)$ and result in $\omega(\infty) \neq 0$.

Undesired zeros of the polynomial $D_L(s)$ are canceled, in cases when the numerator $N_L(s)N_P(s)D_{SC}(s)$ comprises the same zeros. The polynomials N_L and N_P are defined by the disturbance signal $T_L(t)$ and the control object dynamics, respectively. On the other hand, the polynomial $D_{SC}(s)$ can be designed to include $D_L(s)$ zeros, which have to be canceled. In such cases, the load disturbance has no effect on the steady-state value of the output speed ($\omega(\infty) = \omega^*$).

The above conclusions show that the speed controller $W_{SC}(s)$ must comprise an integral action K_I/s in order to reject the load disturbance $T_L(s) = T_{LOAD}/s$. According to Eq. 2.13, the proportional speed controller exhibits a finite steady-state error $\Delta\omega = T_{LOAD}/K_P$. Extending the speed controller with the integral action, $W_{SC}(s)$ is obtained with $N_{SC}(s) = sK_P+ K_I$ and $D_{SC}(s) = s$. Now, $D_{SC}(s)$ cancels $D_L(s) = s$. The result $\omega(\infty) = \omega^*$ is confirmed in Eq. 2.30, obtained in Section 2.2.2, where the load rejection capability of the PI controller is discussed and analyzed.

The need for the speed controller $W_{SC}(s)$ to comprise the load disturbance model $D_L(s)$ in its own denominator $D_{SC}(s)$ is known as the *Internal Model Principle* (IMP) [1].

Note that the speed controller itself cannot contain unstable poles. Therefore, zeros such as $s_{1/2} = \pm j$ cannot be part of the polynomial $D_{SC}(s)$. Hence, load disturbances such as $T_L(s) = T_{LOAD}\sin(t)$ cannot be rejected in the prescribed way. In applications when the load disturbance is known in advance, it can be compensated for by the feedforward control action, discussed in the following section.

2.4 Feedforward compensation

The previous section has shown that certain load torque disturbances, such as $T_L(s) = T_{LOAD}\sin(\omega t)$, cannot be rejected with the speed controller, where the driving torque is generated as $T_{em}(s) = W_{SC}(s)\cdot\Delta\omega(s)$. In Section 2.3.1, it was demonstrated that such a controller cannot track an arbitrary speed reference profile $\omega^*(t)$ without an error $\Delta\omega = \omega^* - \omega$. In this section, the use of feedforward control is considered as a solution capable of suppressing the speed error in cases with known disturbances.

The control structures where the driving force is calculated from the error in the system output are frequently referred to as *feedback controllers.* Such are the systems given in Figs. 2.1, 2.2 and 2.9. The speed controller transfer function $W_{SC}(s)$ can be written as in Eq. 2.55, where the gains K_0 and K_1 correspond to the proportional and integral gains, while the elements K_k/s^k are added to enable error-free tracking of the speed reference profiles comprising t^{k-1}. The implementation of Eq. 2.55 is limited to a rather small number of elements (n), due to practical problems in tuning and stability assurance for $n > 1$. Therefore, zero-free tracking of a number of references, including the sinusoidal profile, is not feasible in cases where the feedback controller is used, with $T_{em} = W_{SC}(s)\cdot\Delta\omega$.

The existence of the error $\Delta\omega = \omega^* - \omega$, while tracking a perpetually changing reference $\omega^*(t)$, follows from the fact that the driving torque $T_{em}(s)$ comes as the product of the speed error and speed controller transfer function ($W_{SC}(s)\cdot\Delta\omega$). In order to provide a dynamically changing T_{em}, required to track the profile $\omega^*(t)$, the speed controller needs an input excitation, making the speed error $\Delta\omega \neq 0$ inevitable. Only in cases when the speed controller structure comprises the complex image of the input disturbance $\omega^*(s)$ can the block $W_{SC}(s)$ generate the desired driving force T_{em} while preserving $\Delta\omega = 0$. As shown in the previous sections, the integral control action ($1/s$) enables zero-free tracking of the step reference (Ω^*/s), while the addition of the control action $1/s^2$ permits the use of the ramp-shaped reference (A^*/s^2). The sinusoidal reference, with $\mathcal{L}(\sin(t)) = 1/(1+s^2)$, and many other relevant reference profiles have complex images $\omega^*(s)$ that cannot be built into the speed controller $W_{SC}(s)$. Therefore, other means need to be envisaged to provide zero-free reference tracking in these cases.

If we consider Eq. 2.56, the task of the speed controller is to provide the driving torque T_{em}, required for the output speed $\omega(t)$ to track the given reference $\omega^*(t)$. In cases where the reference profile $\omega^*(t)$ is known in advance, with the load torque $T_L = 0$ and the initial condition $\omega(0) = \omega^*(0)$, the driving torque required to preserve the error-free tracking $\omega(t) = \omega^*(t)$ for $t > 0$ is given in Eq. 2.57.

$$J\frac{\mathrm{d}\omega}{\mathrm{d}t} = T_{em} - T_L - B\omega \tag{2.56}$$

$$T_{FW} = \hat{J}\frac{\mathrm{d}\omega^*}{\mathrm{d}t} + \hat{B}\omega^* \tag{2.57}$$

In Eq. 2.57, the symbols $\hat{J}$ and $\hat{B}$ are used to indicate the possible discrepancy between the parameters used to calculate the torque T_{FW} and the actual parameters J and B of the mechanical load. For the case where $T_L = 0$ and $\omega(0) = \omega^*(0)$, with the driving torque obtained from Eq. 2.57, and the correct values of J and B assumed to be available, the output speed is calculated from Eq. 2.58. Under these assumptions, the error-free tracking $\omega(t) = \omega^*(t)$ is achieved without using feedback. In 2.57, the driving torque $T_{em} = T_{FW}$ is obtained from the reference $\omega^*(t)$, which requires neither the feedback $\omega(t)$ nor the output error signal $\Delta\omega(t)$.

While the speed control structures in Figs. 2.1 and 2.2 belong to the feedback controllers, the control law given in Eq. 2.57 pertains to feedforward control actions, due to the fact that the input reference $\omega^*(t)$ is processed in Eq. 2.57 and fed forward to the plant input.

$$\omega(t) = \frac{1}{J}\int_0^t \left(\hat{B}\omega(\tau)^* - B\omega(\tau) + \hat{J}\frac{\mathrm{d}\omega(\tau)^*}{\mathrm{d}\tau} \right)\mathrm{d}\tau = \omega^*(t) \tag{2.58}$$

In cases where the load torque is present ($T_L \neq 0$), the driving torque $T_{em} = T_{FW}$ cannot ensure $\omega(t) = \omega^*(t)$. On the other hand, whenever the load torque $T_L(t)$ is known, the driving torque can be calculated as $T_{em} = T_{FW} + T_{FL}$ (Eq. 2.59), where T_{FW} is the feedforward control action calculated from Eq. 2.57, while T_{FL} is the predicted load torque T_L. Given that the driving torque T_{em} equals $T_{FW} + T_{FL}$, the output speed can be calculated from Eq. 2.60. With $\omega(0) = \omega^*(0)$ and the correct values of J and B assumed to be available, the control given in Eq. 2.60 insures a correct error-free tracking of the reference profile (i.e., $\omega(t) = \omega^*(t)$ for $t > 0$).

$$T_{em} = T_{FF} = T_{FW} + T_{FL} = \hat{J}\frac{\mathrm{d}\omega^*}{\mathrm{d}t} + \hat{B}\omega^* + T_{FL} \tag{2.59}$$

$$\omega(t)=\frac{1}{J}\int_0^t\left(T_{FL}(\tau)-T_L(\tau)+\hat{B}\omega^*(\tau)-B\omega(\tau)+\hat{J}\frac{d\omega^*(\tau)}{d\tau}\right)d\tau \quad (2.60)$$
$$=\omega^*(t)$$

When the speed control system is driven in a feedforward manner, several circumstances create the tracking error, and these can be inferred from Eq. 2.60. When the inertia parameter J has an erroneous value, the output speed will track a scaled reference, $\omega(t)=\omega^*(t)\cdot(1+\Delta J/J)$. An error in friction ΔB produces the tracking error that builds up in proportion to ΔB and $\omega(t)$ (Eq. 2.60). According to the same expression, any error in the load torque estimate $\Delta T_{FL}=T_{FL}-T_L$ generates the error $\Delta\omega$ proportional to the integral of ΔT_{FL}. In order to further probe the effects of parameter mismatch, the feedforward controller is modeled and simulated. A Simulink model of the feedforward controller is given in Fig. 2.11 and is contained in the file Fig2_11.mdl. The model implements Eq. 2.59, with the estimates of the load torque, the friction coefficient, and the inertia labeled T_{FL}, B_{FF}, and J_{FF}, respectively. The speed reference profile, supplied on the left side in Fig. 2.11, has a trapezoidal form.

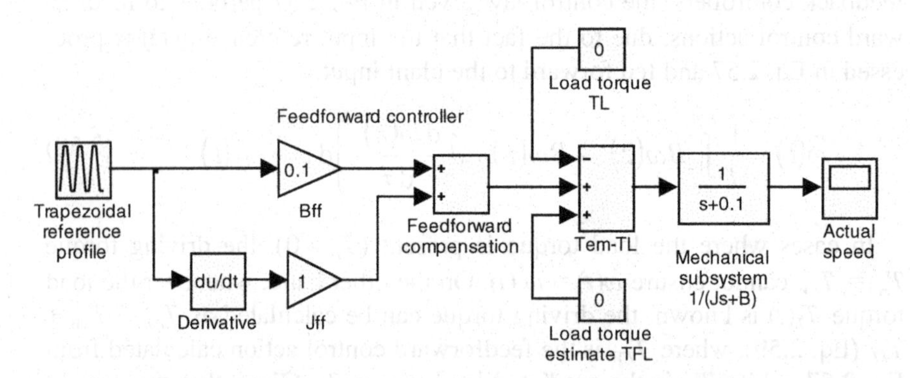

Fig. 2.11. Simulink model of the feedforward controller, given in Eq. 2.59. The output speed is controlled with no feedback. Parameters *Bff*, *Jff*, and the load torque estimate are varied, to give an insight into consequential tracking errors.

Simulation traces obtained from the model in Fig. 2.11 are given in Fig. 2.12. With the proper setting of the friction and inertia and an accurate estimate of the load torque, the output speed tracks the trapezoidal reference with no error (trace A in Fig. 2.12). With an error in inertia J (trace B), the output speed tracks the reference changes with a time delay, converging gradually towards the set speed during the intervals when $\omega^*(t)$ = const. The erroneous friction coefficient (trace C) contributes to significant discrepancies $\omega^*(t) - \omega(t)$, while the error in the load torque estimate T_{FL} (trace D) makes the output speed $\omega(t)$ drift away from the reference profile $\omega^*(t)$.

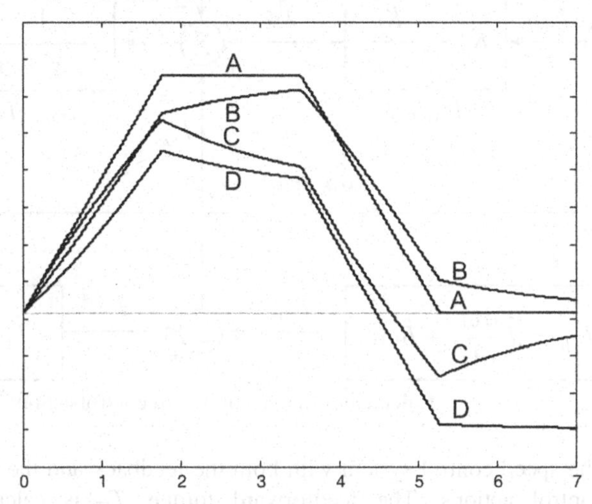

Fig. 2.12. Simulation traces of the output speed obtained from the model in Fig. 2.11. A) Parameters are properly set and the output speed tracks the trapezoidal reference with no error. B) Incorrect inertia setting, Jff = 0.75 J. C) Incorrect setting of the friction, Bff = 0.5 B. D) An error in the load torque estimate, $T_L - T_{FL}$ = 0.05.

In a practical servo system, the friction coefficient depends on the wear on transmission elements, while the inertia is influenced by the weight of the work parts and tools. In complex motion-control systems, where the dynamics of several axes are coupled, the equivalent inertia of one axis depends on the motion of the other axes, making the problem of inertia prediction more difficult. Moreover, the load torque is rarely predictable, since, in most cases, it comprises both deterministic and stochastic disturbances. Therefore, the feedforward controller, summarized in Eq. 2.59, can hardly be used in practice.

The feedback controller and feedforward control action can be beneficially combined to improve the reference tracking capability of the speed-controlled system. With the feedforward control action T_{FF}, obtained from Eq. 2.59, and the feedback controller torque command T_{SC}, obtained from the PI speed controller (Fig. 2.2), the driving torque can be calculated from Eq. 2.61. The speed control system with both the feedback and feed-forward controller is given in Fig. 2.13.

$$T_{em} = T_{FF} + T_{SC} = \left(T_{FW} + T_{FL}\right) + T_{SC} \tag{2.61}$$

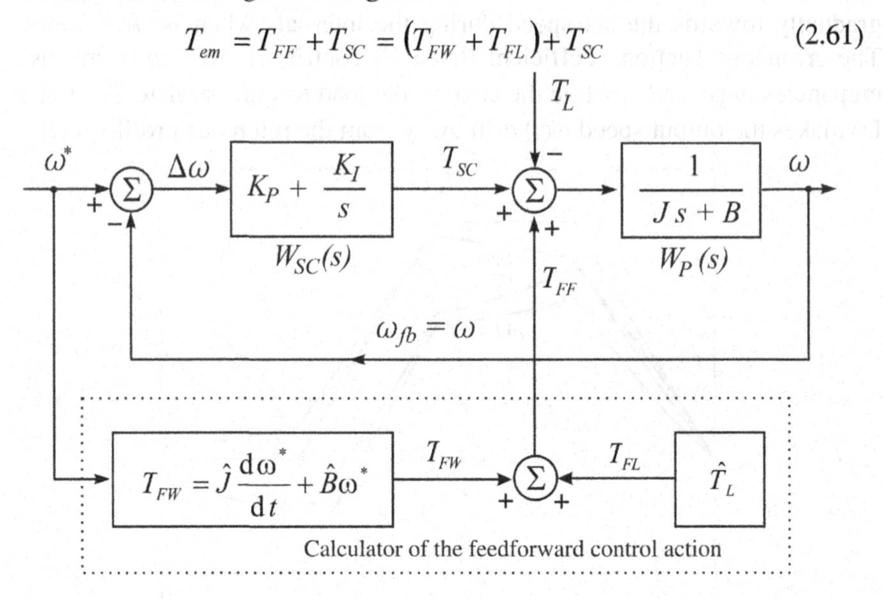

Fig. 2.13. The speed control system with both the feedback and the feedforward control actions. The feedforward torque T_{FF} is calculated from Eq. 2.59.

A brief analysis of the structure in Fig. 2.13 will help in understanding the benefit of using paralleled feedback and feedforward control actions. Consider the case of an ideal feedforward compensation (trace A in Fig. 2.12, with an accurate load torque estimate and exact matching of J and B parameters). In this case, the feedforward torque T_{FF}, alone, will ensure the proper tracking $\omega^*(t) = \omega(t)$, and the resulting speed error $\Delta\omega(t)$ will be zero. Therefore, the torque T_{SC} in Fig. 2.13 will be zero as well. Hence, the addition of the feedback block $W_{SC}(s)$ into the speed controller will make no difference, since the signal T_{SC} is zero. The role of the feedback controller becomes evident in cases when the feedforward parameters are mismatched and/or the load torque estimate T_{FL} is not accurate. Whenever the feedforward torque T_{FF} is inaccurate, the tracking error $\Delta\omega(t)$ emerges. As a consequence, the PI speed controller $W_{SC}(s)$ (Fig. 2.13) obtains the excitation signal at the input, and produces the corrective torque T_{SC}. The feedback control action T_{SC} will reduce the tracking error $\Delta\omega(t)$ in transient conditions. Regarding the steady-state operation, when both the input reference $\omega^*(t)$ and the load torque disturbance $T_L(t)$ are constant, the feedback PI controller, alone, is capable of securing $\Delta\omega(t)$ (Eq. 2.30, Table 2.1).

The addition of the feedforward controller contributes to suppressing the tracking error encountered in the speed-controlled system with feedback action while tracking complex reference profiles. According to the analysis given in Section 2.2.7, error-free tracking requires a rather complex structure of the feedback controller. With profiles $\omega^*(t)$ having a large number of nonzero higher-order time derivatives, the required $W_{SC}(s)$ turns out to be difficult to implement (Eq. 2.55). Practicable feedback controllers can ensure zero-free tracking of the step reference and the ramp-shaped reference (Table 2.1). The proper tracking of the parabolic reference calls for a triple integrator ($1/s^3$) to be part of the $W_{SC}(s)$, aggravating the problems of parameter setting and stability assurance. In practice, the other speed references $\omega^*(t)$, and in particular the sinusoidal profiles, cannot be tracked without an error unless the feedforward action is added to the speed control structure. Ideally, a properly set feedforward controller would drive the tracking error down to zero for an arbitrary reference $\omega^*(t)$ and disturbance $T_L(t)$. Parameter mismatches ΔJ and ΔB will cause a residual tracking error. A set of computer simulations of the structure in Fig. 2.13 has been carried out, in order to investigate the suppression of the tracking error $\Delta\omega$, achieved with a practicable imperfect T_{FF} control action and the PI feedback controller.

The speed control system comprising both the feedback and feedforward control actions (Fig. 2.13) has been modeled in Simulink. The model is contained in the file Fig2_14.mdl, and shown in Fig. 2.14. It comprises the generator of the trapezoidal reference profile (left), the mechanical subsystem with friction $B = 0.1$ and inertia $J = 1$ (right), the proportional-integral feedback controller (top), and the feedforward compensator (bottom). The simulation traces are given in Fig. 2.15.

The simulation traces labeled A in Fig. 2.15 represent the trapezoidal reference profile and the output speed, in the case when the feedforward controller is disconnected and the driving torque is obtained from the feedback controller only ($T_{em} = T_{SC}$). As anticipated in Eq. 2.46, the ramp shaped reference is tracked with an error. When an accurate feedforward compensation T_{FF} is added to the driving torque (Eq. 2.61), with the T_{FF} signal obtained from Eq. 2.59, the reference profile is tracked with no error. Traces B and C in Fig. 2.15 are obtained with both the feedforward and feedback controllers active. In the case labeled B, it is assumed that the inertia J is set with an error of 25%, while in the case labeled C, the friction parameter is set with an error of 50%.

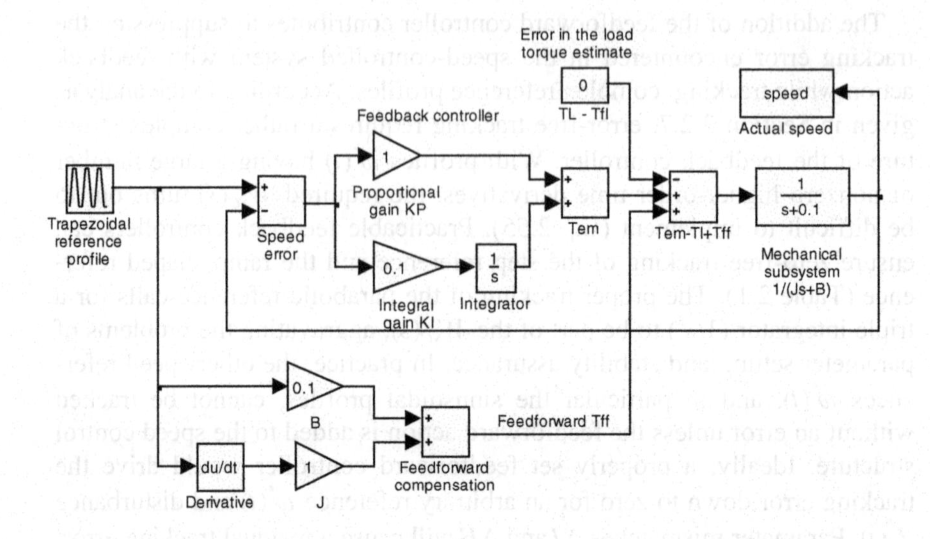

Fig. 2.14. Simulink model of the speed control structure in Fig. 2.13, comprising the feedback control action T_{SC} and the feedforward compensation T_{FF}.

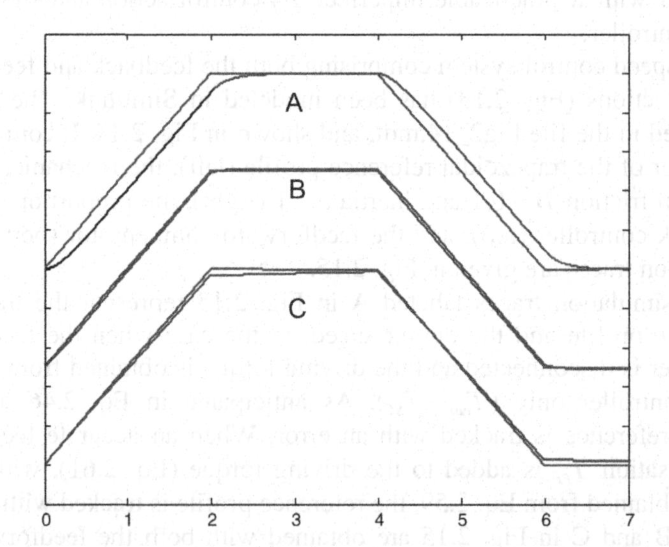

Fig. 2.15. Simulation traces of the output speed obtained from the model in Fig. 2.14. A) The feedforward controller is not activated, $T_{FF} = 0$. Only the feedback controller $W_{SC}(s)$ is active. B) Both the feedforward and the feedback controllers are active, while the inertia setting is incorrect, $Jff = 0.75\ J$. C) Response obtained with an erroneous friction coefficient, $Bff = 0.5\ B$.

It is concluded, from the traces in Fig. 2.15, that the parameter mismatch does introduce a certain tracking error, although this error is much smaller than the one in case A, when the feedforward action is removed altogether. Hence, incorporation of the feedforward compensation T_{FF} into the feedback controller undoubtedly helps to suppress the speed error in tracking intricate speed reference profiles $\omega^*(t)$ with a number of nonzero higher-order time derivatives. In a practical multiaxis motion-control system, the T_{FF} calculation becomes rather involved, as it takes into account cross-coupling effects for a number of concurrently running axes.

Problems

P2.1
The speed-controlled system given in Fig. 2.1 has inertia $J = 0.01$ kgm^2 and proportional gain $K_P = 50$ Nm/(rad/s). Using the Matlab command *step*(), determine the transient response of the output speed to the load step $T_L(t) = T_{LOAD}\, h(t)$, where $T_{LOAD} = 1$ Nm.

P2.2
Consider the speed-controlled system given in P2.1. Under the assumption that the load torque changes as $T_L(t) = T_{RAMP}\, t\, h(t)$, and with $\omega^* = 0$, determine the complex image $T_L(s)$ and the steady-state value of the speed error $\Delta\omega(\infty)$.

P2.3
The speed controller in Fig. 2.2 has the proportional and integral actions in the direct path. The system parameters are $K_M = 1$ Nm/(rad/s), $K_{FB} = 1$, $J = 0.01$ kgm^2, $K_I = 0.1$ Nm/(rad/s), $K_P = 0.03$ Nm/(rad/s), and $B = 0.001$ Nm/(rad/s). Assuming that the load torque changes as $T_L(t) = T_{RAMP}\, t\, h(t)$, and with $\omega^* = 0$, determine the complex image $T_L(s)$ and the steady-state value of the speed error $\Delta\omega(\infty)$.

P2.4
For the system described in P2.3, determine the closed-loop transfer function, the closed-loop zeros and poles, their undamped natural frequency ω_n, and the damping factor ξ.

P2.5(a)
The structure of the PI speed controller is given in Fig. 2.2. It has integral and proportional gains in the direct path. Considering the parameter setting given in P2.3, determine the closed-loop system transfer function, the closed-loop zeros and poles, their undamped natural frequency ω_n, and the damping factor ξ.

P2.5(b)
The structure of the PI speed controller is given in Fig. 2.7. It has the integral gain in the direct path and the proportional gain relocated into the feedback path. Considering the parameter setting given in P2.3, determine the closed-loop system transfer function, the closed-loop zeros and poles, their undamped natural frequency ω_n, and the damping factor ξ.

P2.6
Using the closed-loop system transfer functions obtained in S2.3 and S2.5, and employing the Matlab command *step*(), compare the step responses of the systems in Fig. 2.2 and Fig. 2.7.

P2.7
Given the natural frequency of the closed-loop poles ω_n and their damping $\xi = 1$, determine the closed-loop bandwidth of the system considered in S2.5. The bandwidth frequency f_{BW} is calculated from the condition $|W_{SS}(j2\pi f_{BW})| = 1/\text{sqrt}(2)$.

P2.8
For the speed reference $\omega^*(t) = A^* t$, determine the speed error in the steady-state operating conditions when $T_L = 0$. Consider the speed controller structure in Fig. 2.2, with the proportional action in the direct path, and the structure in Fig. 2.7, with the proportional action repositioned in the feedback path.

P2.9
Consider the speed-controlled system with the control object transfer function $W_P(s) = 1/Js$ and the speed controller $W_{SC}(s) = K_0 + K_1/s + K_2/s^2$. Prove analytically that the system can be brought to instability by reducing the feedback gains.

P2.10
Consider the previous example with $J = 1$, $K_0 = 1$, $K_1 = 2$, and $K_2 = 1$. Use the Matlab function *roots*() to obtain the closed-loop poles. Now reduce the gain K_1 and find the value K_{1MIN} that represents the stability limit.

P2.11

Consider the previous example with $J = 1$, $K_0 = 1$, $K_1 = 2$, and find the maximum value K_{2MAX} for the feedback gain K_2.

P2.12

The output speed of the system is controlled by the feedforward controller given in Fig. 2.11. The driving torque is calculated from the inverse model of the control object. If the load torque T_L and the parameters J and B are known, the torque $T_{em} = J\,d\omega^*/dt + B\omega^* + T_L$ is sufficient to maintain $\omega(t) = \omega^*(t)$. Provided that $\omega(0) = \omega^*(0)$, a properly tuned feedforward controller has the potential to control the output speed without closing the feedback loop. Investigate the transient response of the output speed in the cases when $B = B_{ff} = 0$, $J = J_{ff} = 1$, and the load torque T_L differs from the estimate T_{Lff}. Hint: The Simulink model of the feedforward controller is contained in the model file *P2_12.mdl.* In the model, it is necessary to set $B = B_{ff} = 0$, $J = J_{ff}$, and to extend the simulation *stop time* to 400 units. Introduce $T_L = 0.1$ and $T_{Lff} = 0$, run the model, and observe the changes in the output speed.

P2.13

Consider the speed-controlled system in Fig. 2.14, with the parallel feedforward and feedback control actions. Use the parameter setting defined in P2.12. Investigate the impact of the load torque mismatch $\Delta T_L = T_L - T_{Lff} = 0.1$ in the cases with the feedback gains $KP = 1$ and $KI = 1$. Hint: Use the the Simulink model *P2_13.mdl*

P2.14

Repeat the previous test with $KI = 0$. Repeat the same test with both $KI = 0$ and $KP = 0$.

3 Parameter Setting of Analog Speed Controllers

Practical speed controlled systems comprise delays in the feedback path. Their torque actuators, with intrinsic dynamics, provide the driving torque lagging with respect to the desired torque. Such delays have to be taken into account when designing the structure of the speed controller and setting the control parameters. In this chapter, an insight is given into traditional DC-drives with analog speed controllers, along with practical gain-tuning procedures used in industry, such as the double ratios and symmetrical optimum.

In the previous chapter, the speed controller basics were explained with reference to the system given in Fig. 1.2, assuming an idealized torque actuator ($W_A(s) = 1$). In this chapter, the structure of the speed controller and the parameter settings are discussed for the realistic speed-control systems, including practical torque actuators with their internal dynamics $W_A(s)$. Traditional DC drives with analog controllers are taken as the design example. Delays in torque actuation are derived for the voltage-fed DC drives and for drives comprising the minor loop that controls the armature current. Parameter-setting procedures commonly used in tuning analog speed controllers are reviewed and discussed, including double ratios, symmetrical optimum, and absolute value optimum. The limited bandwidth and performance limits are attributed to the intrinsic limits of analog implementation.

3.1 Delays in torque actuation

The driving torque T_{em}, provided by a DC motor, is proportional to the armature current i_a and to the excitation flux Φ_p. The torque is found as $T_{em} = k_m \Phi_p i_a$, where the coefficient k_m is determined by the number of rotor conductors N_R ($k_m = N_R/2/\pi$). The excitation flux is either constant or slowly varying. Therefore, the desired driving torque T_{ref} is obtained by injecting the current $i_a = T_{ref}/(k_m \Phi_p)$ into the armature winding. Hence, the torque response is directly determined by the bandwidth achieved in controlling the armature current. In cases when the response of the current is faster than the desired speed response by an order of magnitude, neglecting the torque actuator dynamics is justified ($W_A(s) = 1$), and the synthesis of the speed controller can follow the steps outlined in the previous chapter. With reference to traditional DC drives, the current loop response time is moderate. For that reason, delays incurred in the torque actuation are meaningful and the transfer function $W_A(s)$ cannot be neglected.

3.1.1 The DC drive power amplifiers

The armature winding of a DC motor is supplied from the drive power converter. In essence, the drive converter is a power amplifier comprising the semiconductor power switches (such as transistors and thyristors), inductances, and capacitors. It changes the AC voltages obtained from the mains into the voltages and currents required for the DC motor to provide the desired torque T_{em}. In the current controller, the armature voltage u_a is the driving force. The voltage u_a is applied to the armature winding in order to suppress the current error Δi_a and to provide the armature current equal to $T_{ref}/(k_m \Phi_p)$. The rate of change of the torque T_{em} and current i_a are given in Eq. 3.1, where R_a and L_a stand for the armature winding resistance and inductance, respectively; k_m and k_e are the torque and electromotive force coefficients of the DC machine, respectively; Φ_p is the excitation flux; and ω is the rotor speed. Given both polarities and sufficient amplitude of the driving force u_a, it is concluded from Eq. 3.1 that both positive and negative slopes of the controlled variable are feasible under any operating condition. Therefore, any discrepancy in the i_a and T_{em} can be readily corrected by applying the proper armature voltage. The rate of change of the armature current (and, hence, the response time of the torque) is inversely proportional to the inductance L_a. Therefore, for a prompt response of the torque actuator, it is beneficial to have a servo motor with lower values of the winding inductance.

$$\frac{di_a}{dt} = \frac{1}{L_a}\left(u_a - R_a i_a - E\right) = \frac{1}{L_a}\left(u_a - R_a i_a - k_e \Phi_p \omega\right)$$
$$\frac{dT_{em}}{dt} = \frac{k_m \Phi_p}{L_a}\left(u_a - R_a i_a\right) - \frac{k_m k_e \Phi_p^2}{L_a}\omega \tag{3.1}$$

The power converter topologies used in conjunction with DC drives are given in Figs. 3.1–3.3. The thyristor bridge in Fig. 3.1 is line commutated. The firing angle is supplied by the digital drive controller (μP). An appropriate setting of the firing angle allows for a continuous change of the armature voltage. Both positive and negative average values of the voltage u_a

are practicable. With six thyristors in the bridge, the instantaneous value of $u_a(t)$ retains six voltage pulses within each cycle of the mains frequency f_S. Hence, the bridge voltage $u_a(t)$ can be split into the average value, required for the current/torque regulation, and the parasitic AC component, of which the predominant component has the frequency $6f_S$. The AC component of the armature voltage produces the current ripple, inversely proportional to the equivalent series inductance of the armature circuit. Therefore, most traditional thyristorized DC drives make use of an additional inductance installed in series with the armature winding, in order to smooth the $i_a(t)$ waveform. With the topology shown in Fig. 3.1, the current control consists of setting the thyristor firing angle in a manner that contributes to the suppression of the error in the armature current.

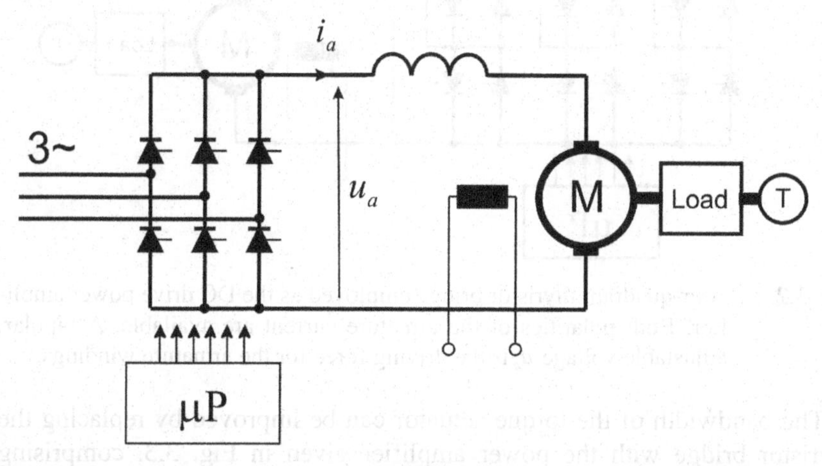

Fig. 3.1. Line-commutated two-quadrant thyristor bridge employed as the DC drive power amplifier. The bridge operates with positive armature currents. The armature winding is supplied with adjustable voltage u_a, controlled by the firing angle. The voltage u_a assumes both positive and negative values.

Each thyristor is fired once within the period $T_S = 1/f_S$ of the mains voltage. Hence, the current controller can effectuate change in the driving force $u_a(t)$ six times per period T_S. In other words, the sampling time of the current controller is $T_S/6$ (2.77 ms or 3.33 ms). A relatively small sampling frequency of practicable current controllers and the presence of an additional series inductance are the main restraining factors for current controllers in thyristorized DC drives. The consequential delays in the torque actuations cannot be neglected and must be taken into account in the speed controller design.

The circuit shown in Fig. 3.1 supplies only positive currents into the armature winding. Therefore, only positive values of the driving torque are feasible. In applications where a thyristorized DC drive is required to supply the torques of both polarities and run the motor in both directions of rotation, it is necessary to devise a power amplifier suitable for the four-quadrant operation. One possibility to supply the four-quadrant DC drive is given in Fig. 3.2.

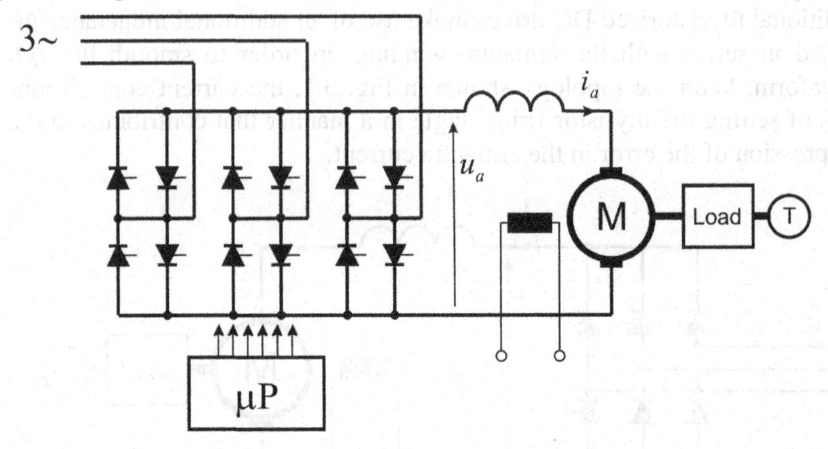

Fig. 3.2. Four-quadrant thyristor bridge employed as the DC drive power amplifier. Both polarities of the armature current are available. A bipolar, adjustable voltage u_a is the driving force for the armature windings.

The bandwidth of the torque actuator can be improved by replacing the thyristor bridge with the power amplifier given in Fig. 3.3, comprising power transistors. While the thyristors (Fig. 3.1) are switched each 2.77 ms (3.33 ms), the switching cycle of the power transistors can go below 100 μs, allowing for a much quicker change in the armature voltage. The transistors Q1–Q4 and the armature winding, placed at the center of the arrangement, constitute the letter *H*. Such an H-bridge is supplied with the DC voltage E_{DC}. The voltage E_{DC} is either rectified mains voltage or the voltage obtained from a battery. The instantaneous value of the armature voltage can be $+E_{DC}$, $-E_{DC}$, or $u_a = 0$. The positive voltage is obtained when Q1 and Q4 are switched on, the negative voltage is secured with Q2 and Q3, and the zero voltage is obtained with either the two upper switches (Q1, Q3) or the two lower switches (Q2, Q4) being turned on. The continuously changing average value (U_{AV}) is obtained by the Pulse Width Modulation (PWM) technique, illustrated at the bottom of Fig. 3.3. Within each period $T_{PWM} = 1/f_{PWM}$, the armature voltage comprises a positive pulse with adjustable width t_{ON} and a negative pulse that completes the period.

The average voltage U_{AV} across the armature winding can be varied in successive T_{PWM} intervals by adjusting the positive pulse width t_{ON}. The PWM pattern can be obtained by comparing the ramp-shaped PWM carrier ($c(t)$ in Fig. 3.3) and the modulating signal $m(t)$.

The pulsed form of the armature voltage obtained from a PWM-controlled H-bridge provides the useful average value $U_{AV}(t_{ON})$ and the parasitic high-frequency component, with most of its spectral energy at the PWM frequency. As a consequence, the armature current will comprise a triangular-shaped current ripple. The PWM frequency $f_{PWM} = 1/T_{PWM}$ can go well beyond 10 kHz. At high PWM frequencies, the motor inductance L_a, alone, is sufficient to suppress the current ripple, and the usage of the external inductance L_m can be avoided. With the H-bridge (Fig. 3.3) being used as the voltage actuator of an armature current controller, the current (torque) response time of several PWM periods can be readily achieved. With T_{PWM} ranging from 50 μS to 100 μS, the resulting dynamics of the torque actuator $W_A(s)$ are negligible, compared with the outer loop transients. Transistorized H-bridges have not been used in traditional DC drives and were made available only upon the introduction of high-frequency power transistors.

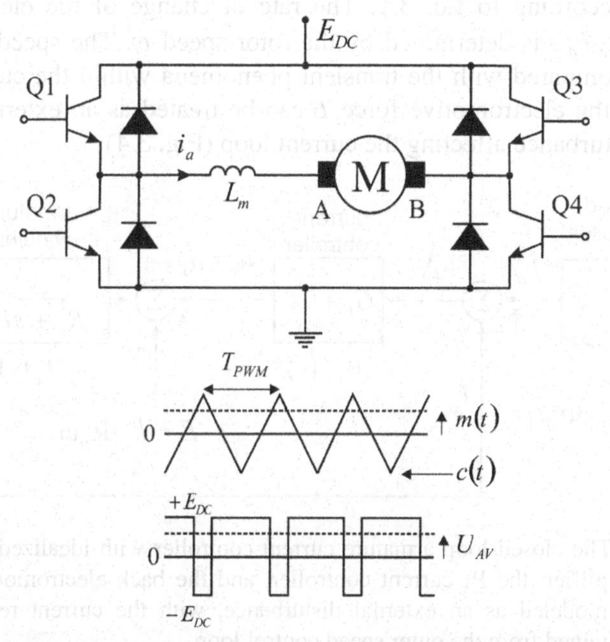

Fig. 3.3. Four-quadrant transistor bridge employed as the DC drive power amplifier. The armature winding is voltage supplied, and both polarities of armature current are available. The average value of the bipolar, adjustable voltage u_a is controlled through the pulse width modulation.

3.1.2 Current controllers

In most traditional DC drives, the driving torque T_{em} is controlled by means of a minor (local) current control loop. The minor loop controls the armature current by adjusting the armature voltage. The power amplifiers used for supplying the adjustable voltage to the armature are outlined in the previous section. The minor current control loop is widely used in contemporary AC drives as well. It is of interest to investigate the current control basics, in order to outline the gain tuning problem and to achieve insight into practicable torque actuator transfer functions.

The simplified block diagram of the armature current controller is given in Fig. 3.4. The current reference i_a^* (on the left in the figure) is obtained from the speed controller $W_{SC}(s)$. With $M_{em} = k_m \Phi_p i_a$, the signal i_a^* is the reference for the driving torque as well. The current controller is assumed to have proportional and integral action, with respective gains denoted by G_P and G_I. Within the drive control structure, the power amplifier feeds the armature winding, with the voltage prescribed by the current controller. In Fig. 3.4, the power amplifier is assumed to be ideal, providing the voltage $u_a(t)$, equal to the reference $u_a^*(t)$ with no delay. The armature current is established according to Eq. 3.1. The rate of change of the electromotive force $E = k_e \Phi_p \omega$ is determined by the rotor speed ω. The speed dynamics are slow, compared with the transient phenomena within the current loop. Therefore, the electromotive force E can be treated as an external, slowly varying disturbance affecting the current loop (Fig. 3.4).

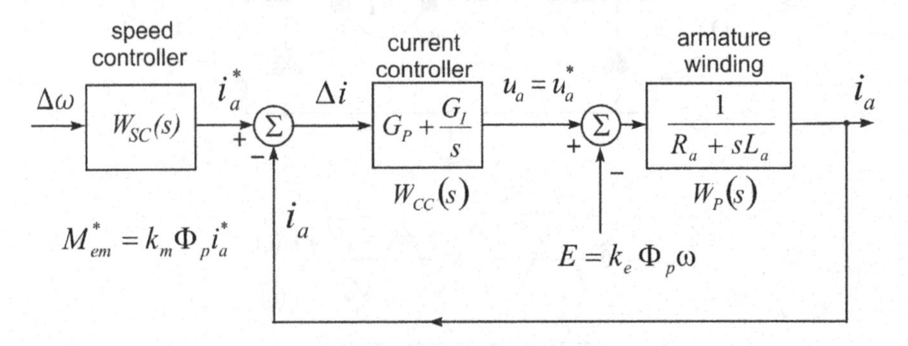

Fig. 3.4. The closed-loop armature current controller with idealized power amplifier, the PI current controller, and the back electromotive force E modeled as an external disturbance, with the current reference obtained from the outer speed control loop.

The analysis of the PI analog current controller is summarized in Eqs. 3.2–3.5. It is based on the assumptions listed in the text above Fig. 3.4.

Minor delays and the intrinsic nonlinearity of the voltage actuator (i.e., the power amplifier) are neglected as well. Practical power amplifiers (Figs. 3.1–3.3) provide the output voltage u_a, limited in amplitude. This situation should be acknowledged by attaching a limiter to the output of the block $W_{CC}(s)$ in Fig. 3.4. At this stage, the analysis is focused on the current loop response to small disturbances. Therefore, nonlinearities originated by the system limits are not taken into account.

The transfer function $W_P(s)$ of the armature winding and the transfer function $W_{CC}(s)$ of the current controller are given in Eq. 3.2. The parameters G_P and G_I are the proportional and integral gains of the PI current controller, respectively. The closed-loop transfer function $W_{SS}(s)$ is derived in Eq. 3.3.

$$W_P(s) = \frac{i_a(s)}{u_a(s) - E(s)} = \frac{1}{R_a + sL_a}, \qquad W_{CC}(s) = \frac{sG_P + G_I}{s} \tag{3.2}$$

$$W_{SS}(s) = \frac{i_a(s)}{i_a^*(s)}\Bigg|_{E=0} = \frac{W_P W_{CC}}{1 + W_P W_{CC}} = \frac{1 + s\dfrac{G_P}{G_I}}{1 + s\dfrac{R_a + G_P}{G_I} + s^2\dfrac{L_a}{G_I}} \tag{3.3}$$

The closed-loop transfer function has one real zero and two poles. The closed-loop poles can be either real or conjugate complex, depending on the selection of the feedback gains. The conjugate complex poles contribute to overshoots in the step response and may result in the armature-current instantaneous value exceeding the rated level. The armature current circulates in power transistors and thyristors within the drive power converter (Figs. 3.1–3.3). The power semiconductors are sensitive to instantaneous current overloads. Therefore, it is good practice to avoid overshoots in the armature current. To this end, the feedback gains G_P and G_I should provide a well-damped step response and preferably real closed-loop poles.

In traditional DC drives, it is common practice to apply feedback gains complying with the relation $G_P/G_I = L_a/R_a$. In this manner, the electrical time constant of the armature winding $\tau_a = L_a/R_a$ becomes equal to $\tau_{CC} = G_P/G_I$ (Eq. 3.4). If we consider $W_{CC}(s)$ in Eq. 3.2, the value τ_{CC} is the time constant corresponding to real zero $z_{CC} = -G_I/G_P$. With $\tau_a = \tau_{CC}$, the zero z_{CC} cancels the $W_P(s)$ pole $p_P = -R_a/L_a$, and the open-loop transfer function $W_S(s) = W_P(s)\ W_{CC}(s)$ reduces to $G_I/(sR_a)$. Consequently, the closed-loop transfer function transforms into the form shown in Eq. 3.5, with only one real pole and no zeros.

$$\tau_a = \frac{L_a}{R_a} = \tau_{CC} = \frac{G_P}{G_I} \tag{3.4}$$

$$W_A(s) = W_{SS}(s) = \frac{i_a(s)}{i_a^*(s)}\bigg|_{E=0} = \frac{1}{1 + s\dfrac{R_a}{G_I}} = \frac{1}{1 + s\tau_{TA}} \tag{3.5}$$

With the parameter setting given in Eq. 3.4 and with the closed-loop transfer function of Eq. 3.5, the transfer function of the torque actuator ($W_A(s)$ in Fig. 1.1) reduces to the first-order lag described by the time constant τ_{TA}. In traditional DC drives, the torque actuator comprises the power amplifier, analog current controller, and separately excited DC motor. In the next sections, the transfer function $W_A(s) = 1/(1+s\tau_{TA})$ is used in considerations related to speed loop-analysis and tuning.

3.1.3 Torque actuation in voltage-controlled DC drives

The torque actuator can be made without the current controller, with the armature winding being voltage supplied. In Fig. 3.5, the speed controller $W_{SC}(s)$ generates the voltage reference u_a^*. Given an ideal power amplifier, the actual armature voltage $u_a(t)$ corresponds to the reference $u_a^*(t)$ without delay. In the absence of the current controller, the armature current $i_a(t)$ is driven by the difference between the supplied voltage and the back electromotive force ($u_a(t) - E(t)$). Since the speed changes are slower compared with the armature current, the electromotive force $E = k_e\Phi_p\omega$ is considered to be an external, slowly varying disturbance. Under these assumptions, the transfer function $W_A(s)$ of the voltage-supplied DC motor, employed as the torque actuator, is given in Eq. 3.6. The transfer function has the static gain $K_M = k_m\Phi_p/R_a$ and one real pole, described by the electrical time constant of the armature winding ($\tau_{TA} = L_a/R_a$).

In the previous section, the transfer function $W_A(s)$ of the torque actuator was investigated for the case when the closed-loop current control is used (Eq. 3.5). In the present section, Eq. 3.6 describes torque generation with voltage-supplied armature winding and no current feedback. In both cases, the function $W_A(s)$ can be approximated with the first-order lag having the time constant τ_{TA}. This conclusion will be used in the subsequent sections in the analysis and tuning of the speed loop.

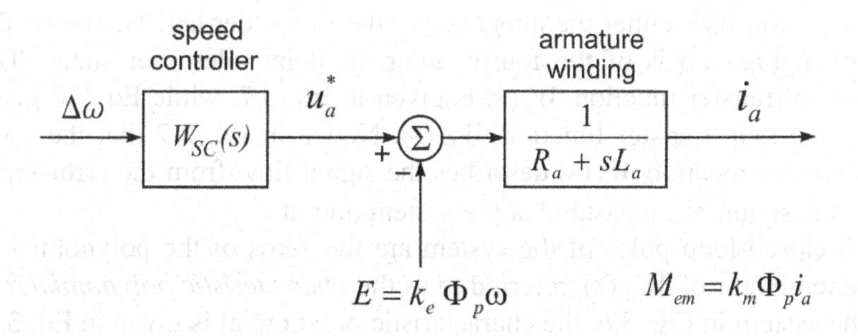

Fig. 3.5. The torque actuation in cases when the speed controller supplies the voltage reference for the armature winding. The current controller is absent, and the actual current $i_a(t)$ depends on the voltage difference $u_a(t) - E(t)$ across the winding impedance $R_a + sL_a$.

$$W_A(s) = \frac{T_{em}(s)}{u_a^*(s)}\bigg|_{E=0} = \frac{k_m \Phi_p}{R_a} \frac{1}{1 + s\dfrac{L_a}{R_a}} = K_M \frac{1}{1 + s\tau_{TA}} \tag{3.6}$$

The transfer function $T_{em}(s)/\Delta\omega(s) = W_{SC}(s)\,W_A(s)$ in Fig. 3.6 can be expressed as $(K_P' + K_I'/s)/(1 + s\tau_{TA})$, where $K_P' = K_P K_M$ and $K_I' = K_I K_M$. Hence, the assumption $K_M = 1$ can be made without lack of generality.

3.2 The impact of secondary dynamics on speed-controlled DC drives

In Fig. 3.6, the speed controlled system employing the DC motor as the torque actuator is shown. The figure includes the secondary phenomena, such as the speed-feedback acquisition dynamics $W_M(s)$ and delays in the torque generation $W_A(s)$. It is assumed that the process of speed acquisition and filtering can be modeled with the first-order lag having the time constant τ_{FB}. The torque actuator is modeled in the previous section (Eqs. 3.5–3.6), with $W_A(s) = 1/(1 + s\tau_{TA})$. It is assumed that the plant $W_P(s)$ is described by the friction coefficient B and equivalent inertia J. The speed controller $W_{SC}(s)$ is assumed to have proportional gain K_P and integral gain K_I.

The presence of four distinct transfer functions within the loop (W_P, W_{SC}, W_M, and W_A) contributes to the complexity of the open-loop and closed-loop transfer functions. Each of the transfer functions W_P, W_{SC}, W_M, and W_A, comprises either the integrator or the first order lag. Therefore, the system in Fig. 3.6 is of the fourth order, as it includes four states. The open-loop transfer function $W_S(s)$ is given in Eq. 3.7, while Eq. 3.8 gives the closed-loop transfer function $W_{SS}(s)$. Notice in Eq. 3.7 that the open loop transfer function $W_S(s)$ describes the signal flow from the error-input $\Delta\omega$ to the signal ω_{fb}, measured at the system output.

The closed-loop poles of the system are the zeros of the polynomial in the denominator of $W_{SS}(s)$, referred to as the *characteristic polynomial* $f(s)$. For the system in Fig. 3.6, the characteristic polynomial is given in Eq. 3.9. The polynomial $f(s)$ is of the fourth order. Therefore, there are four closed-loop poles that determine the character of the closed-loop response. The actual values of the closed-loop poles depend on the polynomial coefficients. The coefficients of $f(s)$ depend on the plant parameters (B, J), time constants (τ_{TA}, τ_{FB}), and feedback gains (K_P, K_I). The plant parameters and time constants are the given properties of the system and cannot be changed. The dynamic behavior of the system can be tuned by adjusting the feedback gains.

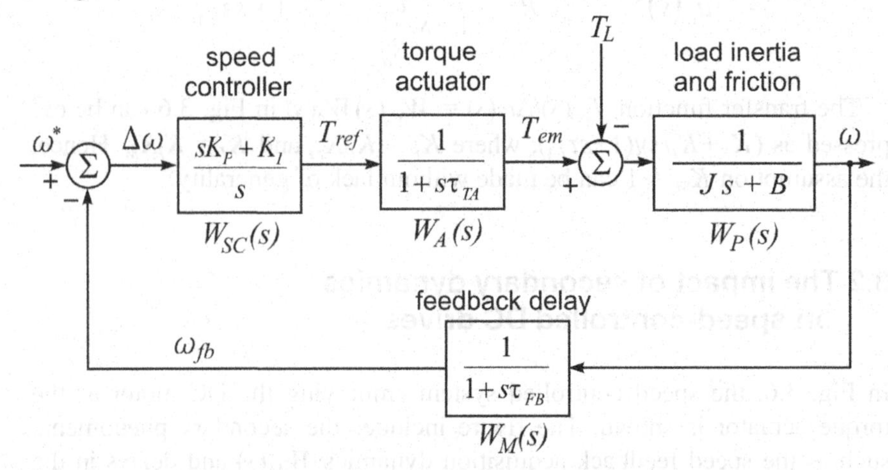

Fig. 3.6. The speed-controlled DC drive system, including the model of secondary dynamic phenomena. The torque generation is modeled as the first-order lag $W_A(s)$. The delays and internal dynamics of feedback acquisition are approximated with the transfer function $W_M(s)$.

$$W_S(s) = W_{SC}(s)W_A(s)W_P(s)W_M(s) \tag{3.7}$$

$$W_{SS}(s) = \frac{W_{SC}(s)W_A(s)W_P(s)}{1+W_{SC}(s)W_A(s)W_P(s)W_M(s)} \tag{3.8}$$

$$f(s) = s^4 + \frac{J(\tau_{TA} + \tau_{FB}) + B\tau_{TA}\tau_{FB}}{J\tau_{TA}\tau_{FB}} s^3 + \frac{B(\tau_{TA} + \tau_{FB}) + J}{J\tau_{TA}\tau_{FB}} s^2 + \frac{B + K_P}{J\tau_{TA}\tau_{FB}} s + \frac{K_I}{J\tau_{TA}\tau_{FB}} \tag{3.9}$$

If we measure that the feedback acquisition system is sufficiently fast, the relevant time constant is presumed to be $\tau_{FB} = 0$, the system reduces to the third order, and the resulting characteristic polynomial is given in

$$f(s) = s^3 + \frac{B\tau_{TA} + J}{J\tau_{TA}} s^2 + \frac{B + K_P}{J\tau_{TA}} s + \frac{K_I}{J\tau_{TA}}. \tag{3.10}$$

In a system of the third order (Eq. 3.10), there are three closed-loop poles and only two adjustable feedback parameters (K_P, K_I). In Eq. 3.9, there are four closed-loop poles (i.e., $f(s)$ zeros) to be tuned by setting the two feedback parameters (K_P, K_I). Under these circumstances, the closed loop cannot be arbitrarily set. An unconstrained placement of the four closed-loop poles requires the state feedback [3], with the driving force being calculated from all four system states. The speed controller transfer function $W_{SC}(s)$ can be enhanced with additional control actions, providing for an implicit state feedback. In such cases, the $W_{SC}(s)$ frequently involves the differentiation of the input signal $\Delta\omega$. Specifically, in order to implement the implicit state feedback, the speed controller in Fig. 3.6 should include the first and the second derivative of the input signal $\Delta\omega$, along with the two associated feedback gains.

Most traditional DC drives do not employ state feedback, nor do they use multiple derivatives within the $W_{SC}(s)$ block. Although the number of relevant closed-loop poles is larger than two, the PI speed controller is commonly used. A number of techniques have been developed and used over the past decades for tuning the PI gains, obtaining a satisfactory placement of multiple poles, and securing a robust, well-damped response. Some of these techniques are discussed in subsequent sections.

3.3 Double ratios and the absolute value optimum

The feedback gains of speed control systems employing traditional DC drives are frequently tuned according to the common design practice called the *double ratios.* The rule is focused on extending the range of frequencies where the amplitude of the closed-loop transfer function remains $|W_{SS}(j\omega)| \approx 1$. As a result, the bandwidth frequency ω_{BW} is increased. The corresponding step response is fast and includes sufficient damping. The *double ratios* design rule is explained in this section.

The closed-loop transfer function can be expressed in the form given in Eq. 3.11, with the numerator *num* (*s*) having *m* zeros and the denominator *f*(*s*) having *n* zeros. The *f*(*s*) is, at the same time, the characteristic polynomial of the system, and its zeros are the closed-loop poles determining the character of the step response. In Eq. 3.11, *out* (*s*) stands for the complex image (i.e., Laplace transform) of the system output, while *ref* (*s*) represents the setpoint disturbance.

$$W_{SS}(s)=\frac{out(s)}{ref(s)}=\frac{a_0+a_1s+a_2s^2+\ldots+a_ms^m}{b_0+b_1s+b_2s^2+\ldots+b_ns^n}=\frac{\sum_{i=0}^{m}a_is^i}{\sum_{i=0}^{n}b_is^i}=\frac{num(s)}{f(s)} \qquad (3.11)$$

The Laplace transform of the system output *out(s)* depends on the input reference *ref(s)* and the transfer function $W_{SS}(s)$: $out(s) = W_{SS}(s)ref(s)$. If we consider the steady-state operation of the closed-loop system with sinusoidal input $ref(t) = \Omega^*\sin(\omega t)$, the Fourier transform of the output can be obtained as $out(j\omega) = W_{SS}(j\omega)ref(j\omega)$. Whatever the input disturbance *ref* (*t*), it is desirable to have the output speed *out* (*t*), which tracks the reference *ref* (*t*) without error in the steady state. Therefore, the closed-loop system transfer function ideally should be $W_{SS}(s) = 1$. With $W_{SS}(j\omega) = 1 + j0$, the system will track the sinusoidal input $ref(t) = \Omega^*\sin(\omega t)$ without errors in amplitude or phase. Hence, it is desirable to have the amplitude characteristic $A(\omega) = |W_{SS}(j\omega)| = a_0/b_0 = 1$ and the phase characteristic $\varphi(\omega) = \arg(W_{SS}(j\omega)) = 0$. The coefficients of the characteristic polynomial $b_0 .. b_n$ and the coefficients of the numerator $a_0 .. a_m$ contribute to changes in amplitude and phase of the closed-loop system transfer function (Eq. 3.12). Therefore, the ideal case of $W_{SS}(j\omega) = 1 + j0$ can hardly be expected, in particular at higher excitation frequencies ω. In Fig. 3.7, the common outline of the amplitude characteristic is shown, with the excitation frequency and the amplitude $A(\omega) = |W_{SS}(j\omega)|$ given in the logarithmic scale. In the middle of the plot, the amplitude characteristic is supposed to have a resonant peak, frequently encountered in systems with conjugate complex poles. Within the frequency range comprising the resonant peak, the most significant closed-loop poles and zeros are found. The frequency ω_{PEAK} is closely related to the bandwidth frequency ω_{BW} (Section 2.1.1).

$$W_{SS}(j\omega) = \frac{\sum_{i=0}^{m} a_i (j\omega)^i}{\sum_{i=0}^{n} b_i (j\omega)^i} = \frac{num(j\omega)}{f(j\omega)} \tag{3.12}$$

As the excitation frequency increases (see the right side of Fig. 3.7), the amplitude $A(\omega)$ reduces towards zero. This reflects the fact that the number of closed-loop poles n in practicable transfer functions (Eq. 3.12) exceeds the number of closed-loop zeros m. Therefore, at very high frequencies, the amplitude characteristic can be approximated by $A(\omega) \approx K/\omega^{n-m}$.

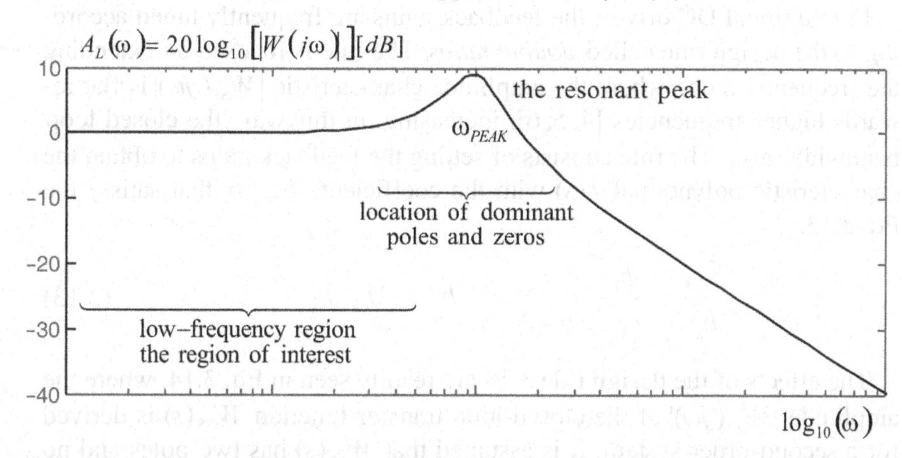

Fig. 3.7. Common shape of the closed-loop transfer function $W_{SS}(j\omega)$ amplitude characteristic. The amplitude characteristic $A(\omega)$ and the excitation frequency ω are given in logarithmic scale.

The low-frequency region extends to the left side of the resonant peak in Fig. 3.7. Within this range, the amplitude characteristic $A(\omega)$ is expected to be close to one. At very low frequencies $\omega \approx 0$, the $A(\omega) = |W_{SS}(j\omega)|$ comes close to $a_0/b_0 = 1$. For sinusoidal reference inputs $\Omega^* \sin(\omega t)$ with excitation frequency ω substantially smaller than ω_{PEAK}, the error in the system output $out(t)$ will be negligible. An insignificant output error can be achieved, as well, with reference signals $ref(t)$ that are not sinusoidal, provided that most of their spectral energy is contained in the low-frequency region, where $A(\omega) = |W_{SS}(j\omega)| \approx 1$. Specifically, in cases when $ref(t)$ comprises a number of frequency components ω_x, these should stay within the frequency range defined as $0 < \omega_x << \omega_{PEAK}$.

When a closed-loop control system is being designed, it is of interest to maximize the range of applicable excitation frequencies $0 < \omega_x << \omega_{PEAK}$. Specifically, it is desirable to extend the range where the amplitude characteristic in Fig. 3.7 is flat ($|W_{SS}(j\omega)| \approx 1$). The frequency ω_{PEAK} and bandwidth frequency ω_{BW} (Section 2.1.1) depend on the closed-loop poles and zeros, which, in turn, are functions of the polynomial coefficients $b_0 .. b_n$ and $a_0 .. a_m$. The coefficients of $f(s)$ and $num(s)$ in Eq. 3.11 are calculated from the plant parameters and control parameters (i.e., feedback gains). The former are given and cannot be changed, while the latter can be adjusted so as to achieve the desired step response and/or the desired amplitude characteristic $|W_{SS}(j\omega)|$.

In traditional DC drives, the feedback gains are frequently tuned according to the design rule called *double ratios.* The rule is focused on extending the frequency range where the amplitude characteristic $|W_{SS}(j\omega)|$ is flat towards higher frequencies [4, 5, 6], increasing, in this way, the closed loop bandwidth ω_{BW}. The rule consists of setting the feedback gains to obtain the characteristic polynomial $f(s)$ with the coefficients $b_0 .. b_n$ that satisfy the Eq. 3.13.

$$\frac{b_{k+1}}{b_k} \le \frac{b_k}{b_{k-1}} \qquad \Rightarrow \qquad b_k^2 \ge 2 b_{k-1} b_k \tag{3.13}$$

The effects of the design rule 3.13 are readily seen in Eq. 3.14, where the amplitude $|W_{SS}(j\omega)|$ of the closed-loop transfer function $W_{SS}(s)$ is derived for a second-order system. It is assumed that $W_{SS}(s)$ has two poles and no zeros ($num(s) = a_0$). Regarding the coefficients of the characteristic polynomial $f(s) = b_0 + b_1 s + b_2 s^2$, it is assumed that $b_1^2 = 2\, b_0 b_2$.

$$\begin{aligned} W_{SS}(j\omega) &= \frac{a_0}{b_0 + b_1 j\omega + b_2 (j\omega)^2} \\ |W_{SS}(j\omega)|^2 &= \frac{a_0^2}{b_0^2 + (b_1^2 - 2b_0 b_2)\omega^2 + b_2^2\omega^4} = \frac{a_0^2}{b_0^2 + b_2^2\omega^4} \end{aligned} \tag{3.14}$$

With $b_1^2 = 2\, b_0 b_2$, the denominator of the amplitude characteristic in Eq. 3.14 reduces to $b_0^2 + b_2^2\omega^4$. The range of frequencies where the amplitude characteristic is flat ($|W_{SS}(j\omega)| \approx 1$) extends towards the corner frequency $\omega_{BW} = (b_0/b_2)^{0.5}$. A similar consideration can be extended to the third order transfer function given in Eq. 3.15, having three closed-loop poles with no finite zeros and with the characteristic polynomial $f(s) = b_0 + b_1 s + b_2 s^2 + b_3 s^3$. The amplitude characteristic $|W_{SS}(j\omega)|^2$ given in Eq. 3.16 includes four factors in the denominator. The coefficients with the second and the fourth power of frequency ω are $(b_1^2-2\,b_0b_2)$ and $(b_2^2-2\,b_1b_3)$, respectively.

$$W_{SS}(j\omega)=\frac{a_0}{b_0+b_1 j\omega+b_2(j\omega)^2+b_3(j\omega)^3} \tag{3.15}$$

$$\left|W_{SS}(j\omega)\right|^2=\frac{a_0^2}{b_0^2+\left(b_1^2-2b_0b_2\right)\omega^2+\left(b_2^2-2b_1b_3\right)\omega^4+b_3^2\omega^6} \tag{3.16}$$

If we apply the *double ratios* setting, the coefficients $(b_1^2-2b_0b_2)$ and $(b_2^2-2b_1b_3)$ in Eq. 3.16 become equal to zero. The amplitude characteristic $A^2(\omega)=|W_{SS}(j\omega)|^2$ reduces to the form shown in Eq. 3.17. In this manner, the frequency range with $|W_{SS}(j\omega)|\approx 1$ spreads towards higher frequencies. The corner frequency ω_{BW}, from where the amplitude characteristic starts to decline, reaches $\omega_{BW}=(b_0/b_3)^{1/3}$. An analogous conclusion can be drawn for the closed-loop systems of the order $n>3$.

$$\left|W_{SS}(j\omega)\right|^2=\frac{a_0^2}{b_0^2+b_3^2\omega^6} \tag{3.17}$$

The *double ratios* extend the range of frequencies where the amplitude characteristic $A(\omega)$ remains $|W_{SS}(j\omega)|\approx 1$. Therefore, this value is frequently referred to as the *absolute value optimum*.

It is interesting to consider the effects of the *double ratios* design rule on the closed-loop poles and, thereupon, the character of the closed loop system step response. In Table 3.1, the closed-loop poles for the second-, third-, and fourth-order systems are derived by calculating the roots of the relevant characteristic polynomials $f_2(s)=b_0+b_1s+b_2s^2$, $f_3(s)$, and $f_4(s)$. Polynomials $f_2(s)$, $f_3(s)$, and $f_4(s)$ are generated by selecting an arbitrary ratio establishing b_0/b_1 and setting the remaining coefficients so as to meet the condition $b_k^2=2b_{k-1}b_{k+1}$. The initial ratio b_0/b_1 determines the natural frequency ω_n of the closed-loop poles in Table 3.1. The damping factor of the closed loop poles ranges from 0.5 to 0.707. The experience in applying the *double ratios* approach [4, 5, 6] provides evidence that the $b_k^2=2b_{k-1}b_{k+1}$ design rule ensures a well damped response, with a reasonable robustness to plant parameter changes. If we apply the rule to characteristic polynomials of the nth order, where n ranges from 5 to 16, the damping coefficients of the resulting conjugate-complex pole remain between 0.64 and 0.66.

Table 3.1. The zeros of the characteristic polynomial and their damping factors for the second-, third-, and fourth-order systems. Polynomial coefficients are adjusted according to the rule of double ratios.

the order	$n = 2$	$n = 3$	$n = 4$
the roots	$s_{1/2} = -\frac{\omega_n}{\sqrt{2}} \pm j\frac{\omega_n}{\sqrt{2}}$	$s_{1/2} = -\frac{\omega_n}{2} \pm j\frac{\omega_n}{2}\sqrt{3}$ $s_3 = -\omega_n$	$s_{1/2} = -\frac{\omega_n}{\sqrt{2}} \pm j\frac{\omega_n}{\sqrt{2}}$ $s_{3/4} = -\frac{\omega_n}{\sqrt{2}} \pm j\frac{\omega_n}{\sqrt{2}}$
damping factor	$\xi = 0.707$	$\xi = 0.5$	$\xi = 0.707$

3.4 Double ratios with proportional speed controllers

The *double ratios* design rule is applied to the speed controlled DC drive, which comprises an imperfect torque actuator $W_A(s)$, with the driving torque T_{em} lagging behind the reference T_{ref}. The block diagram of such a system is given in Fig. 3.8. The torque actuator is modeled as the first-order lag having a time constant of τ_{TA}. In this section, it is assumed that the mechanical load is inertial, with a negligible friction ($B = 0$). The speed controller is supposed to have proportional control action with gain K_P.

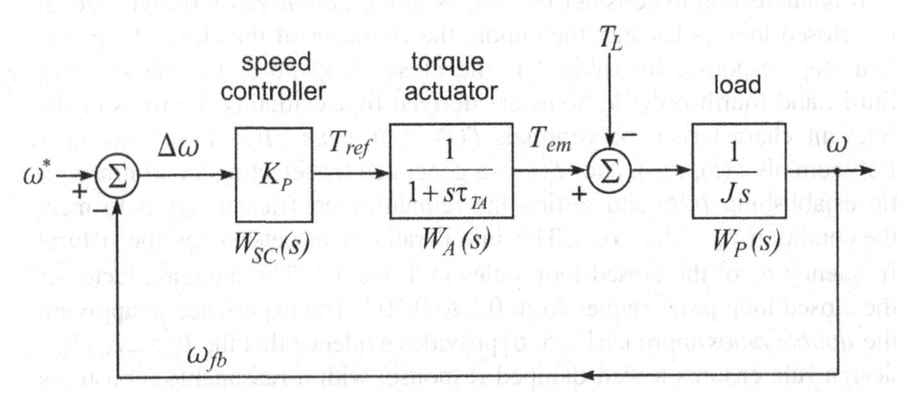

Fig. 3.8. The speed-controlled DC drive system comprising the first-order lag torque actuator $W_A(s)$, inertial load, and the proportional speed controller.

The closed-loop transfer function of the system is given in Eq. 3.18. The zeros of the characteristic polynomial are determined by the coefficients J,

K_P, and τ_{TA}. The feedback gain K_P can be set to meet the *double ratios* relation $b_1^2 = 2b_0b_2$ (Eq. 3.19). With $K_P = J/(2\tau_{TA})$, the absolute value optimum is achieved, as the amplitude characteristic $|W_{SS}(j\omega)|$ remains close to one, in an extended range of frequencies.

$$W_{SS}(s) = \frac{K_P}{J\tau_{TA}s^2 + Js + K_P} \tag{3.18}$$

$$b_1^2 = 2b_0b_2 \quad \Rightarrow \quad J = 2K_P\tau_{TA} \quad \Rightarrow \quad K_P = \frac{J}{2\tau_{TA}} \tag{3.19}$$

The decision 3.19 converts the closed-loop transfer function into the form expressed in Eq. 3.20. The corresponding closed-loop poles are given in Eq. 3.21. The damping of the closed-loop poles is 0.707, as predicted in Table 3.1.

$$W_{SS}(s) = \frac{1}{2\tau_{TA}^2 s^2 + 2\tau_{TA}s + 1} \tag{3.20}$$

$$s_{1/2} = -\frac{1}{2\tau_{TA}} \pm j\frac{1}{2\tau_{TA}} \tag{3.21}$$

The closed-loop step response of the system, shown in Fig. 3.8, subjected to parameter setting 3.19, is given in Fig. 3.9. The output speed reaches the setpoint in approximately five τ_{TA} intervals, where τ_{TA} stands for the time lag of the torque actuator. The output speed overshoots the setpoint by 5%. Following the overshoot, the speed error gradually decays to zero.

The absolute values of the closed-loop poles (Eq. 3.21) are $|s_{1/2}| = 0.707/\tau_{TA}$. At the same time, for the frequency $\omega = 0.707/\tau_{TA}$, the amplitude $A(\omega) = |W_{SS}(j\omega)|$ of the closed-loop transfer function reduces to 0.707 (i.e., to –3 dB). Therefore, the closed-loop bandwidth obtained with the structure in Fig. 3.8, subjected to the parameter setting in Eq. 3.19, is $\omega_{BW} = 0.707/\tau_{TA}$. The bandwidth is inversely proportional to the torque actuator time constant τ_{TA}.

The question arises as to whether the bandwidth ω_{BW} can surpass the value imposed by the internal dynamics of the torque actuator. Preserving the speed controller structure ($W_{SC}(s) = K_P$) and renouncing the design rule $b_1^2 = 2b_0b_2$ by doubling the proportional gain, the step response becomes faster (Fig. 3.10), and the closed loop bandwidth increases. This result is achieved at the cost of a threefold increase in the overshoot. While the optimum gain setting results in an overshoot of 5%, the response obtained with increased K_P gain (Fig. 3.10) exceeds the setpoint by 17%. Therefore, it is concluded that for the system in Fig. 3.8, the absolute value optimum achieved through the *double ratios* design rule secures a well-damped response and provides a reasonable bandwidth.

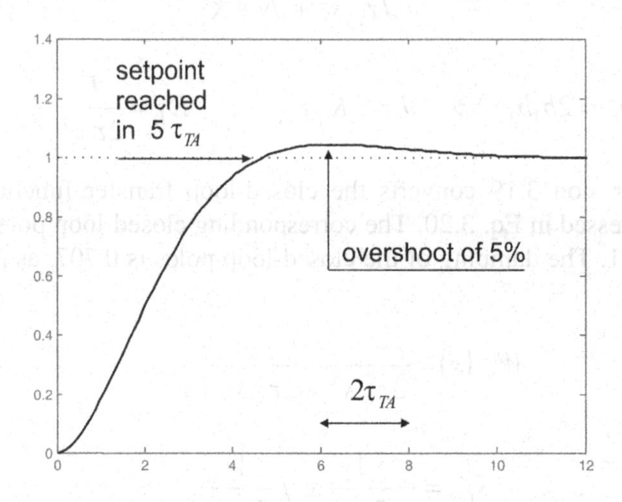

Fig. 3.9. The step response of the second-order speed-controlled DC drive system given in Fig. 3.8, tuned according to the *double ratios* design rule (Eq. 3.19).

Further increase in the closed-loop bandwidth can be achieved by extending the speed controller structure and adding the derivative control action. With the speed controller output T_{ref} augmented by the first derivative of the system output ω, the second-order system in Fig. 3.8 will have an implicit state feedback (i.e., both state variables of the system would have an impact on the driving force). Given the state feedback, the feedback gains can be set to accomplish arbitrary closed-loop poles, resulting in an unconstrained choice of the damping factor, the natural frequency ω_{n}, and the closed-loop bandwidth ω_{BW}. In traditional speed-controlled DC drives, the application of the derivative action is hindered by the presence of high-frequency noise components and by the difficulties of analog implementation and signal processing.

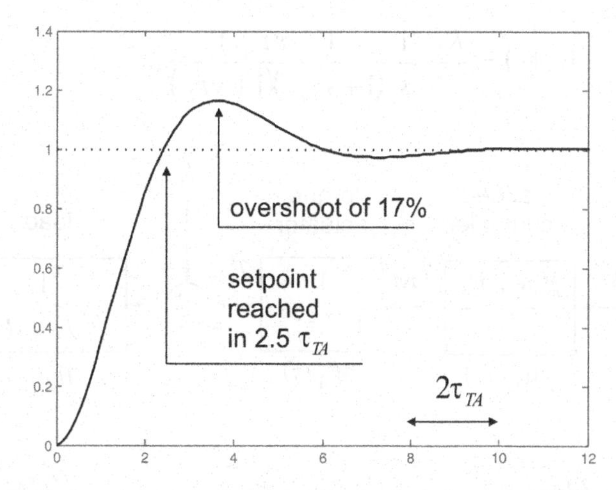

Fig. 3.10. The step response of the second-order speed-controlled DC drive system given in Fig. 3.8. The proportional gain is doubled with respect to the value suggested in Eq. 3.19.

3.5 Tuning of the PI controller according to double ratios

In this section, the *double ratios* rule is applied in setting the feedback gains K_P and K_I for the speed-controlled DC drive with delay in the torque actuator, with friction in the mechanical subsystem, and with a proportional-integral speed controller. The block diagram of the system under consideration is given in Fig. 3.11. The corresponding open-loop transfer function is given in Eq. 3.22.

$$W_S(s) = W_{SC}(s)W_A(s)W_P(s) = \frac{K_I\left(1+s\dfrac{K_P}{K_I}\right)}{Bs(1+s\tau_{TA})\left(1+s\dfrac{J}{B}\right)} \tag{3.22}$$

The ratio $\tau_P = J/B$ represents the time constant of the mechanical system, while the ratio between the proportional and integral gains $\tau_{SC} = K_P/K_I$ stands for the time constant of the speed controller. The values of τ_P and τ_{SC} correspond to the real pole and real zero of the open-loop system transfer function $W_S(s)$. If we introduce τ_P and τ_{SC} in Eq. 3.22, the open-loop system transfer function assumes the following form:

$$W_S(s) = \frac{K_I}{B}\frac{1}{s}\frac{(1+s\tau_{SC})}{(1+s\tau_{TA})(1+s\tau_P)}. \tag{3.23}$$

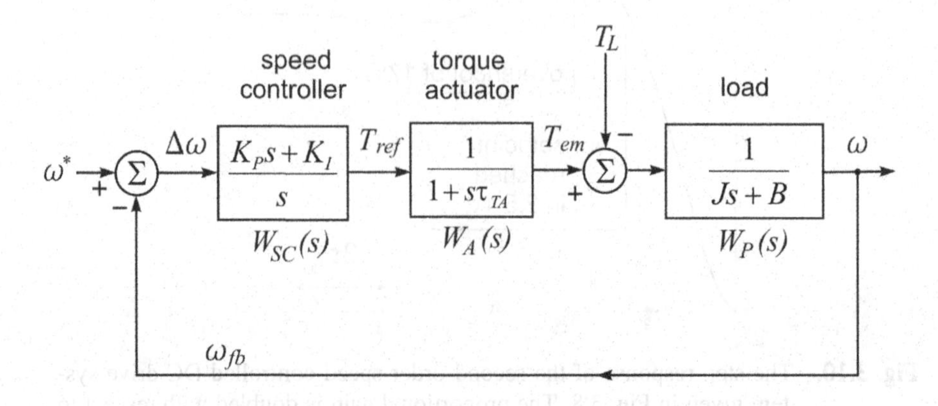

Fig. 3.11. Speed-controlled DC drive with delay τ_{TA} in the torque actuator, with load friction B and load inertia J, and with the PI speed controller.

Two time constants included in the $W_S(s)$ denominator are the plant time constant τ_P (mechanical) and the torque actuator lag τ_{TA} (electrical time constant). In most cases, the mechanical time constant is larger by far. Therefore, the speed controller parameter setting is focused on suppressing the delays brought forward by the mechanical time constant. In traditional DC drives, the feedback gains K_P and K_I are often set with the intent to obtain $\tau_P = \tau_{SC}$ and cancel the pole $-1/\tau_P$ with the speed controllers zero $-1/\tau_{SC}$ [6]. To this end, the K_P and K_I parameters should satisfy Eq. 3.24. Consequently, the open-loop system transfer function $W_S(s)$ reduces to Eq. 3.25.

$$\frac{K_P}{K_I} = \tau_{SC} = \tau_P = \frac{J}{B} \tag{3.24}$$

$$W_S(s) = \frac{K_I}{B}\frac{1}{s(1+s\tau_{TA})} \tag{3.25}$$

The closed-loop transfer function $W_{SS}(s) = W_S(s) / (1 + W_S(s))$ of the system in Fig. 3.11, subjected to decision 3.24, is given in Eq. 3.26. It has a second-order characteristic polynomial in the denominator and no finite zeros:

$$W_{SS}(s) = \frac{K_I}{K_I + sB + s^2 B\tau_{TA}} = \frac{1}{1 + \frac{B}{K_I}s + \frac{B\tau_{TA}}{K_I}s^2} = \frac{1}{b_0 + b_1 s + b_2 s^2}. \tag{3.26}$$

In Eq. 3.26, $b_0 = 1$, $b_1 = B/K_I$, and $b_2 = \tau_{TA}B/K_I$. With application of the *double ratios* design rule $b_1^2 = 2b_0b_2$, the gains of the PI speed controller are obtained as

$$K_P = \frac{J}{2\tau_{TA}}, \quad K_I = \frac{B}{2\tau_{TA}}. \tag{3.27}$$

With the parameter setting given in 3.27, the closed-loop transfer function of the system in Fig. 3.11 becomes essentially the same as the one obtained in Eq. 3.20 in the previous section: it has no finite zeros, while the characteristic polynomial $f(s)$, found in the denominator of the transfer function, takes the form $f(s) = 2\tau_{TA}^2 s^2 + 2\tau_{TA} s + 1$. The values of the closed-loop poles can be found in Eq. 3.21, while Fig. 3.9 presents the step response. Well damped, the step response reaches the setpoint in approximately 5 τ_{TA} and experiences an overshoot of 5%.

The *double ratios* parameter-setting rule, applied to the speed-controlled system in Fig. 3.11, results in the absolute value optimum: that is, the frequency range where the amplitude characteristic $|W_{SS}(j\omega)|$ is flat and close to 0 dB is extended towards higher frequencies. The step response is well damped, while the closed-loop bandwidth ω_{BW} is limited by the time constant τ_{TA}, determined by the internal dynamics $W_A(s)$ of the torque actuator.

An increase of the closed-loop bandwidth can be achieved by adding the derivative action to the structure of the speed controller $W_{SC}(s)$. The application of the derivative action is restricted to the cases where the parasitic high frequency noise is not emphasized. In such cases, the first derivative of the noise-contaminated signal retains an acceptable signal-to-noise ratio. In traditional DC drives with analog implementation of the drive controller, the derivative action is commonly equipped with a first-order low-pass filter, devised to suppress the differentiation noise. In most cases, a practicable derivative action is described by the transfer function $sK_D/(1+s\tau_{NF})$, where the time constant τ_{NF} of the low-pass filter has to be set according to the noise content.

If we assume an ideal noise-free condition, the PID controller can be applied in the form given in Eq. 3.28. Given the system in Fig. 3.11, the third-order characteristic polynomial is obtained (Eq. 3.29). The coefficient b_3 of $f(s)$ is equal to 1, while the coefficients b_2, b_1, and b_0 can be adjusted by selecting an appropriate value for K_D, K_P, and K_I, respectively. With complete control over the coefficients of the characteristic polynomial, the placement of the closed-loop poles is unrestrained. Therefore, the closed-loop bandwidth can exceed the value of $\omega_{BW} = 0.707/\tau_{TA}$, while, at the same time, keeping the damping factor and the overshoot at desirable levels. The practical value of this consideration is restricted by the amount of high-frequency noise encountered in a typical drive environment.

$$W_{SC}(s) = K_D s + K_P + \frac{K_I}{s} \tag{3.28}$$

$$\begin{aligned} f(s) &= s^3 + \frac{J + B\tau_{TA} + K_D}{J\tau_{TA}} s^2 + \frac{B + K_P}{J\tau_{TA}} s + \frac{K_I}{J\tau_{TA}} \\ &= s^3 + b_2(K_D)\, s^2 + b_1(K_P)\, s + b_0(K_I) \end{aligned} \tag{3.29}$$

3.6 Symmetrical optimum

The mechanical subsystem of the speed-controlled DC drive, given in Fig. 3.12, is supposed to have an inertial load with negligible friction. In this section, the use of the *double ratios* rule in setting the K_P and K_I parameters is analyzed and explained. The torque actuator is modeled by the first order low-pass transfer function $W_A(s)$ having time constant τ_{TA}. The corresponding open-loop transfer function is given in Eq. 3.30. The analysis and discussion in this section are focused on deriving the parameter-setting procedure that would result in an acceptable closed-loop bandwidth and a well-damped step response. To begin with, the possibility of simplifying the open-loop function by means of the pole-zero cancellation is discussed briefly.

The parameter τ_{SC} in Eq. 3.30 represents the speed controller time constant K_P/K_I and determines the open-loop zero $-1/\tau_{SC}$ of the transfer function $W_S(s)$. An attempt to cancel out the $W_S(s)$ real pole $-1/\tau_{TA}$ with the zero $-1/\tau_{SC}$ requires the parameters K_P and K_I to satisfy the relation $K_P = \tau_{TA} K_I$. The design decision $\tau_{TA} = \tau_{SC}$ reduces the open-loop system transfer function to $W_S(s) = K_I/(Js^2)$, and the closed-loop characteristic polynomial to $f(s) = s^2 + K_I/J$. The closed-loop poles $s_{1/2} = \pm j(K_I/J)^{0.5}$ result in the damping coefficient $\xi = 0$ and an unacceptable oscillatory response. Therefore, the pole-zero cancellation cannot be used in conjunction with the system in Fig. 3.12. The *double ratios* design rule should be used instead.

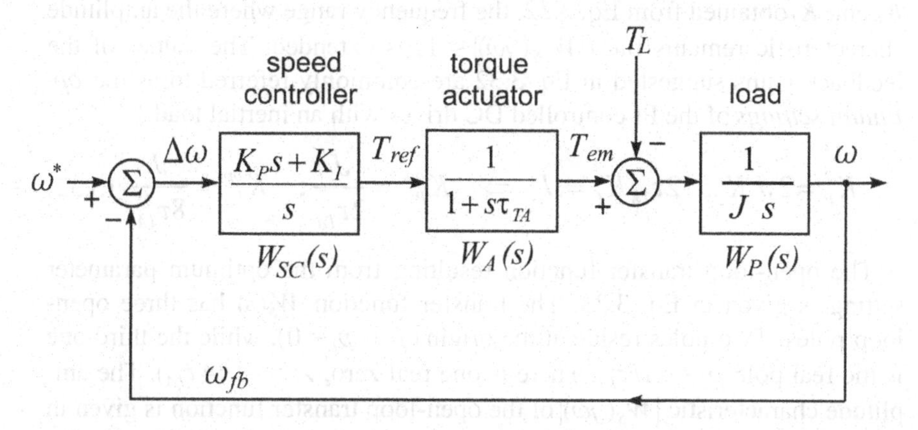

Fig. 3.12. Speed-controlled DC drive with frictionless, inertial load, delay τ_{TA} in the torque actuator, and with the PI speed controller.

$$W_S(s) = W_{SC}(s)W_A(s)W_P(s) = \frac{K_P s + K_I}{s}\,\frac{1}{1+s\tau_{TA}}\,\frac{1}{Js}$$

$$= \frac{K_I}{s}\,\frac{1+s\tau_{SC}}{1+s\tau_{TA}}\,\frac{1}{Js};\qquad \left(\tau_{SC} = \frac{K_P}{K_I}\right) \tag{3.30}$$

The closed-loop system transfer function $W_{SS}(s)$ is given in Eq. 3.31. The closed-loop transfer function has one real zero $(-1/\tau_{SC})$ and three closed-loop poles. The characteristic polynomial coefficients are $b_0 = 1$, $b_1 = \tau_{SC} = K_P/K_I$, $b_2 = J/K_I$, and $b_3 = J\tau_{TA}/K_I$.

$$W_{SS}(s) = \frac{K_I + K_P s}{K_I + K_P s + Js^2 + J\tau_{TA}s^3} = \frac{1 + \frac{K_P}{K_I}s}{1 + \frac{K_P}{K_I}s + \frac{J}{K_I}s^2 + \frac{J\tau_{TA}}{K_I}s^3}$$

$$= \frac{1+\tau_{SC}s}{1+\tau_{SC}s + \frac{J}{K_I}s^2 + \frac{J\tau_{TA}}{K_I}s^3} = \frac{1+\tau_{SC}s}{b_1 + b_1 s + b_2 s^2 + b_3 s^3} \tag{3.31}$$

The *double ratios* design rule requires the coefficients b_0, b_1, and b_2 to satisfy the condition $b_1^2 = b_0 b_2$. The values of b_1, b_2, and b_3 are related by the expression $b_2^2 = b_1 b_3$. The proportional and integral gains that satisfy the conditions above are calculated in Eq. 3.32. Given the feedback gains K_P and K_I obtained from Eq. 3.32, the frequency range where the amplitude characteristic remains flat ($|W_{SS}(j\omega)| \approx 1$) is extended. The values of the feedback gains suggested in Eq. 3.32 are commonly referred to as the *optimum settings* of the PI controlled DC drives with an inertial load.

$$K_P^2 = 2J\,K_I;\;\; 2\tau_{TA}K_I = J \;\Rightarrow\; K_P^{opt} = \frac{J}{2\tau_{TA}};\;\; K_I^{opt} = \frac{J}{8\tau_{TA}^2} \tag{3.32}$$

The open-loop transfer function resulting from the optimum parameter setting is given in Eq. 3.33. The transfer function $W_S(s)$ has three open-loop poles. Two poles reside at the origin ($p_1 = p_2 = 0$), while the third one is the real pole $p_3 = -1/\tau_{TA}$. There is one real zero, $z_1 = -1/(4\tau_{TA})$. The amplitude characteristic $|W_S(j\omega)|$ of the open-loop transfer function is given in Fig. 3.13. Next to the origin, it attenuates at a rate of 40 dB per decade. Passing the open-loop zero z_1, the slope reduces to –20 dB. At the frequency $\omega_0 = 1/(2\tau_{TA})$, the amplitude $|W_S(j\omega)|$ reduces to 1 (0 dB). Due to symmetrical placement of z_1, ω_0, and p_1 ($\omega_0^2 = z_1\, p_1$), the parameter setting given in Eq. 3.32 is known as the *symmetrical optimum*. The closed-loop performance obtained with the symmetrical optimum is discussed later.

$$W_S^{opt}(s) = \frac{1}{8}\frac{1}{(\tau_{TA}s)^2}\frac{1+4\tau_{TA}s}{1+\tau_{TA}s} \tag{3.33}$$

The closed-loop system transfer function $W_{SS}(s)$, obtained with K_P^{opt} and K_I^{opt}, is derived in Eq. 3.34. The pole placement in the *s*-plane is illustrated in Fig. 3.14. The natural frequency $\omega_n = 1/(2\tau_{TA})$ and damping coefficient $\xi = 0.5$ correspond to the values anticipated in Table 3.1. The closed-loop step response is given in Fig. 3.15. Compared with the results obtained in the previous section (Fig. 3.9), the overshoot is increased to approximately 43% due to a lower value of the damping coefficient ξ. Following the overshoot, the oscillation in the step response 3.15 decays rapidly, and the output speed converges towards the reference.

$$W_{SS}^{opt}(s) = \frac{1+4\tau_{TA}s}{1+4\tau_{TA}s+8\tau_{TA}^2s^2+8\tau_{TA}^3s^3} \tag{3.34}$$

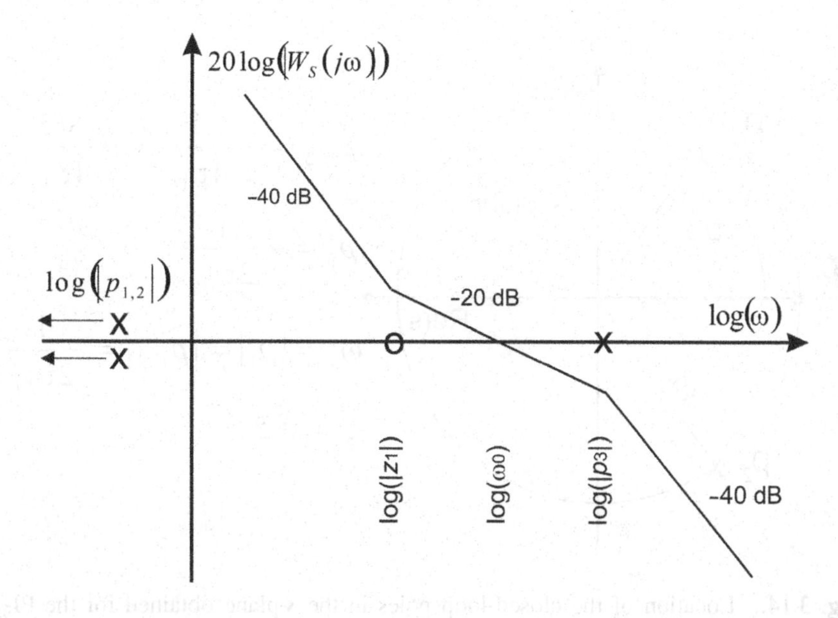

Fig. 3.13. The amplitude characteristic of the open loop transfer function obtained with the gains K_P and K_I calculated from Eq. 3.32. The amplitude $|W_S(j\omega)|$ and frequency ω are given in logarithmic scale. The amplitude $|W_S(j\omega)|$ attenuates to 0 dB at $\omega = \omega_0$. Due to symmetrical placement of z_1, ω_0 and p_1 ($\omega_0^2 = z_1 p_1$), the parameter setting given in Eq. 3.32 is known as the *symmetrical optimum.*

Maintaining the speed controller with the proportional and integral control actions and making further gain adjustments, the step response in Fig. 3.15 can be dampened only at the cost of reducing the closed-loop bandwidth. Likewise, the step response can be made quicker provided that an overshoot in excess of 43% is acceptable. A considerable improvement of the closed-loop performance is feasible in cases where the speed controller $W_{SC}(s)$ can be extended with the derivative control action (Eq. 3.35). A compulsory low-pass filter $1/(1+s\tau_{NF})$ is used in conjunction with the derivative factor in order to suppress the high-frequency noise incited by the differentiation. The time constant τ_{NF} of this first-order filter should be much smaller than the time constants related to the desired closed-loop transfer function. On the other hand, a sufficient value of τ_{NF} is needed to filter out detrimental noise components. In cases when the noise is contained in the high-frequency region, beyond the range comprising the desired closed-loop poles, it is possible to allocate a value of τ_{NF} that meets both requirements.

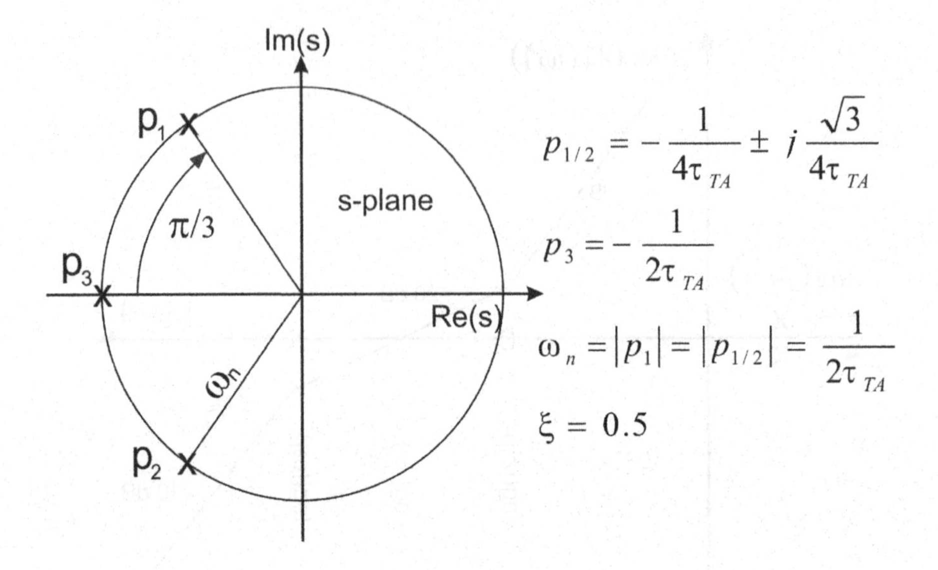

Fig. 3.14. Location of the closed-loop poles in the *s*-plane obtained for the PI-controlled DC drive with inertial load (Fig. 3.12). The feedback gains K_P and K_I are calculated from Eq. 3.32. The damping coefficient and the natural frequency of the conjugate complex poles $p_{1/2}$ are 0.5 and $\omega_n = 1/(2\tau_{TA})$, respectively.

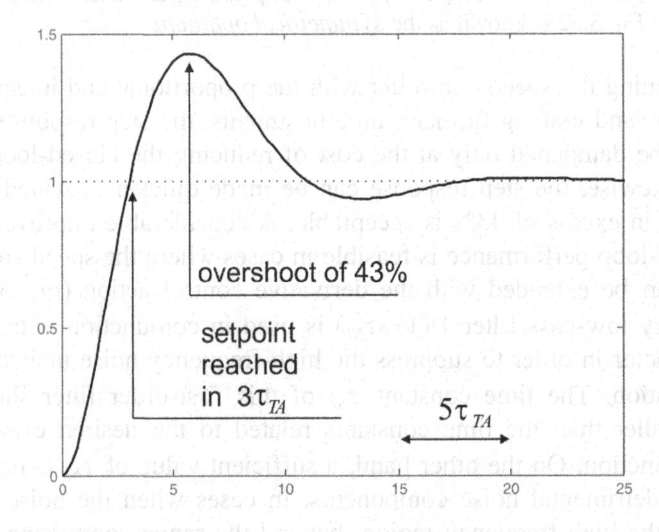

Fig. 3.15. Closed-loop step response of the system in Fig. 3.12, obtained with the adjustable feedback parameters set according to the symmetrical optimum.

With the speed controller transfer function given in Eq. 3.35 and the assumption that the time constant τ_{NF} can be neglected, the resulting third-order characteristic polynomial $f(s)$ is given in Eq. 3.36. With the possibility of altering the three control parameters (K_D, K_P, and K_I), the polynomial coefficients can be controlled in full. Therefore, arbitrary placement of the closed-loop poles is feasible, including the possibility of expanding the closed-loop bandwidth while still preserving the desired damping, and constraining the overshoot.

$$W_{SC}(s) = \frac{K_D s}{1 + s\tau_{NF}} + K_P + \frac{K_I}{s} \tag{3.35}$$

$$\begin{aligned} f(s) &= s^3 + \frac{J + K_D}{J\tau_{TA}} s^2 + \frac{K_P}{J\tau_{TA}} s + \frac{K_I}{J\tau_{TA}} \\ &= s^3 + b_2(K_D)\, s^2 + b_1(K_P)\, s + b_0(K_I) \end{aligned} \tag{3.36}$$

Problems

P3.1

Consider the closed-loop system with the characteristic polynomial $f(s) = b_0 + b_1 s + b_2 s^2 + b_3 s^3 + b_4 s^4$. Assume that $b_0 = 1$ and $b_1 = b_0$. Determine b_2, b_3, and b_4 according to the *double ratios* design rule. Determine the polynomial zeros by using the Matlab function *roots*().

P3.2

(a) Repeat the previous calculation with $b_0 = 10$ and $b_1 = b_0$.
(b) Repeat the previous calculation with $b_0 = 10$ and $b_1 = 10\, b_0$.

P3.3

The speed-controlled DC drive system (Fig. 3.8) comprises a torque actuator that can be modeled as the first-order lag. The friction coefficient B is negligible. The speed controller determines the torque reference $T_{ref} = K_P \Delta\omega$ in proportion to the speed error. The system parameters are $J = 0.1$ kgm^2 and $\tau_{TA} = 10$ ms. Determine the proportional gain K_P so as to obtain the characteristic polynomial in conformity with the *double ratios* design rule. Considering $K_M = 1$ and $K_{FB} = 1$, what are the units of K_P? Calculate the poles of the closed-loop system.

P3.4
For the system in P3.3, determine the closed-loop transfer function. Apply the Matlab command *step*() to obtain the step response to the input disturbance. Estimate the overshoot and the rise time from the plot.

P3.5
The speed-controlled DC drive system (Fig. 3.11) comprises a torque actuator modeled as the first-order lag. The friction coefficient is $B = 0.01$ Nm/(rad/s). The speed controller comprises the proportional and integral action. The system parameters are $J = 0.1$ kgm^2 and $\tau_{TA} = 10$ ms. Determine the gains K_P and K_I so as to obtain the characteristic polynomial in conformity with the *double ratios* design rule. Calculate the poles of the closed-loop system.

P3.6
Determine the closed-loop transfer function of the system in P3.5 and obtain the step response by using the Matlab function *step*().

P3.7
The speed-controlled DC drive system (Fig. 3.12) comprises the torque actuator modeled as the first-order lag. The friction coefficient is $B = 0$. The speed controller comprises the proportional and integral action. The system parameters are $J = 0.1$ kgm^2 and $\tau_{TA} = 10$ ms. Determine the gains K_P and K_I so as to obtain the characteristic polynomial in conformity with the *double ratios* design rule. Calculate the poles of the closed-loop system.

P3.8
For the system given in P3.7, and with the gain setting calculated in S3.7, determine the open-loop transfer function $W_S(s)$, calculate its poles and zeros, and obtain the *Bode* plot by using the appropriate Matlab command. What is the frequency where $|W_S(j\omega)| = 1$? Why is the parameter setting in Eq. 3.32 to be called the *symmetrical* optimum?

P3.9
Determine the closed-loop transfer function of the system in P3.7 and obtain the step response by using the Matlab function *step*(). What are the overshoot and the rise time?

4 Digital Speed Control

Contemporary motion control systems comprise discrete-time speed and position controllers. This chapter restates the basic concepts of discrete-time control, and provides the means to analyze, design, implement, and evaluate discrete-time speed controllers. The sampling process is reviewed and explained. The system dynamics are described in terms of difference equations. The z-transform is introduced as the means of converting the difference equation into an algebraic form, thus simplifying the analysis, design, and performance prediction of discrete-time controllers. The z-transform definition and properties are restated. Before discussing and designing the structure of discrete-time speed controllers, the signal flow between the discrete-time controller and continuous-time control object is explained and detailed. The optimization rule is devised and the parameter setting procedure proposed. In closing sections of the chapter, the large step response is analyzed, explaining the system limits and the wind-up phenomenon. The speed controller structure is enhanced in order to preserve the response character and avoid overshoots and oscillations with large input and load disturbances.

The basic speed control structures have been introduced in previous sections that have addressed the analog implementation, with the relevant continuous-time signals converted into their complex images by means of the Laplace transform. Contemporary speed controllers are implemented in a digital manner. The driving force of the digital speed controller is calculated at discrete, equally-spaced time instants. The calculations within a digital speed controller are initiated by the interrupt events of a microcontroller or a DSP, which executes the speed control algorithm. At each interrupt, the digital controller acquires the speed feedback and calculates a new sample of the driving torque. The interrupt period is known as the *sampling time* or the *sampling interval*. The analysis, synthesis, and tuning of discrete-time or *digital* speed controllers is the subject of this chapter.

4.1 Discrete-time implementation of speed controllers

To begin, a brief comparison of analog and digital implementation is outlined. Prior to the advent of practicable microcontrollers, speed control tasks were implemented in a continuous domain by means of analog electronic circuitry comprising operational amplifiers, resistors, and capacitors. The analog signal, proportional to the revolving speed, was obtained from a tachogenerator attached to the shaft. From an analog amplifier, the reference for the driving torque was obtained in the form of a standard +/–10V analog signal, fed into the torque amplifier. The analog implementation sets the following limits on the closed-loop performance:

- The low-speed operation is impaired by the offset intrinsic to operational amplifiers.
- The feedback gains, defined by the resistors and capacitors, change with temperature and aging.
- The analog shaft sensors, such as the DC tachogenerator, introduce the noise originating from the mechanical commutator, with the noise frequency related to the speed of rotation. The noise in the feedback line reduces the range of applicable gains and, thus, the closed-loop bandwidth.
- While providing a straightforward way of implementing conventional control actions such as $W_{SC}(s) = K_P + K_I/s$, analog implementation encounters great difficulties in implementing nonlinear control laws, backlash compensators, antiresonant filters or finite impulse response filters.
- A speed control task, where adjustable feedback parameters have to be adapted online to specific operating conditions, can hardly be implemented by analog means. A change in parameters or structure of an analog speed controller requires replacement of resistors and capacitors. The implementation of such replacements in real time is hardly practicable with analog tools. Therefore, variable structure controllers and parameter adaptation require implementation on a digital platform.

Digital implementation of the speed-control law began with the introduction of microprocessors and DSP chips that could be employed in the motion-control environment. The development of shaft sensors that provided speed and position information in digital form helped improve the closed-loop performance. Optical encoders and electromagnetic resolvers, equipped with resolver-to-digital converters [7], provided shaft feedback with a signal-to-noise ratio exceeding by far the performance of analog tachogenerators. The numerical capabilities of digital motion controllers allow for the implementation of complex nonlinear compensators, provide for online adaptation of the controller structure and feedback gains, and enable the implementation of FIR filters that were once incompatible with the analog platform.

The role of the digital controller within a speed control system is given in Fig. 4.1. The digital motion controller comprises a microprocessor or a digital signal processor (DSP) equipped with the peripheral units necessary to communicate with the rest of the system. The control algorithm includes the speed controller, which calculates the torque/current reference on the basis of the detected speed error. In addition, the μC/DSP code comprises the local (minor) loop, controlling the motor current (torque). For this reason, the digital controller given on the left in Fig. 4.1 receives speed and current feedback signals and generates the PWM pulses. The latter are used as the firing signals for the power switches within the power converter, thus directing the voltage fed to the motor and constraining the motor current to track the torque reference.

The peripheral module *Counter* in Fig. 4.1 receives a train of pulses coming from an optical encoder attached to the motor shaft. The pulse frequency corresponds to the shaft speed. The *Counter* peripheral counts the pulses and converts the feedback information into a digital word, read and used by the μC/DSP as the speed feedback.

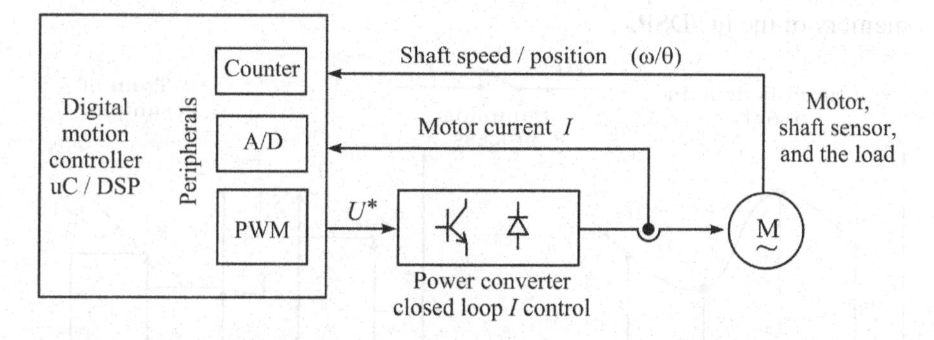

Fig. 4.1. The signal flow between the digital motion controller and the rest of the speed-controlled system. The processor unit on the left outputs the PWM signals, controlling the power converter. It counts the shaft sensor pulses and reads the motor current, by means of an A/D converter.

The A/D peripheral module (Fig. 4.1) receives the analog signal proportional to the motor current. The samples of the motor current are converted into digital words and used as the feedback signal to the current controller. The motor current is controlled by setting the supply voltage to a desired value. The motor voltage is decided by the width of the PWM pulses, generated from the PWM peripheral unit. Within each sampling period, the digital controller evaluates the error in the motor current and calculates the driving force in the form of the voltage reference. The current control algorithm calculates the voltage reference that would reduce the current error in the succeeding sampling intervals, eventually driving it down to zero. Throughout this section, the analysis is simplified by assuming an ideal current controller, wherein the motor current and the delivered torque correspond to reference values.

The digital implementation implies an intrinsic deterioration of signals due to time and amplitude discretization. The loss of information is the inevitable consequence of the sampling process. Indispensable for the digital implementation of control laws, the sampling relates to the conversion of signals from their real-time, continuous domain form ($y(t)$ in Fig. 4.2), into their digital counterparts (Y_n). The sample Y_n is the digital word representing the analog signal $y(t)$ acquired at the instant $t = nT$, converted into a number to be used by the control algorithm. The process of sampling consists of acquiring the analog signal at the sampling instant $t = nT$ and keeping the value $y(nT)$ in a charged capacitor $C_{S/H}$ included in the sample-and-hold circuit. The A/D peripheral then converts the $C_{S/H}$ voltage into a number (Y_n), represented by 12–16 bits and kept within the RAM memory of the μC/DSP.

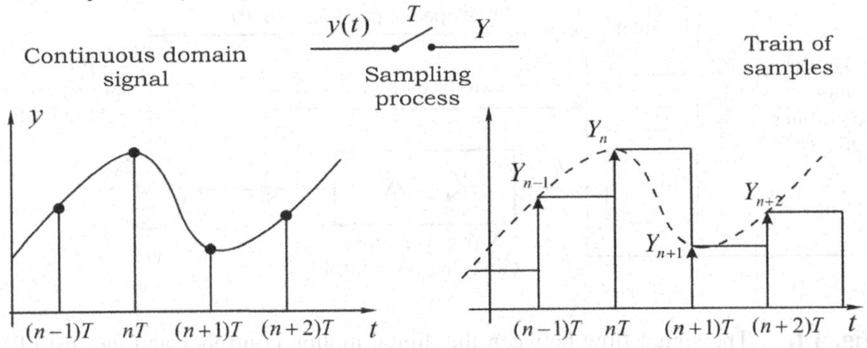

Fig. 4.2. The loss of information in the sampling process: the continuous-time signal $y(t)$ is converted into a train of pulses. The resolution in representing the amplitude of samples Y_n depends on the number of bits in the A/D peripheral.

A very similar process takes place within the *Counter* peripheral (Fig. 4.1), which processes the signals originating from an optical encoder and converts them into samples of the shaft speed or position.

The sampling results in both amplitude and time discretization. The conversion of $y(t)$ into the train of samples Y_n disregards the values of the continuous-time signal in between the sampling instants. Therefore, the reconstruction of the original signal from the train of samples is not always possible. In such a case, the time discretization leads to loss of information. According to the sampling theorem [8, 9], it is possible to sample continuous-time signals without signal deterioration, provided that the input $y(t)$ has a limited bandwidth. If the highest-frequency component of the input does not exceed one half of the sampling frequency $f_S = 1/T$, then $y(t)$ can be recovered in full from the train of samples Y_n. Hence, in such cases, time discretization does not contribute to loss of information. The bandwidth limit $f_{max} = f_S/2 = 1/2T$ is also known as the *Shannon frequency*.

Contemporary microcontrollers and signal processors are equipped with A/D peripheral units capable of performing at rates $f_S > 10^6$. The frequency content of the motor current and the rotor speed does not exceed several tens of kHz. Therefore, the sampling process in digital speed controllers can be organized in such a way that the time discretization does not impair the feedback signals.

The samples Y_n represent the input signal $y(t)$ at instants $t = nT$. The number of data bits in digital words Y_n depends on the resolution of the A/D peripheral unit. Practicable A/D resolutions vary from 10 to 16 bits. Therefore, the samples of the feedback signals are internally represented as digital words having 10–16 bits. When an input signal that has a range of +/–10 V is sampled with a 12-bit A/D peripheral unit, one least significant bit (LSB) of the result corresponds to the quantum $\Delta U = 20\text{V}/2^{12} = 4.88$ mV. As a consequence, the signal amplitude is represented in chunks of ΔU. Fine fluctuations of the signal $y(t)$ within the ΔU boundaries cannot be represented and are therefore lost. Hence, the sampling process involves discretization in amplitude and loss of information caused by the resolution limits of A/D units and finite wordlength.

Within a digital speed control system (Fig. 4.1), the feedback signals contain the parasitic noise components, originating from the power converter commutation processes and other sources. In general, the noise exceeds the quantization step ΔU of the A/D converter, while the errors caused by the amplitude discretization are inferior to the intrinsic noise. Therefore, in most digital speed controllers, the effects of amplitude quantization on feedback signal integrity are negligible.

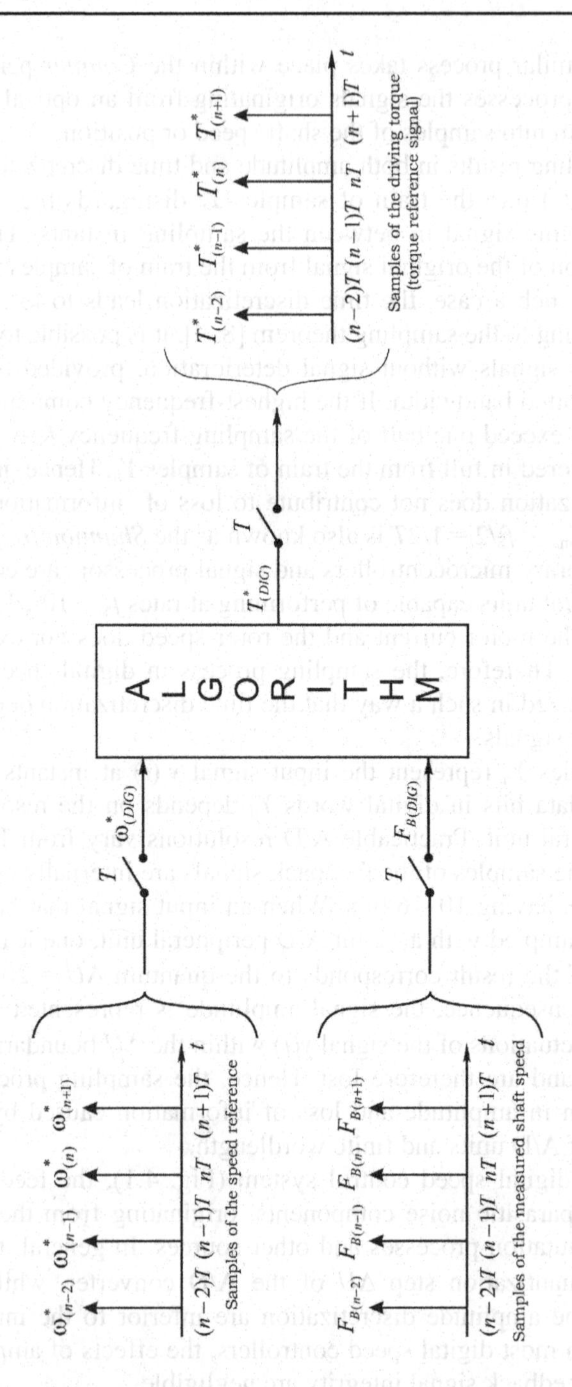

Fig. 4.3. The signal flow within the digital speed controller.

The signal flow within the digital speed controller is given in Fig. 4.3. The speed reference ω^* is available as a train of samples ($\omega^*_{(DIG)}$). At the instant $t = nT$, the speed reference equals $\omega^*_{(n)}$. The train of feedback samples is designated $F_{B\,(DIG)}$. The speed controller acquires the nth feedback sample $F_{B\,(nT)}$, evaluates the speed error $\Delta\omega_{(n)}$, and calculates the torque reference sample $T^*_{(n)}$. The train of torque reference pulses is given in Fig. 4.3 as $T^*_{(DIG)}$. The speed control algorithm may calculate the torque reference value from the present and past samples of the reference and the feedback signals, as well as the past samples of the driving torque:

$$T^*_{(n)} = f\left(F_{B(n)}, F_{B(n-1)}, F_{B(n-2)}, \ldots, \omega^*_{(n)}, \omega^*_{(n-1)}, \omega^*_{(n-2)}, \ldots, T^*_{(n-1)}, T^*_{(n-2)} \ldots\right). \quad (4.1)$$

The speed controller is linear when it calculates the new sample of the driving force as a weighted sum of the relevant samples (Eq. 4.2). In such a case, the weight coefficients G_i^F, G_i^ω, and G_i^T assume the role of the feedback gains. The numbers NF, $N\omega$, and NT in Eq. 4.2 determine the structure of the digital speed controller.

$$T^*_{(n)} = \sum_{i=0}^{NF} F_{B(n-i)} G_i^F + \sum_{i=0}^{N\omega} \omega^*_{(n-i)} G_i^\omega + \sum_{i=1}^{NT} T^*_{(n-i)} G_i^T \quad (4.2)$$

The torque reference $T^*_{(n)}$ is calculated from the input samples acquired at the instant $t = nT$. The new sample $T^*_{(n)}$ is available with a certain delay, required for the μC/DSP to evaluate the expression 4.2. The reference $T^*_{(n)}$ is then supplied to the torque actuator during one whole sampling interval T. In cases when the calculation delay is negligible, the torque reference is set to $T^*_{(n)}$ during the time interval $[nT, (n+1)T\,)$. At the next interrupt instant $t = (n+1)T$, the process repeats, resulting in a new reference $T^*_{(n+1)}$. Hence, the torque reference signal T_{ref} changes in a stepwise manner, assuming a new value $T^*_{(k)}$ at each sampling instance kT.

There are cases when the implementation of the digital speed control algorithm requires a great deal of time, making the calculation delay comparable to the sampling period T. In such cases, the torque reference $T^*_{(n)}$ is applied with a delay of one whole sampling period and remains at the torque actuator input during the time interval $[(n+1)T, (n+2)T\,)$.

Compared with analog implementation, the digital speed controller experiences signal deterioration due to time and amplitude discretization. At the same time, the numerical calculations involved contribute to additional transport delays. Nevertheless, digital speed controllers surpass the performance of their analog counterparts. The quantization effects and calculation delays are suppressed by using specialized, high-throughput motion control DSP platforms [10, 11]. With digital shaft sensors, such as optical encoders, the feedback signals are offset-free and noise-free and maintain their S/N ratio even at very low revolving speeds. A digital speed controller may include nonlinear compensators, an online parameter adaptation, and structural change mechanisms; it may employ finite-impulse response filters and a number of other features that are not available in analog implementation.

In digital controlled electrical drives, the speed control functions may be implemented simultaneously with the current controller (Fig. 4.1). The availability of sampling rates exceeding $f_S = 20$ kHz [10,11] allows for a current-loop bandwidth exceeding 1 kHz. Consequently, the actual torque reaches the reference T_{ref} within 100–200 μs. These dynamics of the torque actuator exceed the time constants of the mechanical subsystem by two orders of magnitude. Therefore, the analysis and design of digital speed controllers can be performed under the assumption that the torque actuator responds instantly to the reference pulses $T^*_{(n)}$. Exceptions to this conclusion need to be considered in cases when the inertia of the mechanical subsystem is extremely low and/or the target bandwidth of the speed loop reaches the 1 kHz range.

4.2 Analysis of the system with a PI discrete-time speed controller

In Fig. 4.1, one can distinguish both the digital and the analog parts of the system. The digital processor in Fig. 4.1 controls the speed of the mechanical subsystem by acquiring samples of feedback signals, processing these through the control algorithm, and obtaining samples of the driving force (Fig. 4.3). The digital part deals with the discrete-time signals, shown in Fig. 4.2 as a train of samples. Such signals reside within the RAM memory of the digital processor in the form of binary words of a finite wordlength. The analog section of the speed control system comprises the torque actuator, the mechanical subsystem, and the analog circuitry associated with the feedback acquisition. The signals involved in the analog part of the system are continuous-time, analog signals.

The analysis of the digital speed controller requires the appropriate modeling of the signal flow in both the analog and digital domains, taking into account the signal conversion between the two. The transition of continuous-time (analog) signals into discrete-time (digital) form is referred to as the *sampling process* and is illustrated in Fig. 4.2. The reconstruction of discrete-time pulse trains into their analog counterparts is discussed later and is given in Fig. 4.5A and Fig. 4.5B [8, 9].

4.2.1 The system with an idealized torque actuator and inertial load

For the purpose of the subsequent analysis, a block diagram of the speed-controlled system with a digital controller is given in Fig. 4.4. The torque actuator dynamics are neglected, assuming that the torque command samples T^* have an immediate effect on the actual driving torque, delivered to the shaft of the electric motor. The signal flow taking part within the digital controller is surrounded by a dashed line, separating the digital section from the rest of the system. The signal transition points between the digital domain (discrete time) and the analog domain (continuous time) are designated with a switch accompanied by the symbol T, standing for the sampling time period.

The speed reference signal $\omega^*(t)$ in Fig. 4.4 is sampled within each sampling interval T and converted into the train of pulses ω^*_{DIG}. In order to evaluate the speed error $\Delta\omega$, it is necessary to acquire the train of feedback pulses ω^{FB}_{DIG}. The speed feedback is obtained from the shaft sensor. In most cases, the continuous-time signals obtained from the sensor are processed in the associated interface circuits, their transfer function being designated in Fig. 4.4 by $W_M(s)$. The digital controller samples the resulting feedback signal F^B and acquires the train of pulses F^B_{DIG}. The feedback signal may correspond to the revolving speed or to the shaft position. When the sensing device is a brushed tachogenerator, the signal F^B is proportional to the speed. With an optical encoder or an electromagnetic resolver, the feedback signal represents the shaft position. In the latter case, the digital controller must evaluate the position-related samples F^B_{DIG} in order to provide digital representation of the speed feedback ω^{FB}_{DIG}. This operation is denoted as block W_{SE} in Fig. 4.4.

The speed error signal error $\Delta\omega$ is fed to the speed regulator (W_{SC}), which calculates the train of torque reference samples T^*_{DIG} intended to drive the error $\Delta\omega$ towards zero. The signal flow within W_{SC} is illustrated in Fig. 4.3, while Eq. 4.2 outlines the algorithm of the T^*_{DIG} calculation for the case of a linear controller. The structure and parameters in Eq. 4.2 determine the character and the bandwidth of the closed-loop step response. The present section is intended to provide the design procedure and the parameter setting for the digital speed controller W_{SC}, and, hence, to supply the proper form of the difference equation 4.2.

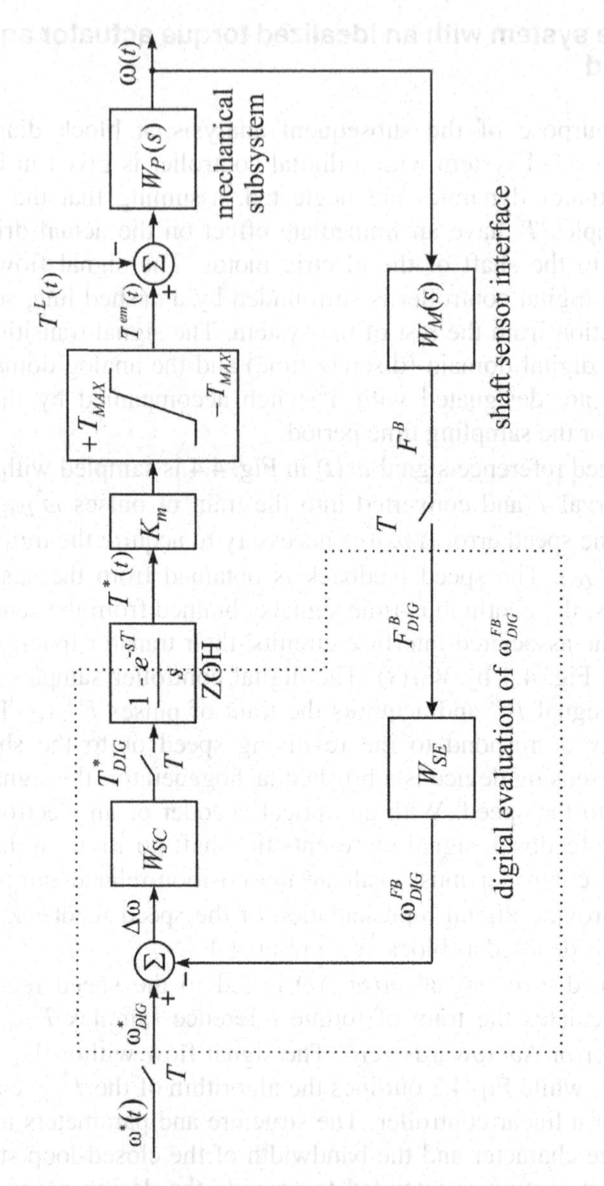

Fig. 4.4. Block diagram of a speed controlled system with digital implementation of the control algorithm. The signal flow taking part within the µC/DSP is surrounded by the dashed line.

The torque reference samples T^*_{DIG} are obtained at the output of the block W_{SC} (Fig. 4.4). In cases when the computation time is negligible with respect to the sampling period T, the torque reference sample $T^*_{(n)}$ is assumed to be available at the sampling instant $t = nT$. With longer computation delays, commensurate with the sampling period T, the torque reference $T^*_{(n)}$ could be put to use at the next sampling instant $t = (n+1)T$. The torque actuator, modeled as the gain block K_m with the associated limiter, requires a continuous-time torque reference $T^*(t)$. Therefore, the train of samples T^*_{DIG} has to be converted into a continuous-time signal. This task is accomplished within the block labeled ZOH in Fig. 4.4, performing the *zero order hold* (ZOH) function. The torque reference $T^*(t)$ is set to $T^*_{(n)}$ during the time interval $[nT, (n+1)T)$. At the instant $t = (n+1)T$, the process repeats, and the new sample $T^*_{(n+1)}$ of the torque reference is calculated and passed forward to the torque actuator. The ZOH function is detailed in Fig. 4.5A: the continuous-time signal $T^*(t)$ changes in a stepwise manner, assuming a new value $T^*_{(k)}$ at each sampling instance kT.

The transfer function of the ZOH block is found to be $W_{ZOH}(s) = T^*(s)/T^*_{DIG}(s)$, where $T^*_{DIG}(s)$ represents the Laplace transform of the pulse train T^*_{DIG}, given on the right in Fig. 4.3, while $T^*(s)$ stands for the Laplace transform of the continuous-time torque reference $T^*(t)$. The complex image $T^*_{DIG}(s)$ is found to be:

$$T^*_{DIG}(s) = \sum_{n=0}^{\infty} T^*_{(n)} \cdot e^{-(nT)s} .$$

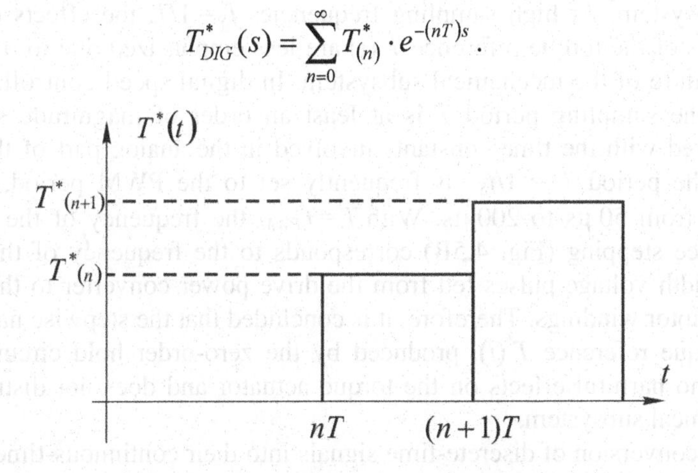

Fig. 4.5A. The zero-order-hold (ZOH) sets the torque reference $T^*(t)$ to $T^*_{(n)}$ during the time interval nT. At the instant $t = (n+1)T$, the new sample $T^*_{(n+1)}$ of the torque reference is calculated and passed forward.

The torque reference pulse applied during the interval $[nT, (n+1)T)$ and shown in Fig. 4.5A can be transformed into the *s*-domain. The complex image $T^*_{PLS(n)}(s)$ of the pulse is calculated as

$$T^*_{PLS(n)}(s) = T^*_{(n)}\left[\frac{e^{-nsT}}{s} - \frac{e^{-(n+1)sT}}{s}\right] = T^*_{(n)}\, e^{-nsT}\left[\frac{1-e^{-sT}}{s}\right].$$

The continuous-time torque reference $T^*(t)$ consists of a train of pulses similar to the one given in Fig. 4.5A. The representative waveform of the reference $T^*(t)$ is given in Fig. 4.5B. The Laplace transform of such a signal is calculated as the sum of $T^*_{PLS(n)}(s)$ elements:

$$T^*(s) = \sum_{n=0}^{+\infty} T^*_{PLS(n)}(s) = \left[\sum_{n=0}^{+\infty} T^*_{(n)}\, e^{-nsT}\right]\left[\frac{1-e^{-sT}}{s}\right] = T^*_{DIG}(s)\left[\frac{1-e^{-sT}}{s}\right].$$

Hence, the transfer function of the ZOH block is found to be:

$$W_{ZOH}(s) = \frac{1-e^{-Ts}}{s}.$$

The question arises as to whether the discontinuous changes in $T^*(t)$ (Fig. 4.5B) have an adverse effect on the torque actuator and the mechanical subsystem. At high sampling frequencies $f_S = 1/T$, the effects of step changes of the torque reference $T^*(t)$ are less emphasized due to the low-pass nature of the mechanical subsystem. In digital speed-controlled systems, the sampling period T is at least an order of magnitude smaller compared with the time constants involved in the analog part of the system. The period $T = 1/f_S$ is frequently set to the PWM period, which ranges from 50 μs to 200 μs. With $T = T_{PWM}$, the frequency of the torque reference stepping (Fig. 4.5B) corresponds to the frequency of the variable-width voltage pulses fed from the drive power converter to the electrical motor windings. Therefore, it is concluded that the stepwise nature of the torque reference $T^*(t)$, produced by the zero-order hold circuit, produces no harmful effects on the torque actuator and does not disturb the mechanical subsystem.

The conversion of discrete-time signals into their continuous-time counterparts can be performed with the *first-order hold* [8, 9]. The first-order hold (FOH) circuit produces continuous-time signals with the ramp-shaped change between the sampling instants. For the aforementioned reasons, the use of the FOH in digital speed controllers is not required.

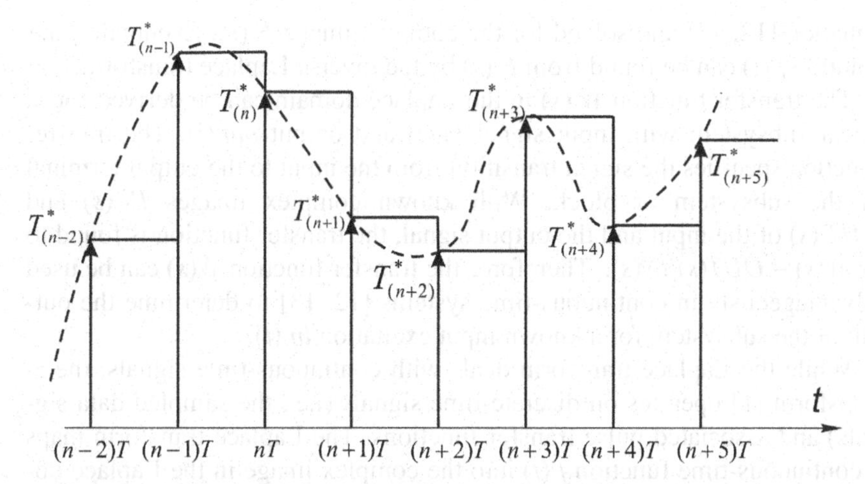

Fig. 4.5B. Reconstruction of the continuous-time torque reference from the train of torque reference samples by means of a zero-order-hold (ZOH).

Derived at the output of the ZOH block in Fig. 4.4, the torque reference $T^*(t)$ is fed to the torque amplifier. Having a negligible delay in the torque actuation, the torque amplifier reduces to a static-gain block K_m followed by a limiter, restricting the driving torque T^{em} to +/– T_{MAX}. The peak driving torque is limited by the permissible current in the semiconductor power switches within the power converter and by characteristics of the electrical motor. Finally, the driving torque $T^{em}(t)$ enters the mechanical subsystem and affects the revolving speed.

4.2.2 The *z*-transform and the pulse transfer function

Analysis of the digital speed controller and design of the proper control algorithm require evaluation of the open-loop transfer function. In Eq. 2.5, the open-loop transfer function $W_S(s)$ was found to be the product of the speed controller transfer function $W_{SC}(s)$ and the transfer function of the plant $W_P(s)$. In Chapter 2, the plant transfer function $W_P(s)$ was defined in the Laplace domain (Eq. 2.4), relating the complex images of the output speed $\omega(s)$ and the driving torque $T_{em}(s)$.

The Laplace transform of the continuous-time signal $f(t)$ results in the complex image $F(s)$. The Laplace transform proved particularly useful in solving linear differential equations. A linear differential equation, involving the continuous-time function $f(t)$ and its derivatives $\mathrm{d}^n f(t)/\mathrm{d}t^n$ with known initial conditions $\mathrm{d}^n f(0)/\mathrm{d}t^n$, can be reduced to an algebraic equation [12, 13] and solved for the complex image $F(s)$. Eventually, the solution $f(t)$ can be found from $F(s)$ by the inverse Laplace transform.

The transfer function $W(s)$ in the Laplace domain can be derived for a linear subsystem with input signal $in(t)$ and output $out(t)$. The transfer function specifies the signal transition from the input to the output terminal of the subsystem or block. With known complex images $IN(s)$ and $OUT(s)$ of the input and the output signal, the transfer function is found to be $W(s) = OUT(s)/IN(s)$. Therefore, the transfer function $W(s)$ can be used advantageously in continuous-time systems [12, 13] to determine the output of the subsystem for a known input excitation $in(t)$.

While the Laplace transform deals with continuous-time signals, the *z*-transform [1] operates on discrete-time signals (i.e., the sampled data signals) and associated pulse transfer functions. The Laplace transform maps a continuous-time function $f(t)$ into the complex image in the Laplace domain, also known as the *s*-domain. The *z*-transform maps the discrete-time signals into their complex representation $F(z)$, residing in the *z*-domain.

The *z*-transform converts the semi-infinite train of pulses $f_{(n)}$ ($n \in [0, +\infty]$), given in Fig. 4.2, into the complex image $F(z)$. The sample values $f_{(n)}$ in semi-infinite trains are equal to zero for $n < 0$. The *z*-transform of the discrete-time signal f_{DIG} is given in Eq. 4.3. In cases when discrete-time samples $f_{(i)}$ are obtained by acquiring the values of the continuous-time function $f(t)$ at the sampling instants $t = iT$ (Fig. 4.2), the *z*-transform $F(z)$ can be found from Eq. 4.4.

$$F(z) = f_{(0)}z^{0} + f_{(1)}z^{-1} + f_{(2)}z^{-2} + \ldots + f_{(i)}z^{-i} + \ldots = \sum_{i=0}^{\infty} f_{(i)}z^{-i} \tag{4.3}$$

$$F(z) = \sum_{i=0}^{\infty} f(iT)z^{-i} \tag{4.4}$$

The *z*-transform is a powerful tool for studying difference equations, such as Eq. 4.2. Linear difference equations involving the sampled data signals reduce to algebraic equations comprising the *z*-transforms of the relevant signals. After solving the algebraic equation and obtaining the $F(z)$, individual samples $f_{(i)}$ can be found by the inverse *z*-transform [1, 12, 13] (Eq. 4.5), wherein the integration contour encircles all the values of the argument *z*, resulting in $F(z)$ singularities.

$$f_{(i)} = f(iT) = \frac{1}{2\pi j} \oint F(z) z^{i-1}\, dz \tag{4.5}$$

Consider the discrete-time subsystem, such as the speed controller in Fig. 4.3. The transition of discrete-time signals from the input in_{DIG} ($\Delta\omega_{DIG}$) to the output out_{DIG} (T^*_{DIG}) can be described by means of the pulse transfer function $W(z)$, also known as the *discrete transfer function.* With known z-transforms of the input $IN(z)$ and the output $OUT(z)$, the discrete transfer function can be derived as $W(z) = OUT(z)/IN(z)$. For the given input and with a known pulse transfer function $W(z)$, the output train of pulses $out_{(n)}$ can be found as an inverse z-transform (Eq. 4.5) of the complex image $OUT(z) = W(z)IN(z)$.

The reference literature [1, 12, 13] *comprises* the tables that include typical continuous-time functions $f(t)$ with their s-domain images $F(s)$ and the z-domain counterparts $F(z)$, obtained by applying the z-transform on the $f(t)$ samples at the sampling instants $t = nT$. Some properties of the z-transform, relevant for further developments, are listed in [1]. Where needed, subsequent analysis throughout this book includes information on z-transform properties and practices. An effort is made to provide the reader with sufficient support to comprehend the analysis, design, and parameter setting of digital speed and position controllers. The reference literature [1, 12, 13, 14] comprises complete information on the z-transform, with rigorous proofs and developments. Some frequently addressed properties of the z-transform are addressed below.

The operators s and z, used in the Laplace and the z-transform, denote the differentiation and the time shift T, respectively. The Laplace operator may indicate the differentiation (s) or integration ($1/s$) [12, 13]. The complex image of the first derivative $df(t)/dt$ is found as $F_1(s) = \mathscr{L}(df(t)/dt) = sF(s) - f(0)$, where $F(s) = \mathscr{L}(f(t))$. The integral of $f(t)$ is transformed into $F(s)/s$. Likewise, the operator z stands for the time advancement by one sampling period T. In the z-domain, the complex image $zF(z)$ corresponds to the z-transform of the $f(t+T)$ samples (Eq. 4.6). Multiplication by the operator z^{-1} denotes the time delay by one sampling period. The z-transform obtained from the samples of the delayed function $f(t-T)$ is found in Eq. 4.7 as $z^{-1}F(z)$, where $f(-T) = 0$, since the train of $f_{(n)}$ pulses is semi-infinite ($f(t) = 0$ for $t < T$).

$$\sum_{i=0}^{\infty} f(iT+T)z^{-i} = z\left(\sum_{k=0}^{\infty} f(kT)z^{-k} - f(0)\right) = z\big(F(z) - f(0)\big) \quad (4.6)$$

$$\sum_{i=0}^{\infty} f(iT-T)z^{-i} = f(-T) + z^{-1}\sum_{k=0}^{\infty} f(kT)z^{-k} = z^{-1}F(z) \quad (4.7)$$

The time shift properties of the z-transform, given in Eq. 4.6 and Eq. 4.7, can be used to obtain the z-domain transfer function $W(z)$ of the subsystem described by a difference equation, such as Eq. 4.8. The values Y_{n+2}, Y_{n+1}, and Y_n, used in the difference equation 4.8, correspond to the samples of the output signal $y(nT+2T)$, $y(nT+T)$, and $y(nT)$, respectively. Likewise, the values X_{n+2}, X_{n+1}, and X_n correspond to the input samples $x(nT+2T)$, $x(nT+T)$, and $x(nT)$. The parameters A_2, A_1, A_0, B_1, and B_0 determine the transition of the discrete-time input pulses $X(z)$ into the output $Y(z)$. If we apply the time shift properties to the difference equation 4.8, the algebraic equation 4.9 is obtained, which relates the z-domain complex images of the input $X(z)$ and output $Y(z)$ signals. Eventually, the transfer function $W(z) = Y(z)/X(z)$ is found in Eq. 4.10. The derivation of the pulse transfer function $W(z)$ described above is used frequently throughout this book.

$$Y_{n+2} = A_2 X_{n+2} + A_1 X_{n+1} + A_0 X_n + B_1 Y_{n+1} + B_0 Y_n \tag{4.8}$$

$$z^2 Y(z) = A_2 z^2 X(z) + A_1 z X(z) + A_0 X(z) + B_1 z Y(z) + B_0 Y[z] \tag{4.9}$$

$$W(z) = \frac{Y(z)}{X(z)} = \frac{A_2 z^2 + A_1 z + A_0}{z^2 + B_1 z + B_0} \tag{4.10}$$

It is of interest to derive the pulse transfer function of the digital (discrete-time) integrator. To this end, recall that the complex image of the Heaviside step $h(t)$ in the s-domain is $H(s) = 1/s$. The samples of $h(t)$ at $t = kT$ instants are all equal to one. The z-transform $H(z)$ of this train of pulses is found in Eq. 4.11. Since $H(s) = 1/s$ maps into $H(z) = z/(z-1)$, the transfer function of the continuous-domain integrator $1/s$ will have the discrete-time counterpart $z/(z-1)$.

$$H(z) = \sum_{i=0}^{\infty} h(iT) z^{-i} = \sum_{i=0}^{\infty} z^{-i} = \frac{1}{1-z^{-1}} = \frac{z}{z-1} \tag{4.11}$$

In cases when the z-domain complex image $F(z)$ of an unknown discrete-time signal is available, the initial value $f(0)$ and the final value $f(\infty)$ of the pulse train can be found from the expressions given in Eq. 4.12 and Eq. 4.13. These expressions are known as the *initial-value theorem* (Eq. 4.12) and the *final-value theorem* (Eq. 4.13) [1]. The final-value theorem holds, provided that the denominator of the expression $(1 - z^{-1})F(z)$ has all of its roots within the unit circle of the z-plane.

$$f(0) = \lim_{z \to \infty} F(z) \tag{4.12}$$

$$f(\infty) = \lim_{z \to 1} \left(\frac{z-1}{z} F(z) \right) \tag{4.13}$$

The step response of a discrete-time system can be accessed from the closed-loop pulse transfer function $W_{SS}(z)$. Recall that the step response of a continuous-time system is determined by the poles and zeros of the relevant s-domain transfer function $W(s)$. The closed-loop bandwidth and the damping factor are observed from the roots of the characteristic polynomial $f(s)$, wherein the $f(s)$ resides in the denominator of the closed-loop system transfer function $W_{SS}(s)$. In the same way, the denominator of the closed-loop system pulse transfer function $W_{SS}(z)$ represents the characteristic polynomial $f(z)$. The roots of $f(z)$ are the closed-loop poles of the pulse transfer function, which determine the character and speed of the step response.

It is helpful to have an understanding of how specific values of the closed-loop poles in the z-domain affect the dynamic behavior of the closed-loop discrete-time system. Meaningful experience in relating the step response of continuous-time systems to the placement of the closed-loop poles in the s-domain can be put to use by mapping the s complex plane into the z counterpart. The poles and zeros in the z-plane can be mapped into their equivalents in the s-domain. The relation between complex variables z and s is given in Eq. 4.14. The function Ln(z) in the equation represents the logarithm function of a complex argument.

$$z = e^{sT}; \quad s = \frac{1}{T} Ln(z) \tag{4.14}$$

$$s_{1/2} = -\sigma \pm j\omega \quad \Rightarrow \quad z_{1/2} = e^{-\sigma} \cos(\omega T) \pm j e^{-\sigma} \sin(\omega T) \tag{4.15}$$

Eq. 4.14 maps the left half-plane of the s-domain into the unit circle in the z-domain. Hence, a discrete-time system is stable if all of its closed-loop poles have their absolute value smaller than one. The conjugate complex poles in the s-domain correspond to the z-domain conjugate complex poles, with their real and imaginary components being related in Eq. 4.15. The damping factor of the conjugate complex pair of poles $s_{1/2}$ determines the ratio between the real and imaginary component of the closed-loop complex poles in the z-domain. Real, stable s-domain poles are located on the negative half of the s-plane real axis ($\sigma \in [-\infty \,..\, 0]$). These poles are mapped to the interval [0 .. 1] of the real axis in the z-plane. The step response of a continuous-time system is faster, inasmuch as the real poles σ move further to the left in the s-plane, towards $-\infty$. The discrete-time system responds quickly, as the closed-loop poles shift towards the origin of the z-plane.

4.2.3 The transfer function of the mechanical subsystem

The torque reference in Fig. 4.4 assumes the form of the train of samples $T^*_{(n)}$. Processed through the zero-order hold block, the torque reference is fed to the torque actuator, affecting, in this way, the output speed and the feedback signal $F^B(t)$. Hence, the samples of the speed feedback $\omega^{FB}{}_{DIG}$ are affected by the torque reference samples $T^*_{(n)}$. Therefore, the plant transfer function of the system in Fig. 4.4 must represent the impact of the discrete-time signal T^*_{DIG} on the discrete-time feedback $\omega^{FB}{}_{DIG}$.

The mechanical subsystem is described by Eq. 4.16. The torque T^L comprises the load torque as well as the friction, the drag in the transmission elements, and other parasitic torque components. Parameter J represents the equivalent inertia of the system. The assumption made in Eq. 4.16 is that the parasitic torque components are decoupled from the internal dynamics of the system. Therefore, the signals summed into T^L can be treated as an external disturbance.

$$J\frac{\mathrm{d}\omega}{\mathrm{d}t} = T^{em} - T^L \tag{4.16}$$

The torque reference $T^*(t)$ in Fig. 4.4 is set to $T^*_{(n)}$ at the instant $t = nT$ and preserves that value until the next sampling instant $t = (n+1)T$, when the speed controller acquires another sample of the feedback and computes the torque reference $T^*_{(n+1)}$. With a negligible time lag in the torque actuator, the driving torque $T^{em}(t)$ in Fig. 4.4 tracks the reference $T^*(t)$ without delays and with no errors. According to Eq. 4.16, the shaft speed $\omega(t)$ changes with the driving torque and is affected by the external disturbance $T^L(t)$. The values of $\omega(t)$ at instants $t = nT$ and $t = (n+1)T$ are denoted by $\omega_{(n)}$ and $\omega_{(n+1)}$. The speed transition from $\omega_{(n)}$ to $\omega_{(n+1)}$ is given in Eq. 4.17.

$$\omega_{(n+1)} = \omega_{(n)} + \frac{1}{J} \cdot \int_{nT}^{nT+T} \left(T^*_{(n)} - T^L(t)\right) \mathrm{d}t$$
$$= \omega_{(n)} + \frac{T}{J} T^*_{(n)} - \frac{1}{J} \cdot \int_{nT}^{nT+T} T^L(t) \mathrm{d}t \tag{4.17}$$

While the driving torque in Eq. 4.17 does not change during the interval $(nT, (n+1)T)$, the generalized load torque $T^L(t)$ is an external disturbance that may fluctuate within the interval in an arbitrary way. This situation hinders the attempt to turn the differential equation 4.16 into a difference equation. Observe that the last factor in Eq. 4.17 represents the average value of the disturbance signals during the time interval $(nT, (n+1)T)$. Hence, the new shaft speed sample $\omega_{(n+1)}$ is uniquely affected by the average value $T^L_{(n)}$ of the signal $T^L(t)$, calculated in the present sampling period T according to the expression given in Eq. 4.18. If we introduce the disturbance average value from Eq. 4.18 into Eq. 4.17, the new speed sample $\omega_{(n+1)}$ is obtained as a weighted sum of the past sample $\omega_{(n)}$, the driving torque $T^*_{(n)}$, and the disturbance $T^L_{(n)}$, resulting in difference equation 4.19.

$$T^L_{(n)} = \frac{1}{T} \cdot \int_{nT}^{nT+T} T^L(t) \mathrm{d}t \tag{4.18}$$

$$\omega_{(n+1)} = \omega_{(n)} + \frac{T}{J} T^*_{(n)} - \frac{T}{J} T^L_{(n)} \tag{4.19}$$

The difference equation 4.19 describes the mechanical subsystem where the signal $T^*_{(n)}$ is the driving force, the speed $\omega_{(n+1)}$ is the output, and $T^L_{(n)}$ is the external disturbance. The transfer function $W_P(z)$ of the mechanical subsystem (plant) can be found to be $W_P(z) = \omega(z)/T^*(z)$, where $\omega(z)$ represents the z-transform of the output samples $\omega_{(n)}$ obtained for the given torque reference $T^*(z)$ in conditions when $T^L(t) = 0$. Applying the time shift properties of the complex operator z (Eqs. 4.6, 4.7), the speed and the torque complex images are related in Eq. 4.20. The plant pulse transfer function $W_P(z)$ is given in Eq. 4.21.

$$z\omega(z) = \omega(z) + \frac{T}{J} T^*(z) \Rightarrow (z-1)\omega(z) = \frac{T}{J} T^*(z) \tag{4.20}$$

$$W_P(z) = \frac{\omega(z)}{T^*(z)}\bigg|_{T^L=0} = \frac{T}{J}\frac{1}{z-1} \tag{4.21}$$

Similarly, the pulse transfer function $W_{PL}(z)$ can be derived by describing the impact of the load disturbance T^L on the output speed of the mechanical subsystem. Given the train of pulses $T^L_{(n)}$, comprising the average values of the signal $T^L(t)$ within individual sampling periods (Eq. 4.18), the z-transform $T^L(z)$ of the disturbance signal can be derived. The pulse transfer function describing the output response to the load disturbance is equal to $-W_P(z)$. The discrete-time transfer function $W_{PL}(z)$ can be found to be $\omega(z)/T^L(z)$ in conditions when $T^*(z) = 0$. From Eq. 4.19 and applying the steps given in Eq. 4.20, it is found that $W_{PL}(z) = -W_P(z)$. Finally, the z-transform of the output speed in conditions when both the load torque and the driving force are present is derived as $\omega(z) = W_P(z)\,(T^*(z)-T^L(z))$.

4.2.4 The transfer function of the speed-measuring subsystem

The speed feedback in digital speed-control systems is mostly obtained from optical encoders or electromagnetic resolvers. In the latter case, continuous time signals obtained from the sensor are processed in resolver-to-digital (R/D) interface circuits. The transfer function of such circuits is denoted by $W_M(s)$ in Fig. 4.4. Both the encoder and the resolver supply information on the shaft position, but not the speed. The former provides the position in the form of pulses. The digital controller (Fig. 4.1) comprises the input peripheral units, which are equipped with counters that accumulate the encoder pulses and provide the position data. With the optical encoder, there are no delays associated with the analog signal processing; therefore, $W_M(s) = 1$. When the electromagnetic resolver with the R/D converter is used, the transfer function $W_M(s)$ achieves bandwidths in excess of 1 kHz [2]. In most digital speed controllers, the R/D circuit dynamics are beyond the frequency range of interest, and the assumption $W_M(s) = 1$ is admissible.

The digital controller samples the shaft position $\theta_{(n)}$ (F^B in Fig. 4.4) and acquires the corresponding train of pulses θ_{DIG} (F^B_{DIG} in the same figure). Position samples must be processed further in order to obtain the desired speed feedback ω^{FB}_{DIG}. In continuous time, the shaft speed is obtained as the first derivative of the position. Discrete-time differentiation involves the calculation of the difference between the neighboring samples. This operation is designated by W_{SE} in Fig. 4.4.

It is of interest to find the pulse transfer function of the block W_{SE}. Consider the instant nT, when the digital controller acquires the sample $\theta_{(n)} = \theta(nT)$. The previously acquired sample $\theta_{(n-1)}$ has to be preserved, and it resides in the RAM memory of the digital controller. At this instant ($t = nT$), the speed feedback $\omega^{FB}{}_{(n)}$ is calculated from the position increment obtained within the past sampling period T (Eq. 4.22). The speed information obtained thusly corresponds to the average speed during the interval of time $[(n-1)T .. nT)$, and not to the actual shaft speed $\omega(nT)$ (Eq. 4.23).

$$\omega_{(n)}^{FB} = \frac{\theta_{(n)} - \theta_{(n-1)}}{T} \tag{4.22}$$

$$\theta_{(n)} = \theta_{(n-1)} + \int_{(n-1)T}^{nT} \omega \, dt \quad \Rightarrow \quad \omega_{(n)}^{FB} = \frac{1}{T} \int_{(n-1)T}^{nT} \omega \, dt \tag{4.23}$$

The relation between the speed feedback signal $\omega^{FB}{}_{(n)}$ and the shaft speed values $\omega_{(n)} = \omega(nT)$ at the sampling instants needs to be established. In Fig. 4.6, the change in the shaft speed between the two sampling instants is given. In cases when both the driving torque and the load disturbance T^L stay constant within the sampling interval T, the speed in Fig. 4.6 changes with a constant slope. In this case, the average speed during the interval T is proportional to the surface S and is found to be the average of the neighboring samples of the actual shaft speed $\omega_{(n-1)}$ and $\omega_{(n)}$ (Eq. 4.24).

The driving torque $T^{em}(t)$ is assumed to be equal to the reference $T^*(t)$. The torque reference is obtained from the zero-order hold circuit in Fig. 4.4. Therefore, the assumption that the driving torque does not change during the sampling interval T holds. In the absence of the load torque T^L, the shaft speed will change from $\omega_{(n-1)}$ to $\omega_{(n)}$ in a linear fashion. A linear speed change such as that shown in Fig. 4.6 is preserved even in the presence of the load disturbance, provided that the $T^L(t)$ stays constant within each individual sampling period. The latter is the case in the majority of practical speed-controlled systems. The possibility and effects of fast $T^L(t)$ fluctuations within the interval are discussed next in Section 4.3.

The difference equation 4.24 is transformed into Eq. 4.25, relating the z-transforms of the shaft speed samples $\omega(z)$ and the speed feedback signal $\omega^{FB}(z)$. Finally, the pulse transfer function $W_{SE}(z)$ of the speed-measuring system is given in Eq. 4.26.

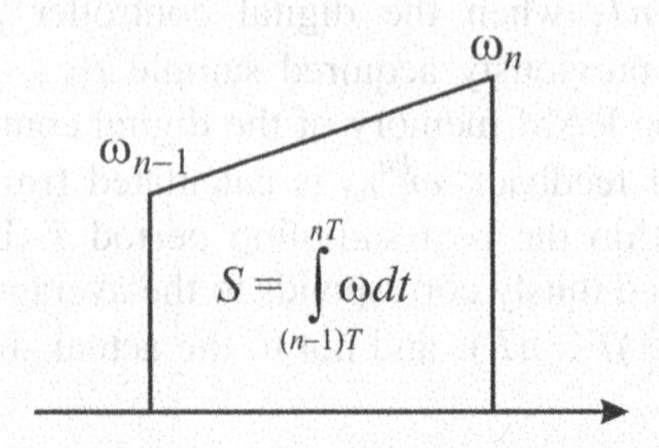

Fig. 4.6. The change in shaft speed between two sampling instants. When the load torque stays constant during the interval $[(n-1)T\ ..\ nT)$, the speed change is linear. The average value of the speed, proportional to the surface S, is found to be the average value of the neighboring samples $\omega_{(n)}$ and $\omega_{(n-1)}$.

$$\omega_{(n)}^{FB} = \frac{\omega_{(n)} + \omega_{(n-1)}}{2} \tag{4.24}$$

$$\omega^{FB}(z) = \frac{1}{2}\omega(z) + \frac{z^{-1}}{2}\omega(z) \tag{4.25}$$

$$W_{SE}(z) = \frac{\omega^{FB}(z)}{\omega(z)} = \frac{z+1}{2z} \tag{4.26}$$

4.3 High-frequency disturbances and the sampling process

The load torque $T^L(t)$ is an external disturbance that may take an arbitrary form. It is possible to envisage the frequency contents of T^L with spectral components that go beyond the sampling rate $f_S = 1/T$. In such cases, the load disturbance may change along the interval $[(n-1)T\ ..\ nT)$.

In Section 4.2.3, in the derivation of the transfer functions $W_P(z)$ and $W_{PL}(z)$ of the mechanical subsystem, the load disturbance was taken into account as the train of samples $T^L_{(n)}$, each sample being the $T^L(t)$ average value of the load within the individual sampling interval (Eq. 4.18). The conclusions drawn in this section, in particular the result $W_{PL}(z) = -W_P(z)$ and Eq. 4.21, are valid for an arbitrary form of $T^L(t)$.

On the other hand, Fig. 4.6 and the resulting Eq. 4.21 in Section 4.2.4 are based on the assumption that the shaft speed experiences a linear change from $\omega_{(n-1)}$ to $\omega_{(n)}$. The assumption does not hold in cases when the $T^L(t)$ changes are fast, which places into question the result obtained in Eq. 4.26. As indicated in Fig. 4.6 and given a parabolic change in speed, the surface S and the feedback signal $\omega^{FB}{}_{(n)}$ are not proportional to the average value of the neighboring samples $\omega_{(n)}$ and $\omega_{(n-1)}$. Therefore, it is necessary to discuss the possibility of experiencing high-frequency load disturbances and to evaluate their effects on the speed-controlled system and its model. The present section draws the conclusion that input signals at frequencies in excess of the Shannon frequency $f_{SH} = 1/2T$ cannot pass through the sampling circuit and the associated prefilters. High-frequency disturbances in a digital controlled system cannot be sampled, nor can they be suppressed by deliberate control action. Such signals are attenuated by the low-pass nature of the plant and are frequently referred to as the *unmodeled dynamics* [1].

The choice of sampling time T is the vital step in designing a digital speed controller. It is widely discussed in the reference literature [1, 12, 13] and will be addressed in subsequent chapters. In many practical cases, the speed loop sampling period T ranges from 50 μs to 200 μs. Hence, for the load disturbance $T^L(t)$ to assume meaningful changes within the interval T, it should comprise the spectral components at frequencies in excess of 1 kHz. In a practical speed-controlled system, the disturbance $T^L(t)$ is passed from the load to the motor by means of mechanical transmission elements. The inertia of the motor, the load mass, and the inertia of transmission elements act as a low-pass filter ($1/Js$), attenuating the high-frequency contents of the disturbance signal. Therefore, it is unlikely for the shaft speed signal to assume a form other than the one shown in Fig. 4.6.

From a more general point of view, it is interesting to distinguish between the low-frequency disturbance (LFD) signals and the high-frequency disturbance (HFD) signals. The effects of the former on the system output are to be removed by appropriate control action. The latter are beyond the capabilities of the control system and cannot be attenuated by deliberate control action. On the other hand, most plants have a low-frequency nature and do not react to a high-frequency excitation. The above discussion briefly summarizes the error-suppression strategy of most controllers. The output errors caused by LFD signals are put down in an active way, through the control effort, while the errors caused by HFD signals are suppressed in a passive way, relying on the low-pass nature of the plant.

The separation of LFD and HFD signals is related to the closed-loop bandwidth ω_{BW} of the controller, discussed in Section 2.1.1. With input and load disturbances at frequencies lower than ω_{BW}, the closed-loop controller is capable of suppressing the output error. For excitation frequencies higher than ω_{BW}, the controller and/or the torque actuator are not capable of providing the necessary drive for the plant, and the output signal departs from the reference value. Hence, the LFD signals are comprised within the bandwidth, while the HFD signals reside above the ω_{BW} level.

With a discrete-time controller, the choice of sampling time is conditioned by the frequency separating the LFD and HFD signals and the desired closed-loop bandwidth. The sampling process (Fig. 4.2) converts the analog signals into their discrete-time equivalent, losing some of the information due to time and amplitude discretization. According to the sampling theorem [1, 8, 9], the bandlimited signals with their highest frequency (f_{max}) components residing below the Shannon frequency $f_{SH} = f_S/2 = 1/2T$ are sampled without losing information. In other words, a bandlimited continuous-time signal $y(t)$ can be recovered in full from the train of samples Y_n. Moreover, the sampling of signals not complying with $f_{max} \leq f_{SH}$ introduces severe errors. With $y(t)$ comprising the frequency components at $f_{SH} + \Delta f$, the spectral content of the resulting samples Y_n would comprise alias (false) spectral components at $f_{SH} - \Delta f$ [8, 9]. If we consider the example where the sampling frequency equals 10 kHz ($T = 100$ μs), the presence of a 9 kHz component in $y(t)$ produces a false (*alias*) component at 1 kHz in the spectrum of the output pulse train Y_n. Ultimately, a false signal in the feedback path prevents the proper control of the system output variable.

In order to avoid alias components and ensure that the condition $f_{max} \leq f_{SH}$ is respected, the analog signals brought to the sampling circuits and A/D converters are processed through analog low-pass filters, intended to remove the frequency content above the Shannon frequency $f_{SH} = f_S/2 = 1/2T$. Such filters are referred to as *anti-aliasing filters*. Their goal is to remove any frequency component above one half of the sampling frequency. The anti-aliasing filters may be implemented as passive low-pass networks employing series resistors and parallel capacitors. Larger attenuation of high- frequency components requires active filters, comprising operational amplifiers. However, a complete removal of all the spectral content above the Shannon frequency can hardly be achieved. Recall at this point that the anti-aliasing filter feeds the signal to the A/D converter, which turns the sampled data into digital words having a finite wordlength.

In practice, the attenuation of an anti-aliasing filter is considered sufficient when the residual high-frequency content does not alter the digital representation of the sample Y_n by more than one least significant bit (LSB). With a 10-bit A/D converter having an input range of +/– 10V, the anti-aliasing filter must reduce the amplitude of residual high-frequency components below +/– 10mV. With an N-bit A/D converter, the suggested attenuation ratio of 2^{-N} is feasible in most cases.

The conclusion drawn from prior considerations is that the high-frequency feedback signals cannot pass through the sampling chain. The anti-aliasing filter suppresses the frequency content above the Shannon frequency $f_{SH} = 1/2T$. Moreover, the sampling block is incapable of processing inputs at frequencies higher than f_{SH}. Therefore, any input signal or load disturbance cannot enter the digital controller and does not have an impact on the reference torque. The aforesaid proves the consistency of results given in Fig. 4.6. and Eq. 4.24.

High-frequency dynamic processes are often called *unmodeled dynamics*. They correspond to HFD signals and are beyond the control capabilities of the closed-loop system. Although they do have an effect on the analog part of the system, the consequences are negligible due to the low-pass nature of the plant under control. An example of HFD signals is the PWM voltage pulses in three-phase inverters feeding AC motors and the associated ripple in motor currents.

4.4 The closed-loop system pulse transfer function

The closed-loop pulse transfer function $W_{SS}(z)$ represents the signal transition from the reference input $\omega^*(z)$ to the output speed $\omega(z)$. Poles and zeros of $W_{SS}(z)$ define the character of the step response and the closed-loop bandwidth. In this section, the closed-loop transfer function is derived for the system with a digital speed controller, given in Fig. 4.4. For the purpose of calculating the transfer functions, the system in Fig. 4.4 can be considered for small signals, and the torque limiter can be neglected. The torque actuator gain K_m multiplies the output of the PI speed controller, resulting in the effective proportional and integral gains being multiplied K_m times. Without lack of generality, the block K_m can be merged into W_{SC}, assuming that the torque actuator has a gain of one. A simplified block diagram of the closed-loop system is given in Fig. 4.7.

The mixed-signal block diagram in Fig. 4.7 comprises both continuous-time and discrete-time signals. The continuous-time output $\omega(t)$ can be turned into a train of samples $\omega_{(n)}$, the z-transform of such samples being

$\omega(z)$. The transfer function $W_P(z)$ of the mechanical subsystem is given in Eq. 4.21. This function relates the input samples of the driving torque to the output samples $\omega_{(n)}$. The pulse transfer function $W_{SE}(z)$ in Eq. 4.26 describes the calculations that result in the feedback signal $\omega^{FB}{}_{DIG}$. If we introduce $W_P(z)$ and $W_{SE}(z)$, the block diagram assumes the form in Fig. 4.8, comprising discrete-time signals.

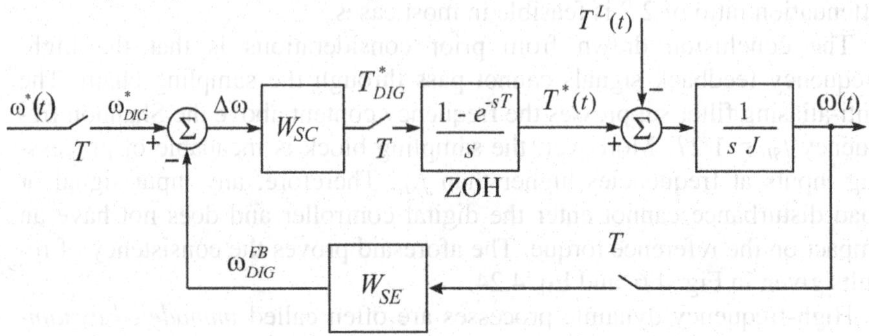

Fig. 4.7. Digital speed controller with ideal torque actuator with $K_m = 1$. The speed controller deals with discrete-time signals. The mechanical subsystem is described in terms of continuous domain variables.

In order to derive the open-loop transfer function $W_S(z) = W_{SC}(z)$ $W_P(z)W_{SE}(z)$, the pulse transfer function $W_{SC}(z)$ of the speed controller is required. The digital speed controller W_{SC} in Fig. 4.8 calculates the samples of the reference torque from the train of speed error pulses $\Delta\omega_{DIG}$. Equation 4.2 provides the driving force $T^*_{(n)}$ as a weighted sum of the past $\Delta\omega$ and T^* samples. The number of past samples involved and the corresponding weight coefficients determine the structure and actions of the digital speed controller. Section 2.2, which deals with continuous time controllers, shows that the proportional and integral control actions ensure rejection of step-shaped input and load disturbances. Therefore, it is assumed that the pulse transfer function $W_{SC}(z)$ comprises a discrete-time version of the proportional and integral action.

A discrete-time speed controller with proportional and integral action is given in Eq. 4.27. The new output sample $T^*_{(n)}$ is calculated at the sampling instant $t = nT$. The integral action in its discrete-time form comprises the sum of $\Delta\omega_{(j)}$ samples, starting with $\Delta\omega_{(0)}$ and ending with $\Delta\omega_{(n)}$. The feedback gains K_P and K_I correspond to proportional and integral gains. The gain setting should result in control actions that drive the speed error down to zero along the consecutive sampling periods. Affecting the pulse transfer function $W_{SC}(z)$, the feedback gains define the open-loop transfer function $W_S(z)$ and the closed-loop performance of the system.

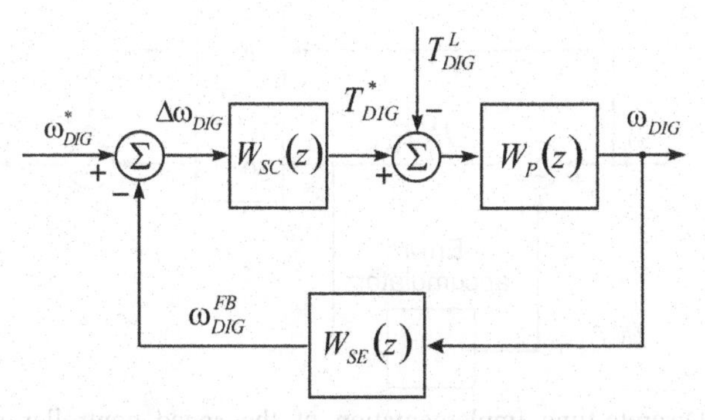

Fig. 4.8. The block diagram of a digital speed-controlled system comprising discrete-time signals. The output ω_{DIG} represents the train of samples $\omega_{(n)}$ of the shaft speed $\omega(t)$, acquired at sampling instants.

A straightforward implementation of the previous equation requires all of the speed error samples $\Delta\omega_{(j)}$ acquired in the interval [0 .. nT]. Storage of such samples in the internal memory of a digital controller cannot be achieved. A more practical approach consists of assigning a dedicated memory location for the sum of the speed errors ($INT_{(n)}$ in Fig. 4.9). At the instant $t = nT$, such an error accumulator INT is incremented by $\Delta\omega_{(n)}$. Multiplied by K_I, the error accumulator provides the integral action of the speed controller. In Fig. 4.9, the torque reference $T^*_{(n)}$ is calculated as K_I $INT_{(n)} + K_P\,\Delta\omega_{(n)}$.

$$\begin{aligned} T^*_{(n)} &= K_P\left(\omega^*_{(n)} - \omega^{FB}_{(n)}\right) + K_I \sum_{j=0}^{j=n}\left(\omega^*_{(j)} - \omega^{FB}_{(j)}\right) \\ &= K_P\,\Delta\omega_{(n)} + K_I \sum_{j=0}^{j=n} \Delta\omega_{(n)} \end{aligned} \tag{4.27}$$

If we introduce the transfer function $(1-z^{-1})^{-1}$ of the discrete time integrator, the pulse transfer function of the speed controller in Fig. 4.9 is given in Eq. 4.28.

$$W_{SC}(z) = \frac{T^*(z)}{\Delta\omega(z)} = K_P + K_I \frac{1}{1-z^{-1}} \tag{4.28}$$

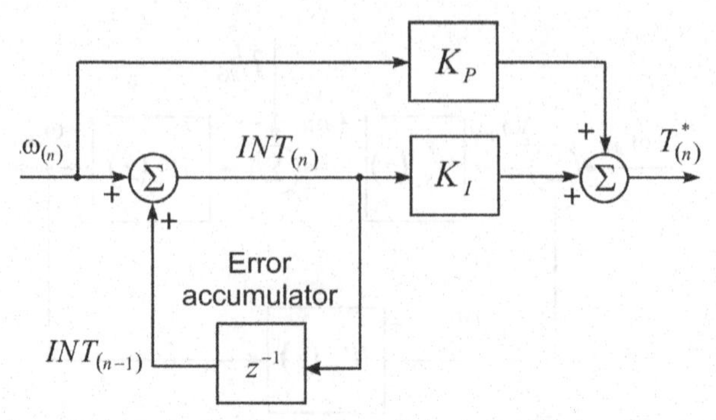

Fig. 4.9. Discrete-time implementation of the speed controller with proportional and integral action. The output $INT_{(n)}$ of the discrete-time integrator represents the sum of the past samples.

An alternative implementation of a discrete-time PI controller is shown in Fig. 4.10. The torque increments $\Delta T^*_{(n)}$ are calculated over each sampling interval and accumulated within the discrete-time integrator, given on the right in the figure. The error increment is calculated as $\Delta\omega_{(n)} - \Delta\omega_{(n-1)}$ and multiplied by the proportional gain, providing the increment of the proportional action. The increment of the integral action is found to be $K_I \Delta\omega_{(n)}$.

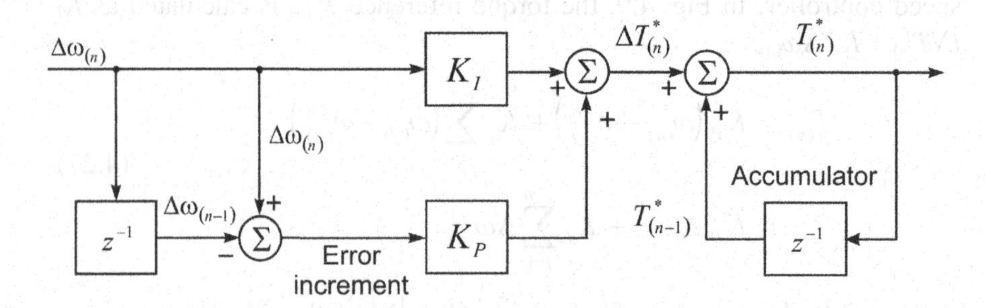

Fig. 4.10. The speed controller with proportional and integral action implemented in incremental form. The torque increments $\Delta T^*_{(n)}$ are accumulated within the discrete-time integrator on the right.

The operation of the structure in Fig. 4.10 is described by the difference equation 4.29. Using the properties of the operator z, the algebraic equation 4.30 is derived, relating the z-transforms of the speed error and torque reference. From Eq. 4.30, the transfer function of the speed controller $W_{SC}(z)$ is found, and is essentially the same as the one in Eq. 4.28. Hence, the two implementations of the speed controller outlined in Fig. 4.9 and

Fig. 4.10 result in the same pulse transfer function. The incremental form has some advantages in handling the operating conditions where the driving torque reaches the limits of the system, as will be explained later in this chapter.

$$T^*_{(n)} - T^*_{(n-1)} = K_P\left(\Delta\omega_{(n)} - \Delta\omega_{(n-1)}\right) + K_I\,\Delta\omega_{(n)} \tag{4.29}$$

$$T^*(z)\left(1 - z^{-1}\right) = K_P\Delta\omega(z)\left(1 - z^{-1}\right) + K_I\,\Delta\omega(z) \tag{4.30}$$

If we introduce $W_P(z)$ from Eq. 4.21, and $W_{SE}(z)$ from Eq. 4.26 and consider the pulse transfer function of the speed controller (Eq. 4.28), the closed-loop transfer function $W_{SS}(z)$ of the system in Fig. 4.8 is derived in Eq. 4.31. The closed-loop system transfer function is obtained as $\omega(z)/\omega^*(z)$ in conditions when $T^L_{DIG} = 0$. It has three closed loop poles and one zero.

The signal T^L_{DIG} in Fig. 4.8 represents the train of samples $T^L_{(n)}$, wherein each sample stands for the average value of the load torque within the sampling interval T (Eq. 4.18). The z-transform of this pulse train is the load torque complex image $T^L(z)$. With $\omega^*_{DIG} = 0$ and in the presence of the load disturbance, the output speed can be found as $\omega(z) = W_{LS}(z)T^L(z)$, with $W_{LS}(z)$ being the disturbance transfer function, reflecting the output susceptibility to load disturbances (Eq. 4.32).

$$W_{SS}(z) = \frac{\omega(z)}{\omega^*(z)}\bigg|_{T_L=0} = \frac{W_{SC}(z)W_P(z)}{1 + W_{SE}(z)W_{SC}(z)W_P(z)}$$

$$= \frac{z\left(K_P(z-1) + K_I z\right)\frac{T}{J}}{z(z-1)^2 + \frac{T}{2J}\left(K_P(z-1) + K_I z\right)(z+1)}$$

$$= \frac{\left(K_P + K_I\right)\frac{T}{J}z^2 - K_P\frac{T}{J}z}{z^3 - \left(2 - K_P\frac{T}{2J} - K_I\frac{T}{2J}\right) + \left(1 + K_I\frac{T}{2J}\right)z - K_P\frac{T}{2J}} \tag{4.31}$$

$$W_{LS}(z) = \frac{\omega(z)}{T^L(z)}\Big|_{\omega^*=0} = \frac{-W_P(z)}{1+W_{SE}(z)W_{SC}(z)W_P(z)}$$

$$= \frac{-\frac{T}{J}z^2 + \frac{T}{J}z}{z^3 - \left(2 - K_P\frac{T}{2J} - K_I\frac{T}{2J}\right) + \left(1 + K_I\frac{T}{2J}\right)z - K_P\frac{T}{2J}} \tag{4.32}$$

The denominator of $W_{SS}(z)$ and $W_{LS}(z)$ is the characteristic polynomial $f(z)$ of the system. The roots of the equation $f(z) = 0$ represent the closed-loop poles. Note in Eq. 4.31 and Eq. 4.32 that the feedback gains K_P and K_I appear multiplied by the factor $T/2J$. Therefore, it is convenient to introduce the normalized gains p and i, defined in Eq. 4.33. If we introduce p and i into the previous expressions, the closed-loop system transfer functions $W_{SS}(z)$ and $W_{LS}(z)$ assume the form given in Eq. 4.34.

$$p = K_P\frac{T}{2J},\; i = K_I\frac{T}{2J} \tag{4.33}$$

$$W_{SS}(z) = \frac{2(p+i)z^2 - 2pz}{z^3 - (2-p-i)z^2 + (1+i)z - p}$$
$$W_{LS}(z) = -\frac{T}{J}\frac{z^2 - z}{z^3 - (2-p-i)z^2 + (1+i)z - p} \tag{4.34}$$

4.5 Closed-loop poles and the effects of closed-loop zeros

The characteristic polynomial $f(z)$ resides in the denominator of Eq. 4.34. The three polynomial zeros σ_1, σ_2, and σ_3 are, at the same time, the closed-loop poles, determining the character and speed of the step response. The poles σ_1, σ_2, and σ_3 depend on the normalized gains p and i. In Eq. 4.35, $f(z)$ is expressed in terms of its zeros. If we equate the coefficients multiplying z^n on both sides of Eq. 4.35, a set of three equations is derived (Eq. 4.36), which relates the closed-loop poles to the normalized feedback gains p and i. Note that an arbitrary pole placement is not feasible, since the three poles are to be tuned by only two adjustable parameters.

$$\begin{aligned} f(z) &= z^3 - (2 - p - i)z^2 + (i+1)z - p \\ &= (z - \sigma_1)(z - \sigma_2)(z - \sigma_3) \end{aligned} \tag{4.35}$$

$$\begin{aligned} &\sigma_1 + \sigma_2 + \sigma_3 = 2 - p - i \\ &\sigma_1\sigma_2 + \sigma_2\sigma_3 + \sigma_3\sigma_1 = i + 1 \\ &\sigma_1\sigma_2\sigma_3 = p \end{aligned} \tag{4.36}$$

The numerator of $W_{SS}(z)$ is given in Eq. 4.37. It is a polynomial of the second order, having two zeros. The roots of the equation $num(z) = 0$ are the closed-loop zeros of the transfer function. The zero z_2 resides in the origin and designates the time shift of one sampling interval T. Hence, the presence of z_2 is beneficial, as it reduces delays introduced by the third-order polynomial $f(z)$ in the denominator. The closed-loop zero z_1 is positive and real and it lies within the unit circle. Therefore, it maps onto the negative side of the real axis in the s-plane.

$$num(z) = 2(p+i)z^2 - 2pz, \qquad z_1 = \frac{p}{p+i}, \quad z_2 = 0 \tag{4.37}$$

The closed-loop zero z_1 contributes to the derivative action of the transfer function. This differential nature of $W_{SS}(z)$ emphasizes the rising edge of the input and the load disturbances. Abrupt changes in the driving torque, produced by the derivative action, have an adverse effect on the mechanical subsystem and contribute to wear on the transmission elements. In cases when the load consists of distributed masses with elastic coupling, sudden torque changes give rise to mechanical resonance.

The closed-loop zero z_1 contributes to an overshoot in the step response. This effect has been discussed in Section 2.2.4 for a continuous-time speed-controlled system. For a discrete-time transfer function $W_{SS}(z)$ and its closed-loop zero z_1, the presence of an overshoot is illustrated in Fig. 4.11. The waveforms in the figure represent the step response obtained from the pulse transfer function $W_{SS}(z)$, given in Eq. 4.34. The closed-loop poles (σ_1, σ_2, and σ_3) and zeros (z_1 and z_2), obtained with the sample parameters $p = 0.15$ and $i = 0.01$, are given in the figure. The values of the closed-loop poles and zeros and the step response are obtained from Matlab, by entering the set of commands shown in Table 4.1 at the Matlab command prompt.

Table 4.1. The Matlab command sequence used to calculate the closed-loop poles and zeros and to obtain the step response for the transfer function $W_{SS}(z)$.

```
>> p = 0.15, i = 0.01                     % Setting the p and i gains
>> den = [1  -(2-p-i)  (1+i)  -p]         % Polynomial f(z) defined as den (Eq. 4.35)
>> num =  [2*(p+i)  -2*p   0 ]            % Numerator num(z) defined as num
>> roots(den)                             % Calculates zeros of f(z) (closed-loop poles)
>> roots(num)                             % Will calculate zeros of num(z) (c.l. zeros)
>> response = dstep(num,den)              % The step response samples will be
>>                                        % obtained from the Matlab function dstep
>>                                        % and stored in the array response
>> stairs(response)                       % Plotting the step response
```

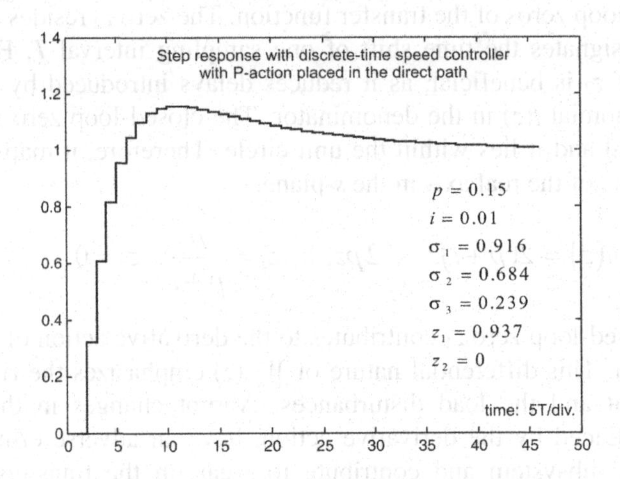

Fig. 4.11. The step response obtained from the closed-loop system transfer function $W_{SS}(z)$ given in Eq. 4.34 for the feedback parameters $p = 0.15$ and $i = 0.01$.

All the closed-loop poles obtained in Fig. 4.12 are real. In the absence of conjugate complex poles, an aperiodic step response is to be expected. However, the presence of real zero z_1 contributes to the overshoot. In numerous applications of servo drives, overshooting the setpoint is not acceptable. The overshoot may result in bringing the mechanical load and its vital parts, such as the tools, into a position where they may collide with other objects and eventually break. If we consider the response in Fig. 4.11, the driving torque must turn negative towards the end of the transient, resulting in the deceleration phase required to dissipate the excess speed. Frequent changes in the sign of the driving torque emphasize the backlash in the transmission elements, which gives rise to wear, and reduced controllability of the load speed and position.

To probe the relation between z_1 and the overshoot, the step response is derived for the system with the same characteristic polynomial (Eq. 4.35), yet without the real zero z_1. The resulting waveform is given in Fig. 4.12 and is aperiodic, without any overshoot. The sequence of Matlab commands required to obtain the step response in Fig. 4.12 is given in Table 4.2.

Table 4.2. The Matlab command sequence used to obtain the step response for the system with two real poles and no finite zeros.

>> p = 0.15, i = 0.01	% Setting the *p* and *i* gains
>> den = [1 -(2-p-i) (1+i) -p]	% Polynomial *f*(*z*) defined as in Eq. 4.35
>> num = [2*i 0 0]	% Numerator *num*(*z*) with zeros $z_1 = z_2 = 0$
>> roots(num)	% Will calculate zeros of *num*(*z*)
>> response = dstep(num,den)	% Obtaining the step response
>> stairs(response)	% Plotting the array response

A comparison of the step responses obtained in Fig. 4.11 and Fig. 4.12 indicates that real, closed loop zeros may contribute to the overshoot in the step response, even in cases when all of the closed-loop poles are real. Therefore, to avoid the overshoot, the speed controller design has to provide a closed-loop system transfer function without zeros. The solution to the problem is the relocation of the proportional gain, as explained in Section 2.2.4, for the continuous-time controller and analyzed in the next section for discrete-time implementation of the controller.

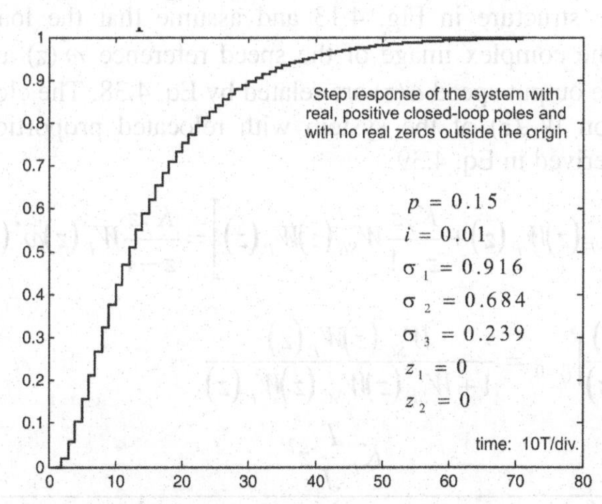

Fig. 4.12. The step response obtained from the closed-loop system transfer function with $num(z) = 2iz^2$ and with the characteristic polynomial $f(z)$ given in Eq. 4.35.

4.6 Relocation of proportional gain

The proportional gain of the speed controller (Fig. 4.9) multiplies the speed error. That is, the proportional action is located in the direct path. Alternatively, the structure of the speed controller can be changed into the form given in Fig. 4.13, where the gain K_P resides in the feedback path and multiplies the speed feedback. A similar operation is performed in Section 2.2.4, where the gain relocation affects the closed-loop zeros and leaves the closed-loop poles unaltered.

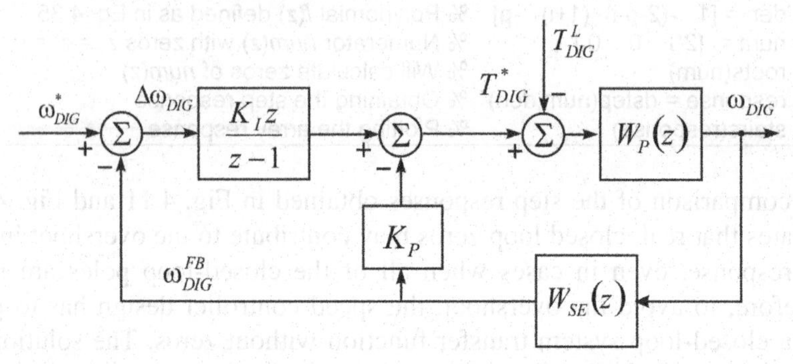

Fig. 4.13. Discrete-time speed controller with the proportional gain relocated in the feedback path.

Consider the structure in Fig. 4.13 and assume that the load torque $T^L = 0$. Then, the complex image of the speed reference $\omega^*(z)$ and the z-transform of the output speed $\omega(z)$ are related by Eq. 4.38. The closed-loop transfer function $W_{SS}(z)$ of the system with relocated proportional gain (Fig. 4.13) is derived in Eq. 4.39.

$$\omega(z)\left[1 + K_P W_{SE}(z) W_P(z) + \frac{K_I z}{z-1} W_{SE}(z) W_P(z)\right] = \frac{K_I z}{z-1} W_P(z) \omega^*(z) \quad (4.38)$$

$$W_{SS}(z) = \left.\frac{\omega(z)}{\omega^*(z)}\right|_{T_L=0} = \frac{W_{SC}(z) W_P(z)}{1 + W_{SE}(z) W_{SC}(z) W_P(z)}$$

$$= \frac{K_I \frac{T}{J} z^2}{z^3 - \left(2 - K_P \frac{T}{2J} - K_I \frac{T}{2J}\right) + \left(1 + K_I \frac{T}{2J}\right) z - K_P \frac{T}{2J}} \quad (4.39)$$

With the proportional gain in the feedback path, the input disturbance affects the driving torque through the integral action. Hence, the input pulsations will result in much smaller fluctuations of T^*, and the stress exercised on the mechanical subsystem will be reduced. The practical implementation of such a controller is given in Eq. 4.40. In comparison with Eq. 4.29, the only difference found is the proportional action, with K_P multiplying the shaft speed samples instead of the samples of the speed error.

$$\begin{aligned} \Delta T^*_{(n)} &= T^*_{(n)} - T^*_{(n-1)} = K_P\left(\omega_{(n-1)} - \omega_{(n)}\right) + K_I\,\Delta\omega_{(n)} \\ T^*_{(n)} &= T^*_{(n-1)} + \Delta T^*_{(n-1)} \end{aligned} \tag{4.40}$$

If we introduce the normalized gains $p = K_P \cdot (T/2J)$ and $i = K_I \cdot (T/2J)$, the closed-loop system transfer function is presented in the form given in Eq. 4.41. This proves that K_P relocation does not change the characteristic polynomial, which retains the form given in Eq. 4.35. Compared with the previous equation (4.34), the numerator becomes $num(z) = 2iz^2$. Hence, the real zero $z_1 = p/(p + i)$ is removed, and $W_{SS}(z)$ in Eq. 4.41 has two zeros at the origin. The relevant closed-loop step response is shown in Fig. 4.12. The output speed changes in an aperiodic way and does not overshoot the set speed. Depending on the specific requirements, both implementations of discrete-time speed controllers (Eq. 4.29 and Eq. 4.40) are applied in the field.

$$W_{SS}(z) = \frac{2iz^2}{z^3 - (2-p-i)z^2 + (1+i)z - p} \tag{4.41}$$

4.7 Parameter setting of discrete-time speed controllers

In this section, the setting of the adjustable feedback parameters is discussed. The objective is to achieve the aperiodic response of the shaft speed with the fastest response practicable. The objective is first formulated in a criterion function and expressed in terms of the feedback parameters p and i. Thereupon, the optimized values of the feedback parameters are found, resulting in the extreme value of the criterion function.

4.7.1 Strictly aperiodic response

The parameter setting delineated in this section provides the aperiodic character of the step response: namely, the closed-loop poles are real, and the output speed reaches the setpoint without overshoots. In many servo-system applications, the conjugate complex poles (Eq. 2.34) with the associated damped oscillations (Fig. 2.3) in the output speed and the driving torque give rise to some detrimental consequences.

A sample step response of a discrete-time speed controller with conjugate complex poles is given in Fig. 4.14. Given the closed-loop transfer function $W_{SS}(z)$ in Eq. 4.34 and the normalized gains of $p = 0.2$ and $i = 0.1$, the system has one pair of conjugate complex poles, $s_{1/2} = 0.70 \pm j\ 0.44$, and one real pole. The step response of this system is given in Fig. 4.14. Along with the output speed, the waveform corresponding to the driving torque is plotted as well. The traces in Fig. 4.14 are derived by using the sequence of Matlab commands listed in Table 4.1. In addition to the speed trace, it is interesting to observe the change in the driving torque during the transient. With $T^L = 0$, the driving torque is proportional to the first derivative of the output speed. In order to obtain the waveform representing the driving torque, the array *response*, obtained with dstep(num,den), is differentiated using the Matlab command *filter* (see Table 4.3). In this way, discrete-time differentiation is performed, as described by the transfer function $W_F(z) = (1 - z^{-1})$.

Table 4.3. The Matlab command sequence used to reconstruct the torque signal from the response of the output speed.

```
>> response = dstep(num,den)          % obtains the output speed response
>> torque = filter([1 -1],1,response) % torque(z) = (1 – z^-1) response(z)
```

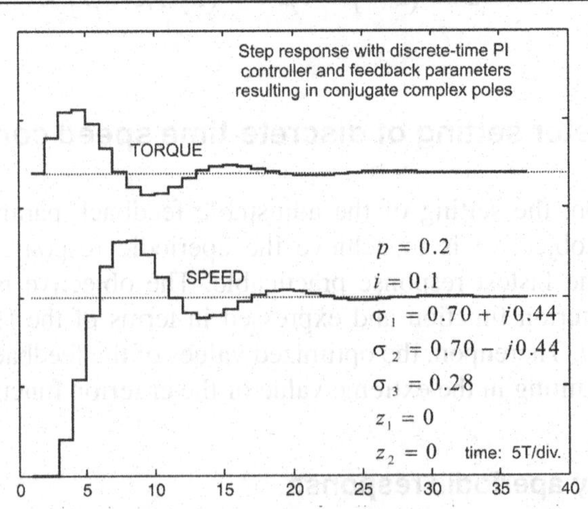

Fig. 4.14. The speed and torque response of a discrete-time speed controller with conjugate complex poles in the *z*-domain.

In Fig. 4.14, the driving torque exhibits damped oscillations. During the transient, the driving torque changes sign several times. The changes affect the stress exercised on transmission elements, providing the mechanical coupling between the motor and the load. In cases when the transmission includes gears, the backlash effect is noticed at each zero crossing of the torque/force. Due to a finite tolerance of mechanical parts, the load can make a small movement while the motor is stopped, and vice versa. Within the closed loop, this free travel is experienced as a tiny position step, modeled as $\Delta\theta = K_{BL}$ sign(T^*), with K_{BL} dependent on the tolerance of the gears. Hence, the torque oscillations emphasize the backlash effects and worsen the load position controllability.

The waveform corresponding to the shaft speed in Fig. 4.14 overshoots the setpoint and results in a negative speed error. During the transient, the speed error oscillates, changing in sign and reducing in amplitude. In the case when the system runs at a positive speed and decelerates towards a standstill, the motor shaft will stop and turn briefly in the negative direction. The resulting speed error will be dissipated in damped oscillations, similar to those shown in Fig. 4.14. In many cases, such an overshoot brings the mechanical parts into an undesirable position, with potentially harmful effects and risks. Therefore, in a number of applications, it is desirable to have the speed step response without an overshoot and the driving torque transient with no oscillations.

It is of interest to point out that the presence of conjugate complex poles is not uniquely related to the overshoot. The example given in Fig. 4.11 shows that a system with positive and real closed-loop poles may have an overshoot in cases when the closed-loop transfer function has positive, real zeros. On the other hand, it is possible to feature a system with conjugate complex poles and with no overshoot. This example is given in Fig. 4.15, where the step response of the output speed is given for the system with one pair of conjugate complex poles ($\sigma_{1/2} = 0.2 \pm \text{j}\ 0.8$) and one real, positive pole ($\sigma_3 = 0.8$). The real pole is dominant: that is, the time constant corresponding to $\sigma_3 = 0.8$ has the prevailing impact on the step response. The oscillations introduced by the conjugate complex pair $\sigma_{1/2}$ are damped before the speed approaches the setpoint. Therefore, the speed does not overshoot the reference. On the other hand, oscillations are observed in the driving torque, which has several zero crossings during the transient.

The results shown in Fig. 4.15 suggest that the aperiodic nature of the driving torque cannot be secured by achieving the step response without overshoot. Instead, all the closed-loop poles have to be real, residing on the interval (0 .. 1) of the real axis in the z-plane. The systems with this kind of pole placement are strictly aperiodic.

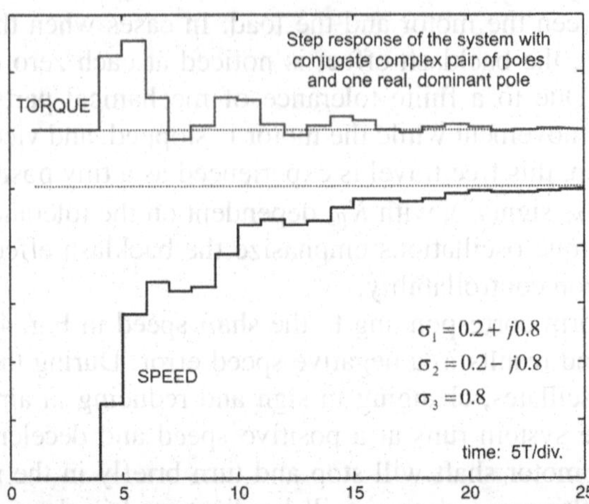

Fig. 4.15. The step response of the system with two conjugate complex poles and one real pole. The real pole $\sigma_3 = 0.8$ is dominant, and the speed step response does not have an overshoot.

The parameter setting of the PI speed controller consists of determining the values of the normalized gains p and i in such a manner that the closed-loop poles σ_1, σ_2 and σ_3 have the desired placement within the unit circle of the z-plane. Being the zeros of the characteristic polynomial $f(z)$ (Eq. 4.35), the closed-loop poles are related to the feedback gains by Eq. 4.36. A strictly aperiodic response imposes the following constraints on the closed loop poles:

$$\begin{aligned} &\mathrm{Im}(\sigma_1) = 0,\ \mathrm{Im}(\sigma_2) = 0,\ \mathrm{Im}(\sigma_3) = 0 \\ &0 < \mathrm{Re}(\sigma_1) < 1,\ 0 < \mathrm{Re}(\sigma_2) < 1,\ 0 < \mathrm{Re}(\sigma_3) < 1 \end{aligned} \tag{4.42}$$

Under the constraints in Eq. 4.36 and Eq. 4.42, the feedback gains are to be chosen so as to provide the fastest response possible and the maximum closed-loop bandwidth. A large bandwidth enables successful suppression of the speed fluctuations caused by the load torque disturbances. In Section 2.1.1, the bandwidth frequency ω_{BW} is related to the closed-loop poles in the s-domain. On the other hand, the z-domain equivalent z_m of the given s-domain pole s_m is obtained from $\sigma_m = \exp(s_m T)$. Hence, for the given sampling interval T, the bandwidth frequency ω_{BW} is determined by σ_1, σ_2, and σ_3. Higher bandwidth frequencies are obtained in cases when the z-domain poles are closer to the origin.

Note at this point that the sampling time has a considerable effect on ω_{BW}. For the given z-domain poles σ_1, σ_2, and σ_3, the closed-loop bandwidth ω_{BW} is proportional to the sampling rate $f_S = 1/T$.

4.7.2 Formulation of criterion function

The objectives of the parameter-setting procedure designed in this section must be formulated in a criterion function and expressed in terms of the feedback parameters p and i. The feedback gains are to be found from the optimization procedure, driven by the criterion function, while respecting at the same time the constraints given in Eq. 4.36 and Eq. 4.42.

Fig. 4.16 presents a strictly aperiodic step response of the shaft speed $\omega(t)$. At instant $t = 0$, the speed reference $\omega^*(t)$ steps to the new setpoint ω^*. The speed-error samples $\Delta\omega_{(n)} = \omega^* - \omega(nT) = \Delta\omega(nT)$ are strictly positive, as the speed does not overshoot the setpoint. The shaded area in Fig. 4.16 can serve as an indicator of the response speed. A faster response results in a smaller surface of the shaded area. The surface S is defined by the integral in Eq. 4.43. Given the discrete-time nature of the controller, the surface S can be expressed in terms of the speed error samples $\Delta\omega_{(n)}$. Hence, the integral S turns into the sum Q, given in Eq. 4.44. The value of Q will serve as the criterion function. The choice of the adjustable feedback parameters should drive Q down to the smallest possible values for the given constraints.

$$S = \int_{0}^{+\infty} \Delta\omega(t)\,dt \tag{4.43}$$

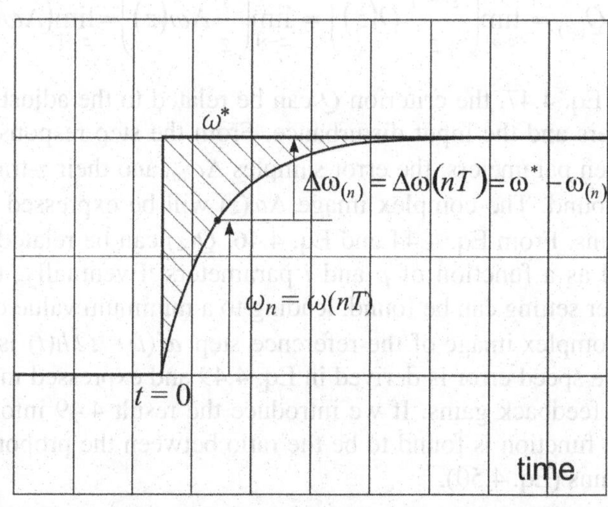

Fig. 4.16. A strictly aperiodic step response. The shaded surface corresponds to the speed error integral. The smaller the shaded area, the faster the step response.

$$Q = \sum_{k=0}^{\infty} \Delta\omega(kT) \tag{4.44}$$

It is necessary to express the criterion function Q in terms of p and i. To this end, consider the train of samples $Q_{(n)}$, defined in Eq. 4.45. Each element $Q_{(n)}$ of the series is the sum of the speed error samples for the interval $[0..nT]$. With $n \rightarrow \infty$, the series $Q_{(n)}$ converges to the criterion function Q. The discrete-time signal comprising the sample $Q_{(n)}$ has its complex image $Q(z)$ in the z-domain. The z-transform of the series $Q_{(n)}$ is given in Eq. 4.46, relating $Q(z)$ to the speed error $\Delta\omega(z)$.

$$Q_{(n)} = \sum_{k=0}^{n} \Delta\omega(kT) \tag{4.45}$$

$$Q(z) = \sum_{n=0}^{\infty} Q_{(n)} z^{-n} = \frac{1}{1-z^{-1}} \Delta\omega(z) \tag{4.46}$$

The criterion function can be found as the final value of the sample train, namely, the value of $Q_{(n)}$ obtained for $n \rightarrow \infty$. According to the final value theorem (Eq. 4.13), the criterion Q can be calculated from the complex image $Q(z)$ and hence, from the z-transform of the speed error:

$$Q = Q_{(\infty)} = \lim_{z \to 1}\left(\frac{z-1}{z} Q(z)\right) = \lim_{z \to 1}\left(\frac{1}{z} \Delta\omega(z)\right) = \lim_{z \to 1}\left(\Delta\omega(z)\right) \tag{4.47}$$

From Eq. 4.47, the criterion Q can be related to the adjustable feedback parameters and the input disturbance. From the step response of a system with given parameters, the error samples $\Delta\omega_{(n)}$ and their z-transform $\Delta\omega(z)$ can be found. The complex image $\Delta\omega(z)$ will be expressed in terms of p and i gains. From Eq. 4.44 and Eq. 4.46, $Q(z)$ can be related to $\Delta\omega(z)$ and obtained as a function of p and i parameters. Eventually, the optimized parameter setting can be found, leading to a minimum value of Q.

The complex image of the reference step $\omega^*(t) = \Omega^* h(t)$ is given in Eq. 4.48. The speed error is derived in Eq. 4.49 and expressed in terms of normalized feedback gains. If we introduce the result 4.49 into Eq. 4.47, the criterion function is found to be the ratio between the proportional and integral gains (Eq. 4.50).

$$\omega^*(z) = \frac{\Omega^*}{1-z^{-1}} \tag{4.48}$$

$$\begin{aligned}
\Delta\omega(z) &= \omega^*(z)-\omega(z)=\omega^*(z)(1-W_{SS}(z)) \\
&= \frac{\Omega^*}{1-z^{-1}}\left(1-\frac{2iz^2}{z^3-(2-p-i)z^2+(i+1)z-p}\right) \\
&= \frac{z\Omega^*}{z-1}\left(\frac{z^3+z^2(p-i-2)+z(1+i)-p}{z^3-(2-p-i)z^2+(i+1)z-p}\right) \\
&= \frac{z\Omega^*}{z-1}\,\frac{(z-1)\left(z^2+z(p-i-1)+p\right)}{z^3-(2-p-i)z^2+(i+1)z-p} \\
&= \frac{z\Omega^*\left(z^2+z(p-i-1)+p\right)}{z^3-(2-p-i)z^2+(i+1)z-p}
\end{aligned} \tag{4.49}$$

$$Q=Q_{(\infty)}=\lim_{z\to 1}(\Delta\omega(z))=\omega^*\left(\frac{p}{i}-\frac{1}{2}\right) \tag{4.50}$$

The criterion function Q can be minimized by applying the feedback gains p and i with the minimum possible ratio p/i. The gain selection is constrained by requirement 4.42, imposing the real and positive closed-loop poles positioned within the unit circle. From the above, the optimized parameter setting for the discrete-time speed controller with integral action in the direct path and proportional action in the feedback path can be formulated. For the fastest strictly aperiodic step response, the feedback gains should provide the minimum ratio p/i, respecting at the same time the constraint in Eq. 4.42. In the subsequent developments, the optimized values of normalized gains are found in a procedure searching for the maximum value of $Q_1 = i/p = 1/Q$.

4.7.3 Calculation of the optimized values for normalized gains

In this section, the optimized values of p and i are found, resulting in the fastest possible strictly aperiodic step response. The relation between the closed-loop poles σ_1, σ_2, and σ_3, and the values of normalized proportional and integral gains is given by the three relations in Eq. (4.36). Summing the three expressions, one obtains the constraint 4.51, restricting the pole placement. Namely, with only two adjustable feedback parameters, two out of three closed-loop poles can be set at will, while the third is set as a consequence and is calculated from Eq. 4.51.

$$(\sigma_1 + \sigma_2 + \sigma_3) + (\sigma_1\sigma_2 + \sigma_2\sigma_3 + \sigma_3\sigma_1) + (\sigma_1\sigma_2\sigma_3) = 3 \tag{4.51}$$

The proportional and integral gain can be expressed in terms of the closed-loop poles, by using the second and the third expressions in Eq. 4.36. The criterion function $Q_1 = i/p$ can be expressed in terms of the closed-loop poles as well:

$$Q_1 = \frac{i}{p} = \frac{\sigma_1\sigma_2 + \sigma_2\sigma_3 + \sigma_3\sigma_1 - 1}{\sigma_1\sigma_2\sigma_3}. \tag{4.52}$$

Further steps are facilitated by introducing the reciprocal values of the closed- loop poles, namely, $x = 1/\sigma_1$, $y = 1/\sigma_2$ and $v = 1/\sigma_3$. Given the constraint in Eq. 4.42, the values of x, y, and v are positive numbers larger than one ($x > 1$, $y > 1$, $v > 1$). Expression 4.51 assumes the form

$$xy + yv + vx + x + y + v + 1 = 3xyv$$

and therefore the third variable v can be expressed in terms of x and y,

$$v(x, y) = \frac{xy + x + y + 1}{3xy - x - y - 1} \tag{4.53}$$

resulting in the criterion function $Q_1(x, y)$ reformulated as follows:

$$Q_1 = \frac{i}{p} = x + y + v(x, y) - xyv(x, y) \tag{4.54}$$

$$= x + y + (1 - xy)\frac{xy + x + y + 1}{3xy - x - y - 1}.$$

At this point, it is necessary to find the values of x and y that result in the maximum possible $Q_1(x, y)$. The positive, real arguments x and y assume their values on the interval [1 .. +∞]. It can be proved that the function $Q_1(x, y)$ does not have a maximum at the boundaries of the interval. From Eq. 4.51, the following is concluded:

- With $x = y = +\infty$, the closed-loop poles $\sigma_1 = 1/x$ and $\sigma_2 = 1/y$ are at the origin, while the third pole remains outside the unit circle ($\sigma_3 = 3$), causing the instability.
- When $x = y = 1$, the poles $\sigma_1 = 1/x$ and $\sigma_2 = 1/y$ are at the unit circle in the z-plane. They are mapped in the s-domain as $s_{1/2} = 0$, introducing the double integrator into the closed-loop function and a step response that does not converge towards the setpoint.

- With either $x = 1$, $y = +\infty$, or $y = 1$, $x = +\infty$, the third pole $\sigma_3 = 1/v$ settles on the unit circle ($\sigma_3 = 1$), resulting in the s equivalent $s_3 = 0$ and a lack of output speed convergence towards the reference Ω^*.

Hence, the criterion function $Q_1(x, y)$ has its extremum within the region in the x–y plane defined by $x\in [1 .. +\infty]$ and $y\in [1 .. +\infty]$. Since the extremum does not reside on the region boundaries, its coordinates x_{OPT} and y_{OPT} can be found by equating the first derivates of $Q_1(x, y)$ to zero:

$$f_1(x,y) = \frac{\partial Q_1(x,y)}{\partial x} = 0, \quad f_2(x,y) = \frac{\partial Q_1(x,y)}{\partial y} = 0 .$$

If we apply partial differentiation, the functions $f_1(x, y)$ and $f_2(x, y)$ are obtained as

$$f_1(x,y) = \frac{(y-1)^2\left(2xy-3yx^2+y+x^2+2x+1\right)}{(3xy-x-y-1)^2}$$

$$f_2(x,y) = \frac{(x-1)^2\left(2yx-3xy^2+x+y^2+2y+1\right)}{(3xy-x-y-1)^2} .$$

Disregarding the solution ($x = 1$, $y = 1$), it is necessary to find the values $x > 1$ and $y > 1$ that satisfy the following equations:

$$2xy-3yx^2+y+x^2+2x+1=0, \quad 2yx-3xy^2+x+y^2+2y+1=0 .$$

From the first equation, the variable y can be expressed in terms of x:

$$y(x) = \frac{x^2+2x+1}{3x^2-2x-1} . \tag{4.55}$$

Introducing $y = y(x)$ into the second equation, one obtains:

$$4x\frac{\left(-1-4x-6x^2+3x^4\right)}{\left(3x^2-2x-1\right)^2} = 0 .$$

The solution $x = 0$ leads to unstable zeros of the characteristic polynomial $f(z)$. The remaining solutions are found as roots of the following equation:

$$H(x) = 3x^4 - 6x^2 - 4x - 1 = 0 .$$

The equation $H(x) = 0$ is of the fourth order. However, the coefficient next to x^3 is equal to zero. This facilitates solving $H(x) = 0$ for x. The four roots of the equation are found to be

$$x_1 = 1.7024; \; x_{2/3} = -0{,}3512 \pm j\,0{,}2692; \; x_4 = -1.$$

Given $\sigma_1 = 1/x$ and the strict aperiodicity constraint, the roots x_2, x_3, and x_4 are rejected. Introducing $x_1 = 1.7024$ in Eq. 4.53 and Eq. 4.55, one obtains

$$x_{OPT} = y_{OPT} = v_{OPT} = 1.7024\,.$$

Consequently, in order to obtain the fastest strictly aperiodic step response, the characteristic polynomial $f(z)$ has to assume the following form:

$$f(z) = (z-\sigma_1)(z-\sigma_2)(z-\sigma_3) = (z-\sigma)^3\,. \tag{4.56}$$

The optimized values of the closed-loop poles σ_1, σ_2, and σ_3 and the corresponding values of the normalized feedback gains p_{OPT} and i_{OPT} are given in Eq. (4.57):

$$\begin{aligned} &\sigma_1 = \sigma_2 = \sigma_3 = \sigma = 0.587 \\ &p_{OPT} = \sigma^3 = 0.2027 \\ &i_{OPT} = 3\sigma^2 - 1 = 0.03512\,. \end{aligned} \tag{4.57}$$

Note at this point that the values of the normalized gains p_{OPT} and i_{OPT} hold for any and all discrete-time PI controllers, whatever the plant and the torque actuator parameters, provided that the mechanical system can be represented by the inertial load and that delays in the torque actuations are negligible. The absolute gains K_P and K_I depend on the sampling period T and the inertia J:

$$K_{POPT} = 0.2027\frac{2J}{T}; \; K_{IOPT} = 0.03512\frac{2J}{T}\,. \tag{4.58}$$

In cases when the inertia J is altered at run time, Eq. 4.58 can be used to adjust the absolute gains according to the inertia changes.

At this point, it is worthwile to discuss some practical aspects in applying result 4.58 in setting the feedback parameters. In the simplified block diagram in Fig. 4.7 of the closed-loop speed controlled system, the speed reference, the feedback signal and the speed error are expressed in [rad/s]. The torque reference output in the same figure is expressed in [Nm]. As a consequence, the optimized gains given in Eq. 4.58 are expressed in [Nm/(rad/s)] units, namely, in terms of [Nm s/rad]. Without lack of generality, it is assumed in Fig. 4.7 that the torque actuator gain K_m can be merged into the W_{SC} block, supposing that the torque actuator has a gain of one.

In practical applications of digital speed controllers, signals such as the speed reference, the speed feedback, and the error signal, as well as the digital representation of the torque reference, are binary words residing in the RAM of the DSP controller. In most cases, the signals are represented as 16-bit signed integers. Such integers do not have dimension: that is, the digital representation of the torque reference of 10 hardly ever corresponds to the torque reference of 10 Nm. The ratio between the actual speed, expressed in [rad/s] and the digital word representing the speed is to be decided by the programmer. This decision is governed by the requirement to maximize the resolution in representing the speed and to minimize the quantization noise produced by the finite wordlength. One least significant bit (LSB) of the digital representation of the speed should correspond to one as-small-as-possible quantum of the actual speed, expressed in [rad/s]. At the same time, the digital representation of the shaft speed must be organized in such way that the top speed ω_{MAX} can be properly digitized and expressed in the form of an N-bit integer. In a system with 16-bit representation of the speed signal and with maximum speed of ω_{MAX}[rad/s], the ratio between the actual speed and its digital representation is to be set in such a way that 1 LSB corresponds to the quantum $\omega_Q = \omega_{MAX}/2^{15}$. The coefficient of proportionality between the digital and the actual speed can be designated as K_{FB}.

In a like manner, the ratio between the actual driving torque and its digital image residing in RAM can be denoted by K_M, corresponding to the overall gain of the torque controller. With 16-bit representation of the torque reference, and with maximum driving torque of T_{MAX}[Nm], the ratio between the actual, and the digitized torque has to be set for 1 LSB to represent the quantum $T_Q = T_{MAX}/2^{15}$. In this way, the 16-bit wordlength will be exploited for the best resolution and minimum quantization noise.

For the designer to apply the conclusions given above, it is necessary to analyze the signal flow in both the analog and digital domain. Said analysis requires information about the shaft sensor, the analog and digital interface circuits, and the peripheral units of the DSP controller, such as the A/D unit and the pulse detection/pulse generation peripherals.

In Fig. 4.17, the block diagram of a speed-controlled system is given, which introduces the scaling coefficients K_{FB} and K_M explained above. Said coefficients multiply the open-loop transfer function and affect the ratio between the optimized values of the feedback gains K_P and K_I and their normalized counterparts p and i. Expression 4.59 provides the means for the optimized setting of the feedback parameters. In order to set the absolute gains, it is necessary to provide the values for the sampling time T, the inertia J, and the scaling coefficients K_{FB} and K_M.

$$K_{P\,OPT} = 0.2027\frac{2J}{T}\frac{1}{K_M K_{FB}},\; K_{I\,OPT} = 0.03512\frac{2J}{T}\frac{1}{K_M K_{FB}}. \quad (4.59)$$

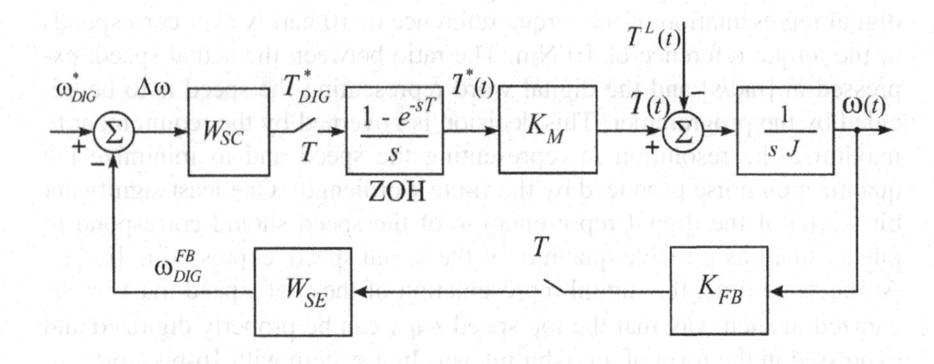

Fig. 4.17. A practical digital speed controller indicating the scaling coefficients K_{FB} and K_M and establishing the ratio between the actual speed and torque signals and their finite-wordlength digital representations T^*_{DIG} and ω^*_{DIG}, residing in RAM.

4.8 Performance evaluation by means of computer simulation

To verify the findings from the previous section and further probe the dynamic performance of discrete-time speed controllers, the system in Fig. 4.17 has been modeled and simulated using the Matlab and Simulink tool. In the sample speed-controlled system, the inertia is set to $J = 0.11$ kgm^2, the sampling time to $T = 0.001$ s, and the scaling coefficients to one. The model, shown in Fig. 4.18, takes into account the limited resolution of the shaft sensor (note the block titled *limited resolution of position reading* at the bottom right in the Figure). For proper use of the model, the parameters K_P, K_I, K_{FB}, K_M, and J must be entered to provide gains for the relevant Simulink blocks. The sampling time needs to be defined by entering $T_S = 0.001$ at the Matlab command prompt. A summary of the parameters that have to be set for the model to run is given in the center of Fig. 4.18.

The speed controller is implemented according to the block diagram given in Fig. 4.13. The integral control action processes the speed error, while the proportional gain multiplies the speed feedback. Within the model, the speed reference block and the load torque block provide the input excitation and the disturbance to the model. Note in Fig. 4.18 that the speed feedback signal is estimated from the shaft position sample in the manner prescribed in Eq. 4.22. Therefore, the speed feedback corresponds to the average speed within the past sampling period and differs slightly from the running speed (Eq. 4.22). The block *measured speed* and the block *torque reference* capture samples of the speed feedback and the torque reference and store them in the arrays *speed* and *torque* for plotting and further processing. In order to suppress the errors in simulating the mechanical subsystem, the simulation step is to be set to a value of $T_S/10$ or smaller.

In Fig. 4.19, the simulation traces are plotted for the speed feedback and the torque reference. The traces are obtained from the model in Fig. 4.18. In the simulation, the speed reference is stepped up at the beginning of the session. In the second half, the load step is applied to the system. The feedback gains of the system are set according to Eq. 4.59, while the effects of the finite resolution are neglected. In Fig. 4.19, the time division is set to five sampling periods per division, leading to a total simulation time of 50 sampling periods.

The simulation model in Fig. 4.18 does not take into account nonlinearities such as the system limits imposed on the driving torque. Therefore, the simulation traces obtained correspond to the operating conditions when the input and the load disturbances are limited. Such disturbances do not involve the driving torque transients reaching the torque limit of the servo drive. Consequently, the traces in Fig. 4.19 correspond to the linear regime of the speed-controlled system.

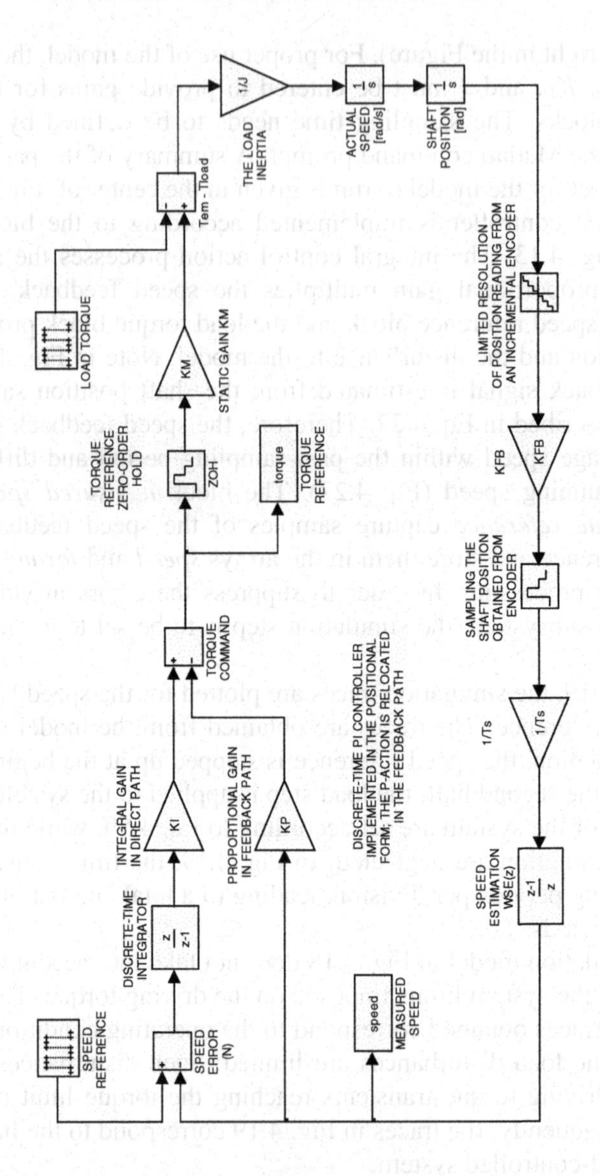

Fig. 4.18. Simulink model of a speed controlled system with a digital PI controller. The proportional gain is relocated in the feedback path. The model takes into account the limited resolution of the shaft sensor.

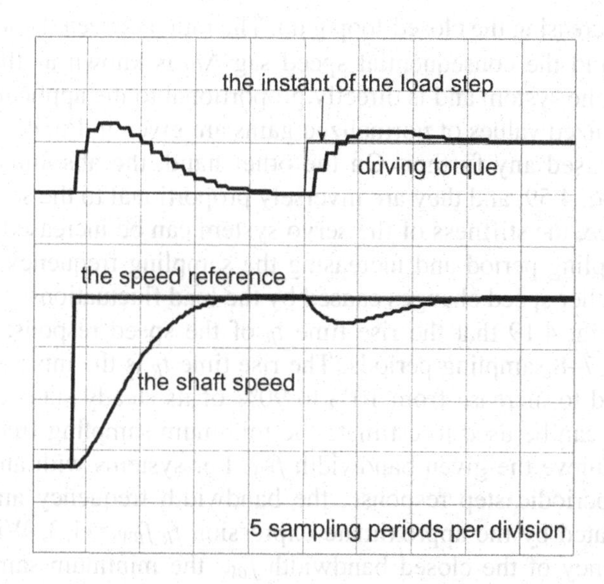

Fig. 4.19. Response of a discrete-time speed controller to the input and the load step disturbance. The PI controller has proportional gain relocated into the feedback path. The feedback gains are set according to Eq. 4.59, while the shaft sensor resolution is assumed to be infinite.

The trace corresponding to the shaft speed in Fig. 4.19 is strictly aperiodic and does not overshoot the setpoint. While the speed approaches the setpoint, the driving torque does not change its sign and remains strictly positive. Within the load step response, shown on the right in Fig. 4.19, the driving torque increases in an attempt to balance for the load step, while the speed experiences a sag. In the final stage of the transient, the driving torque exceeds the load torque, in order to provide for a brief acceleration required to bring the speed up to the reference value. Hence, the driving torque exceeding the load is not a manifestation of an overshoot, nor does it contradict the strict aperiodicity. Instead, an inevitable excess torque is required to suppress the speed sag.

The load rejection capability of the speed-controlled system depends on the feedback gains. According to Eq. 2.28, the speed sag encountered upon the load step is inversely proportional to the absolute value of the feedback gain K_I. The principal task of the speed controller is to keep the actual speed on the reference profile and to suppress the speed errors caused by the load fluctuations. Therefore, it is of interest to minimize the load impact by increasing the closed-loop gain. The ratio between the load disturbance ΔT^L and the consequential speed sag $\Delta\omega$ is known as the *stiffness* (rigidity) of the system and is directly proportional to the applicable gain.

The optimized values of normalized gains are given in Eq. 4.57 and cannot be increased any further. On the other hand, the absolute gains are defined in Eq. 4.59, and they are inversely proportional to the sampling period T. Hence, the stiffness of the servo system can be increased by reducing the sampling period and increasing the sampling frequency. This will result in smaller speed changes caused by the load fluctuations.

Note in Fig. 4.19 that the rise time t_R of the speed response takes approximately 7–8 sampling periods. The rise time t_R is the interval required for the speed to increase from 10% to 90% of its steady state-value. This information can be used to estimate the minimum sampling frequency required to achieve the given bandwidth f_{BW}. For systems with an aperiodic or nearly aperiodic step response, the bandwidth frequency and the rise time are related by the approximate expression $t_R\, f_{BW} = 1/3$. With the desired frequency of the closed bandwidth f_{BW}, the minimum sampling frequency $f_S = 1/T$ can be estimated as $f_S = 21 f_{BW}$. As an example, for the closed-loop bandwidth $f_{BW} = 200$ Hz, the sampling frequency must be at least $f_S = 4200$ Hz, and the sampling period has to be smaller than $T = 238$ μs. In an attempt to increase the stiffness of the systems, the sampling frequencies are often set to even higher values than indicated in the example.

The simulation results explained above assume that the shaft sensor reads the position with infinite resolution. In Fig. 4.20, the input step and the load step responses are given for the case when the shaft sensor has a finite resolution. It is assumed that 1 LSB of the position reading corresponds to the quantum of 0.1[mrad]. Note in Fig. 4.20 that both the torque reference and the speed estimate signal contain a parasitic component produced by quantization effects. The noise spectral content is related to the sampling frequency. In a practical servo system, such parasitic components of the driving torque have an adverse effect. They increase the tracking error, create an audible noise, accelerate wear, and may give rise to mechanical resonance phenomena. Therefore, every effort should be made to increase the resolution of the position reading and the speed estimation and to alleviate the quantization effects.

The number of encoder pulses per mechanical turn of the motor is limited by the minimum width of the dark and transparent windows printed on the circumference of the glass disk. The resolution of resolver-based systems is restricted by the characteristics of R/D converters [2] and by the precision of the resolver construction. An increase in resolution of speed reading can be achieved by using advanced features of the pulse detection peripherals built into motion-control DSP chips [10, 11]. Parallel reading of both the width of the encoder pulses and the number of pulses detected within each sampling period provides a resolution increase of an order of magnitude. A new generation of optical encoders [15] provides both traditional digital signals with N pulses per turn and their analog, sinusoidal counterparts with N periods of the sine wave within each turn. Such devices are known as *sincoders*. The presence of analog signals provides the means for position interpolation within the one-pulse boundaries, thus increasing significantly the effective resolution of position reading.

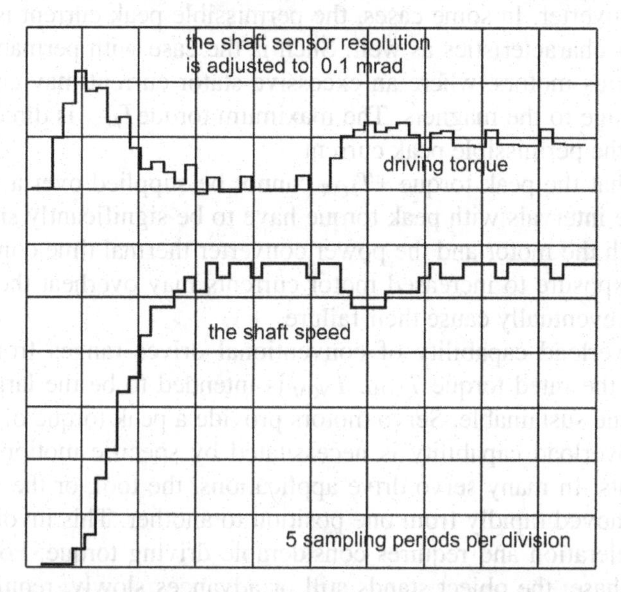

Fig. 4.20. The step response obtained with the resolution of the shaft-sensor position reading of 0.1 mrad. The torque reference and the measured speed contain the quantization noise.

4.9 Response to large disturbances and the wind-up phenomenon

The step response plotted in Fig. 4.19 corresponds to the linear operating mode, where the shaft speed and the driving torque do not reach the system limits. When we consider the response to the step input, the amplitude of the torque transients is proportional to the disturbance. While the system remains in linear mode, the changes in amplitude of the speed step will not affect the rise time and the character of the response. Instead, the speed and torque traces will change in proportion to the input step. At a certain point, the torque transient required will reach the maximum torque available. A further increase in the input excitation will drive the system into a nonlinear operating mode, as the driving torque will be limited to $\pm T_{MAX}$ boundaries.

The maximum torque T_{MAX} available at the shaft of the servo motor is limited by the characteristics of the drive power converter and the motor itself. The peak current available in the motor windings is limited by the characteristics of the semiconductor power switches constituting the drive power converter. In some cases, the permissible peak current is limited by the motor characteristics as well. Such is the case with permanent magnet synchronous motors, where an excessive stator current may cause permanent damage to the magnets. The maximum torqueT_{MAX} is directly proportional to the permissible peak current.

Note that the peak torque $\pm T_{MAX}$ cannot be supplied over a longer time span. The intervals with peak torque have to be significantly smaller compared with the motor and the power converter thermal time constants. Prolonged exposure to increased motor currents may overheat the vital drive parts and eventually cause their failure.

The overload capability of conventional drives ranges from 120% to 200% of the rated torque T_{NOM}. T_{NOM} is intended to be the largest steady-state torque sustainable. Servo motors provide a peak torque of 5–10 T_{NOM}. A high overload capability is necessitated by specific motion-control requirements. In many servo drive applications, the tool, or the work piece, is to be moved rapidly from one position to another. This involves significant acceleration and requires considerable driving torque. Following the motion phase, the object stands still or advances slowly, requiring only a small amount of the torque to account for friction and gravity.

The block diagram of the system, accounting for the torque limit $\pm T_{MAX}$, is given in Fig. 4.21. The torque limit is modeled as a nonlinear block placed at the output of the speed controller. It is valuable to investigate the speed step response obtained for a large input disturbance, which can drive the torque to the limit. In order to simulate the large step response, the Simulink model of the speed-controlled system from Fig. 4.18 is modified to include the torque limit. Fig. 4.22A illustrates the necessary changes. The simulation results given in Fig. 4.23 present the speed and the torque traces obtained in conditions when a large input step drives the torque into saturation. Namely, the amplitude of the input disturbance is such that the torque requirement exceeds the limit $\pm T_{MAX}$. With the torque limit activated, the system shown in Fig. 4.21 operates in a nonlinear mode, resulting in a transient response dissimilar to the linear case (Fig. 4.19).

Note in Fig. 4.23 that the driving torque quickly reaches the limit T_{MAX} and remains on the same level for approximately 100 sampling intervals.

Due to the constant acceleration, the shaft speed increases linearly. It reaches the setpoint in 60 sampling periods, taking much longer than the rise time of 7–8 T in Fig. 4.19.

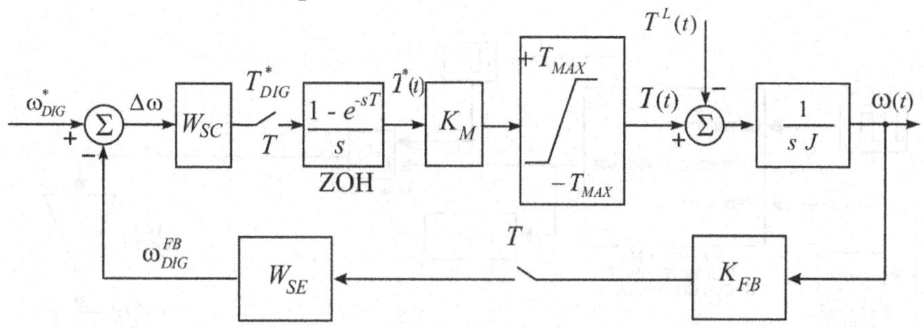

Fig. 4.21. The speed-controlled system with limited torque capability of the actuator. The driving torque is limited to $\pm T_{MAX}$, thus introducing the nonlinear block into the direct path.

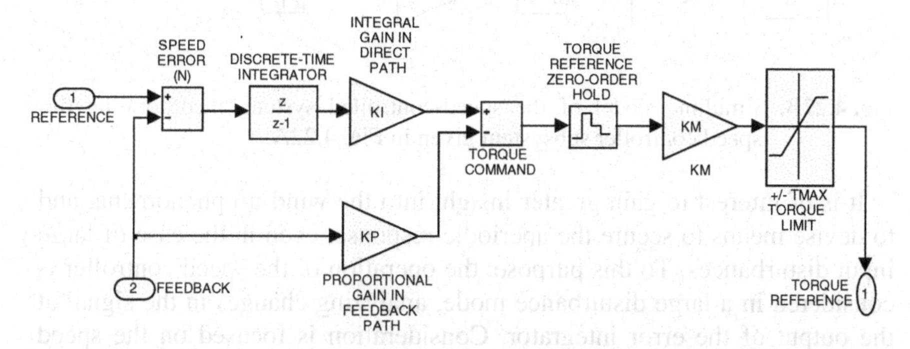

Fig. 4.22A. Simulink subsystem comprising the speed controller, with the speed reference and the speed feedback at the input terminals (left), and the torque reference at the output (right). Simulink model from Fig. 4.18 is modified to account for the limited torque capability of the torque actuator. This subsystem is used in the Simulink model in Fig. 4.22B.

Note in Fig. 4.23 that the driving torque remains at the positive limit even after the speed reaches the setpoint. Therefore, the speed overshoots the reference by almost 60–70 %. During the transient, the torque retains the absolute value of T_{MAX}, while changing the sign. The speed error oscillations are triangular shaped, with a gradual decay in amplitude. As the oscillations reduce in amplitude, their frequency increases, giving evidence that the system is nonlinear. Namely, the oscillations experienced in linear systems, caused by poorly damped conjugate complex poles, do not change in frequency. Instead, the oscillating phenomena in Fig. 4.23 originate from an interaction between the system nonlinear elements, such as the torque limiter and the speed error integrator, contained within the speed controller. Such a detrimental interaction is known as the *wind-up*.

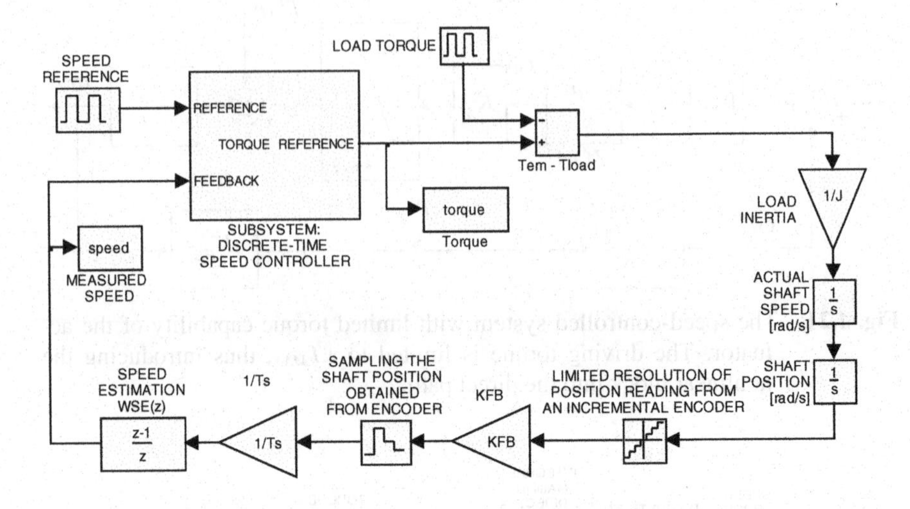

Fig. 4.22B. Simulink model of the speed-controlled system, incorporating the speed controller subsystem given in Fig. 4.22A.

It is of interest to gain greater insight into the wind-up phenomena, and to devise means to secure the aperiodic response, even in the case of large input disturbances. To this purpose, the operation of the speed controller is considered in a large disturbance mode, analyzing changes in the signal at the output of the error integrator. Consideration is focused on the speed controller structure given in Fig. 4.13 and represented by the Simulink model, plotted in Fig. 4.22A. In both diagrams, the error integrator accumulates speed error samples and multiplies them by the integral gain K_I. The wind-up is illustrated in Fig. 4.24, which presents the traces of the speed reference, the shaft speed, the error integrator output, and the torque reference obtained in the case of a large input step.

Note in Fig. 4.24 that the error integrator reaches the level of T_{MAX} at the instant t_1, immediately as the input step cuts in. In cases when the proportional action is in the direct path, the torque reference will reach T_{MAX} at instant t_0 due to the $K_P \Delta\omega$ component of the driving torque. With the proportional action in the feedback path, the maximum torque T_{MAX} will be reached after the instant t_1.

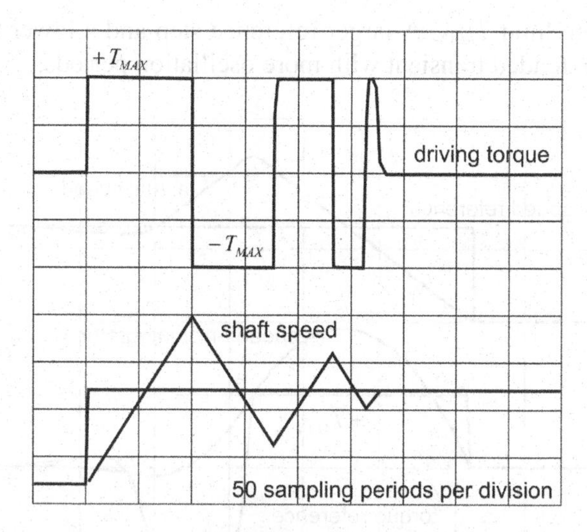

Fig. 4.23. Response to a large input step, driving the torque into saturation. Although the closed-loop poles are strictly aperiodic, a nonlinear element such as the torque limiter gives rise to oscillations.

In the interval $[t_1 .. t_2]$, the input to the error integrator is positive due to $\Delta\omega = \omega^* - \omega > 0$. Therefore, the integrator output keeps building until the instant t_2, when the speed reaches the setpoint and the speed error becomes zero. During the same interval, the torque remains at the limit T_{MAX}. With $\Delta\omega(t_2) = 0$, it would be convenient to drive the torque down to zero, thus keeping the speed at the reference value. With the speed controller structure as shown in Fig. 4.13 and Fig. 4.22A, this outcome cannot be achieved. The error integrator is well beyond the T_{MAX} level. In other words, it is charged or *wound up*. For the torque to decay, the error integrator must be discharged first. Within the interval $[t_2 .. t_3]$ in Fig. 4.24, the torque remains at T_{MAX}, resulting in a continued acceleration. The speed overshoots the reference and keeps increasing in a linear manner. The speed error becomes negative, and the error integrator discharges.

At the instant t_3, the integrator output drops below the limit value and the torque gradually decreases. However, the steady-state condition cannot be reached, as the shaft speed $\omega(t_3)$ is well beyond the reference and the speed error is negative. Now, a negative torque is generated ($t > t_3$), and the system decelerates. Nonlinear oscillations around the setpoint (Fig. 4.23) may last several cycles, diminishing in amplitude and increasing in frequency. The number of cycles and the time required for the system to reach the steady state depends upon the amplitude of the input disturbance and the torque limit T_{MAX}. A larger reference step and a lower torque limit produce an extended transient with more oscillation periods.

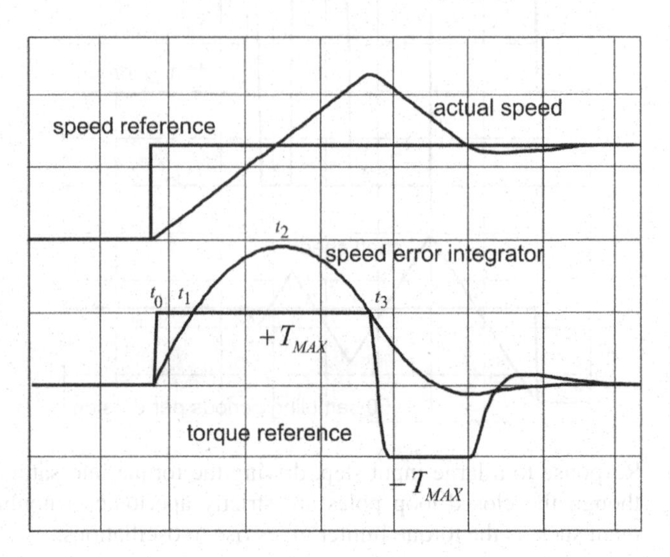

Fig. 4.24. Illustration of the wind-up phenomenon. The change in the error integrator output in response to a large input step goes well above T_{MAX}. While the integrator is discharged, the speed overshoots the setpoint.

The effects caused by the wind-up are not acceptable. Therefore, the speed controller has to include measures devised to suppress the wind-up in the error integrator. Such measures are known as the *Anti-Wind-Up* (AWU). The implementation of AWU measures with PI controllers in positional form (Fig. 4.9, Fig. 4.13) becomes quite involved. The AWU implementation, in conjunction with the incremental form of the PI controller (Fig. 4.10, Eq. 4.29), is discussed in the next section.

Consider the controller structure in positional form (Fig. 4.22A). The proportional and integral actions are calculated in separate blocks, added at the summation point, and then limited to $\pm T_{MAX}$. The wind-up occurs when the error integrator output goes well beyond the limit. To avoid the wind-up of the error integrator, the following AWU steps are to be taken:

- Within each sampling period, new values of both the proportional and the integral action are calculated separately.
- The sum of the two is compared with the torque limit.

- In cases when the sum exceeds the limit by ΔT, the error integrator is decreased by the same amount.
- The sum of the proportional action and adjusted integral action will set the torque reference at the $\pm T_{MAX}$ limit.

This procedure is rather involved and has several disadvantages in specific operating regimes. In speed-controlled systems operating in noisy environments and/or having low resolution of the shaft sensor, noise contributes to sporadic pulses that may trigger the AWU mechanism and alter the contents of the error integrator. In this way, the integrator output will not correspond to the sum (integral) of the error samples. Therefore, the principal function of the error integrator will be impaired and the speed controller will be unable to secure the suppression of the speed error in the steady state. In addition, the AWU measures applied to the structure in Fig. 4.22A may reduce the average torque available from the servo drive. In an operation with a considerable environmental and/or quantization noise and a torque reference adjacent to $+T_{MAX}$, positive noise spikes go beyond the limit, triggering the AWU mechanism and limiting the driving torque to $+T_{MAX}$. On the other hand, negative excursions contributed by noise will result in a decrease of the resulting torque reference. On average, the peak torque available will be lower than the limit, to the extent proportional to the amplitude of parasitic noise signals.

The application of AWU measures to the speed controller, as implemented in its incremental form, eliminates some of the above-mentioned problems and alleviates the others.

4.10 Anti-Wind-Up mechanism

The introduction of the AWU mechanism into the positional form of the PI speed controller (Fig. 4.9, Fig. 4.13) becomes complex and gives rise to the problems listed in the previous section. In this section, a quite straightforward implementation of the AWU in the incremental form of the PI controller (Fig. 4.10, Eq. 4.29) is discussed in detail.

The increments of the proportional and integral action are given in Eq. 4.29. In cases when the integral action of the controller resides in the direct path and processes the speed error, while the proportional gain is relocated into the feedback path, the torque increment is expressed as

$$\Delta T^*_{(n+1)} = T^*_{(n+1)} - T^*_{(n)} = K_P\left(\omega_{(n)} - \omega_{(n+1)}\right) + K_I\,\Delta\omega_{(n+1)}, \tag{4.60}$$

while the sample $T^*_{(n+1)}$ of the driving torque is obtained as the sum of the torque increments accumulated on the interval $t \in [0 .. (n+1)T]$:

$$T^*_{(n+1)} = \sum_{k=0}^{n+1} \Delta T^*_{(k)} . \tag{4.61}$$

In its incremental form, the speed controller structure should calculate the increments of the proportional and integral action according to Eq. 4.60, and feed them into the output integrator, such as the one shown on the right in Fig. 4.10. The integrator accumulates the torque increments according to Eq. 4.61 and provides the torque reference sample $T^*_{(n+1)}$.

The structure of the discrete-time speed controller implemented in its incremental form is given in Fig. 4.25. The proportional action is repositioned into the feedback path. The increments of the proportional and integral action are summed in the junction S1. The delay block D1 provides the most recent sample of the speed $\omega^{FB}_{(n)}$, required for calculation of the increment in the proportional action $K_P(\omega^{FB}_{(n+1)} - \omega^{FB}_{(n)})$. The summation point S3 provides the increment in the feedback signal between the two successive sampling instants. The increment of the integral action is obtained by multiplying the gain K_I and the newly-acquired sample of the speed error.

The torque increment $\Delta T^*_{(n+1)}$ is fed from S1 into the summation junction S1, where it is added to the most recent value of the torque reference $T^*_{(n)}$, obtained at the previous sampling instant $t = nT$. The most recent torque reference is obtained from the delay block D2 in Fig. 4.25. Note in the figure that the delay line D2 takes into account the torque reference value processed through the torque limiter, ensuring that the actual torque command never exceeds the capabilities of the servo drive. The new torque reference signal $T^*_{(n+1)} = T^*_{(n)} + \Delta T^*_{(n+1)}$, obtained from the summation point S2 may exceed the limit T_{MAX}. At this point, the torque limiter will cut in, securing a limited driving torque and ensuring that delay line D2 receives the signal within $\pm T_{MAX}$ boundaries.

In conditions when the driving torque stays within the limits, the torque limiter in Fig. 4.25 acts as the unity gain block. In such cases, and in conjunction with D2 and S2, the group in the upper right in the figure constitutes a discrete-time integrator, accumulating torque increments according to Eq. 4.61 and providing a ready-to-use torque reference $T^*_{(n+1)}$. The integrator mentioned is the only one used within the controller. In other words, the process of integrating the speed error and, thus, carrying out the integral action is implicitly provided in the discrete-time integrator comprising the limiter, D2, and S2. The torque limiter placement prevents the integrator wind-up and suppresses its negative consequences, listed in the preceding section.

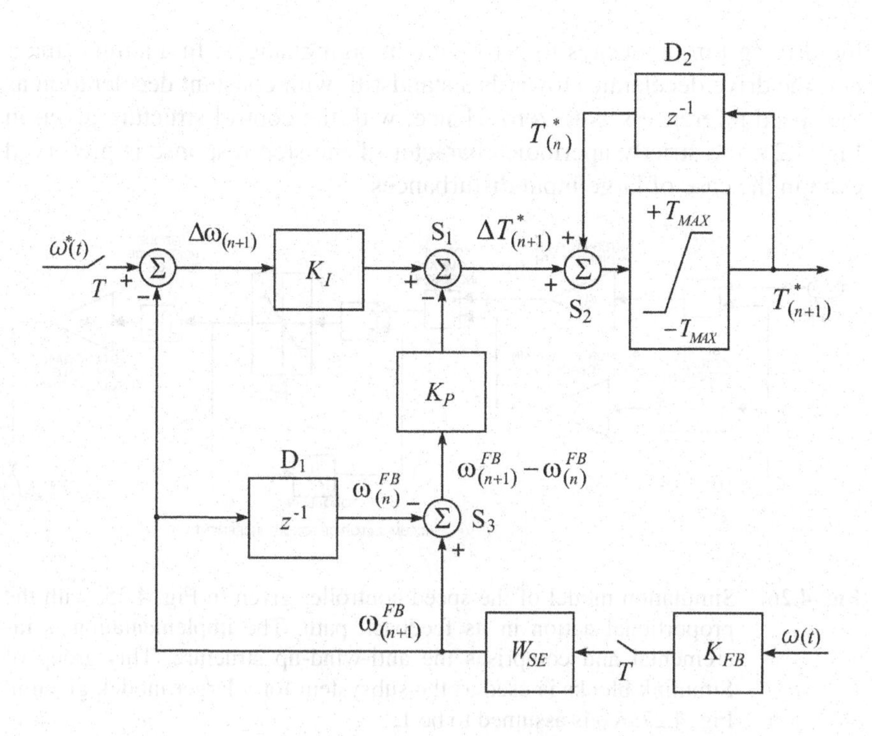

Fig. 4.25. Incremental form of the digital speed controller with proportional action relocated into the feedback path. The discrete-time integrator D2 accumulates the torque increments. The torque limiter is positioned within the integrator D2, suppressing, in this way, the wind-up effects.

The operation of the control structure given in Fig. 4.25 has been investigated by means of computer simulations. To this purpose, the Simulink model plotted in Fig. 4.18 is modified so as to comprise the incremental speed controller with an anti-wind-up structure from Fig. 4.25. In Fig. 4.26, the Simulink subsystem is shown, comprising the implementation of the speed controller structure summarized in Fig. 4.25. This speed controller is integrated in the Simulink model of the overall speed-controlled system, given in Fig. 4.27. For simplicity, it is assumed that the discrete-time integrator deals with the torque signal expressed in [Nm], leading to $K_M = 1$.

The simulation traces obtained from the model are shown in Fig. 4.28, which gives the large step response of the speed and the driving torque over a time span of 500 sampling periods. After 50 periods, the speed reference exhibits a large positive step. It remains at the setpoint for 300 T and then steps back to zero. The shaft speed follows the reference, with acceleration $a^0_{max} = T_{MAX}/J$, defined by the peak torque T_{MAX} and the load inertia J. The speed reaches the setpoint without an overshoot. At this point, the driving torque decays to zero without sign changes. In a similar manner, the drive decelerates towards a standstill, with constant deceleration as the speed reference sets to zero. Hence, with the control structure given in Fig. 4.25, the strictly aperiodic character of the step response is preserved even in the case of large input disturbances.

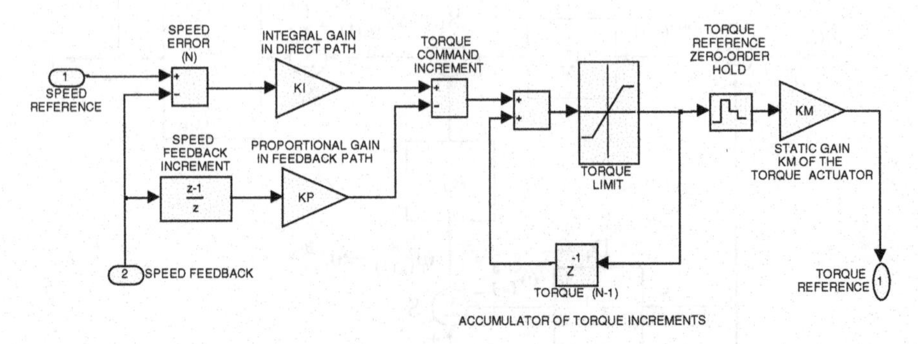

Fig. 4.26. Simulation model of the speed controller given in Fig. 4.25, with the proportional action in its feedback path. The implementation is incremental and comprises the anti-wind-up structure. This group of Simulink blocks is used as the subsystem for a larger model, given in Fig. 4.27. K_M is assumed to be 1.

It is important to discuss the signal flow within the controller during a large step transient. Given a large reference step, a speed error of considerable amplitude will quickly drive the torque signal $T^*_{(n)}$ into the limit T_{MAX}. The speed will accelerate towards the setpoint at a constant slope T_{MAX}/J. Prior to the speed reaching the setpoint, the speed error samples will be positive. Therefore, the summation junction S1 in Fig. 4.25 will keep generating positive increments $\Delta T^*_{(n+1)}$. The limiter will reject any further increase in the driving torque over the T_{MAX} limit, and the increments $\Delta T^*_{(n+1)}$ will not have an effect on $T^*_{(n+1)} = T_{MAX}$.

With the shaft speed approaching the setpoint, the servo drive delivers the torque T_{MAX} and the acceleration $a = \mathrm{d}\omega/\mathrm{d}t$ is constant. In such a case, the shaft speed increases by $\omega^{FB}_{(n+1)} - \omega^{FB}_{(n)} = aT$ within each sampling interval. Therefore, the torque increment $\Delta T^*_{(n+1)}$, obtained from S1 in Fig. 4.25, becomes negative prior to the speed reaching the reference. As indicated by Eq. 4.60, the torque increment changes sign when the residual speed error becomes smaller than $\Delta\omega = (K_P/K_I)\cdot aT$. Beyond this point, the content of the D2–S2 integrator diminishes. With the torque limiter inactive, the system operates in linear mode. The decay of the speed error and the remaining torque are determined by the closed-loop poles. Therefore, the speed error reduces from $\Delta\omega = (K_P/K_I) \cdot aT$ towards zero in a strictly aperiodic manner, leading to the absence of any overshoot. When the speed sets to the reference, the driving torque reaches zero without changing sign, and the system enters the steady-state condition, interrupted only by the next step in the speed reference (Fig. 4.28). This proves that the structure in Fig. 4.25 eliminates the wind-up in the error integrator and provides a strictly aperiodic step response, even in cases with large input disturbances. With reference to the speed control problem, further AWU measures are not required.

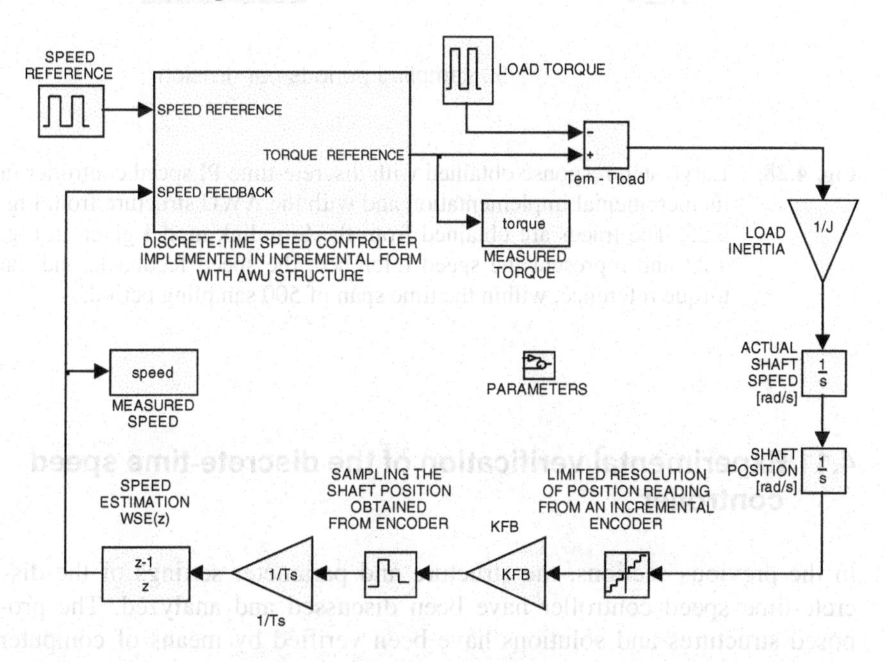

Fig. 4.27. Simulink model of the speed-controlled system with the speed controller given in Fig. 4.25. The speed controller is represented as the subsystem, given in detail in Fig. 4.26.

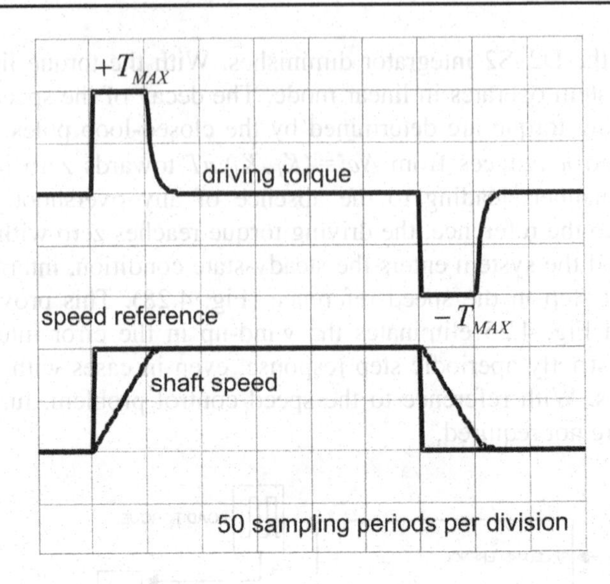

Fig. 4.28. Large step response obtained with discrete-time PI speed controller in its incremental implementation and with the AWU structure from Fig. 4.25. The traces are obtained from the Simulink model given in Fig. 4.27 and represent the speed reference, the speed feedback, and the torque reference, within the time span of 500 sampling periods.

4.11 Experimental verification of the discrete-time speed controller

In the previous sections, the structure and parameter settings of the discrete-time speed controller have been discussed and analyzed. The proposed structures and solutions have been verified by means of computer simulations. In this section, the results are applied to an experimental setup and verified in several experimental runs.

The speed-controlled system under consideration is the test bed in the Laboratory for Digital Control of Electrical Drives at the University of Belgrade, comprising a 0.75kW three-phase induction motor. The motor current is controlled from a Current Regulated PWM inverter (CRPWM). The flux and torque of the induction motor are vector controlled. The implementation details related to the induction motor control are given in [16], along with the control algorithms, coded in ANSI C. The code is executed on a PC platform, comprising the I/O boards for communication with the CRPWM inverter and the shaft encoder. The sampling period T for the speed loop is set to 10 ms. The motor is coupled to an inertial disk and a single-phase synchronous generator providing the load torque. The rated speed of the motor is $\omega_{nom} = 145$ rad/s, while the equivalent inertia of the system comprising the motor, the load, and the inertial disk amounts to $J = 0.032$ kgm^2. An incremental encoder with $N = 1250$ pulses per turn acquires the shaft position. The peak torque capability of the torque actuator, comprising the motor and the CRPWM inverter, is 13.6 Nm.

In Fig. 4.29, the experimental traces are given for the shaft speed (the upper trace) and the driving torque (the lower trace) in cases when the speed references changes from –300 rpm to + 300 rpm. Within the setup, the driving torque is adjusted by altering the i_q current of the vector-controlled induction motor. A similar experiment is conducted with a speed reference of 600 rpm (Fig. 4.30) and 1000 rpm (Fig. 4.31). Note that the responses of the shaft speed and the driving torque are strictly aperiodic, and their character does not depend on the amplitude of the reference step. In Fig. 4.31, when the driving torque remains at the limit for more than 500 ms due to a large reference step, the transient phenomena end within 70–80 ms from the instant when the torque leaves the limit and the system enters linear operating mode. This behavior is consistent with the simulation traces in Fig. 4.19, where the response time amounts to 7–8 sampling periods T.

The experimental traces given in Fig. 4.30 correspond to the speed reversal at 600 rpm. It is of interest to notice that the driving torque fluctuations have a frequency of 10 Hz. While the induction motor controls the running speed at +600 rpm, the two-pole single-phase synchronous generator provides load torque pulsations at approximately 10 Hz. The load pulsations are at a relatively low frequency, staying within the closed-loop bandwidth of the speed controller. Therefore, the speed controller detects the error caused by the load pulsations and generates the torque command, which provides the necessary compensation. In turn, the driving torque suppresses the 10 Hz disturbance and keeps the running speed unaltered.

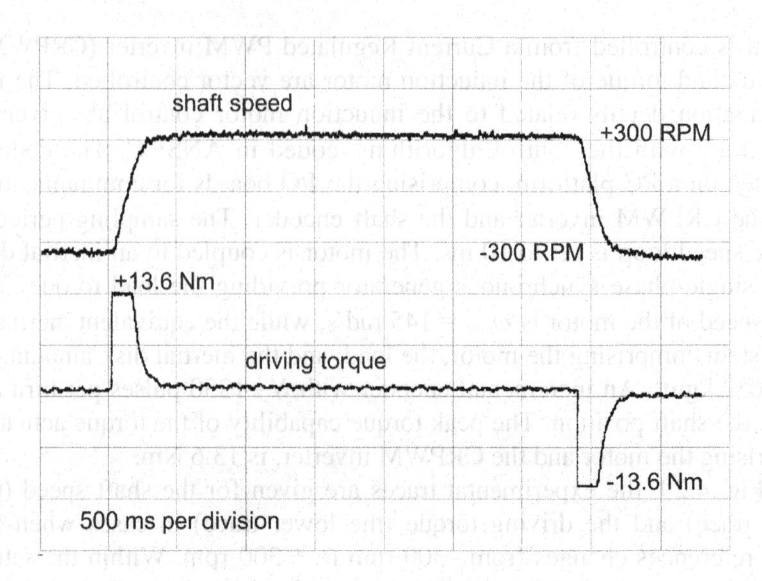

Fig. 4.29. Experimental traces of the shaft speed and the driving torque. The speed-controlled system with discrete-time controller is exposed to a speed reversal from –300 rpm to + 300 rpm.

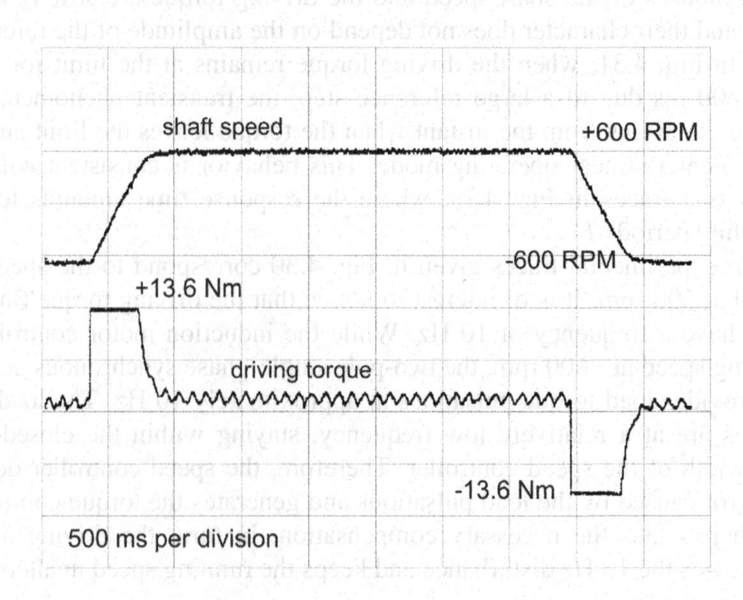

Fig. 4.30. Experimental traces of the shaft speed and the driving torque. The speed-controlled system with discrete-time controller is exposed to a speed reversal from –600 rpm to + 600 rpm.

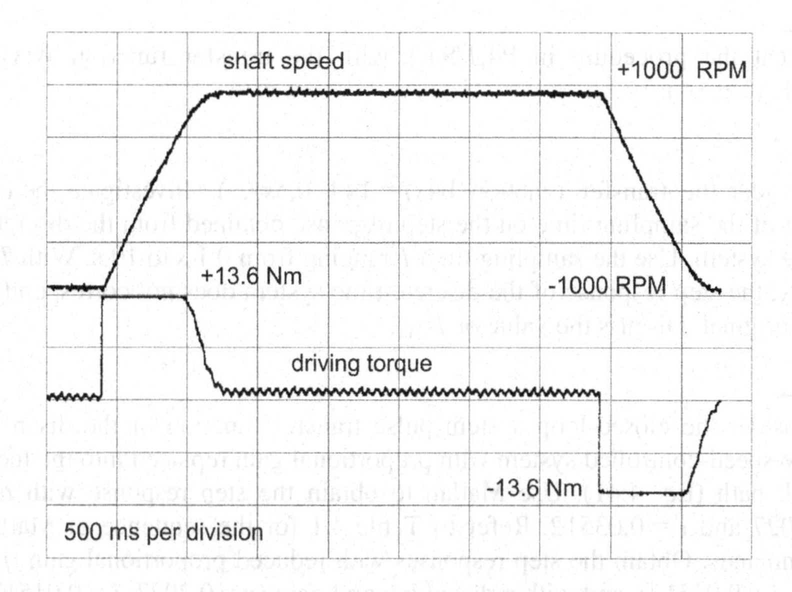

Fig. 4.31. Experimental traces of the shaft speed and the driving torque. The speed-controlled system with discrete-time controller is exposed to a speed reversal from –1000 rpm to +1000 rpm.

In the traces shown in Fig. 4.31, obtained at a higher speed (1000 rpm), the frequency of the load disturbance increases in proportion to the shaft speed and becomes 1000/60 = 16.66 Hz. As the disturbance frequency increases and comes closer to the bandwidth frequency, the capability of the speed controller to provide corrective action diminishes. Therefore, the amplitude of the driving torque oscillations is smaller compared with Fig. 4.30. On the other hand, hardly any oscillations in the running speed are noticed. Due to the low-pass nature of the plant ($1/(Js)$), disturbances at higher frequencies have a minor impact on the output speed.

Problems

P4.1
Use Matlab to obtain the step response from the transfer function $W(s) = 1/(1+s\tau)$ with τ=1s. Calculate the z-domain equivalent $W(z)$ by using the function *c2d* and assuming that the sampling time is T = 1s. Plot the step response obtained with the pulse transfer function $W(z)$ by means of the function *dstep*, and compare this plot to the previous one.

P4.2
Repeat the procedure in P4.1/S4.1 with the transfer function $W(s) = (s+1)/(1+s+s^2)$.

P4.3
Consider the transfer function $W(s) = 1/(1+0.5s+s^2)$. Investigate the impact of the sampling time on the step response obtained from the discrete-time system. Use the sampling time T ranging from 0.1 s to 10 s. With $T > T_{MAX}$, the step response of the discrete-time system does not correspond to the original. Discuss the value of T_{MAX}.

P4.4
Consider the closed-loop system pulse transfer function of the discrete-time speed-controlled system with proportional gain replaced into the feedback path (Eq. 4.41). Use Matlab to obtain the step response with $p = 0.2027$ and $i = 0.03512$. Refer to Table 4.1 for the sequence of Matlab commands. Obtain the step responses with reduced proportional gain ($p = 0.1$, $i = 0.03512$), and with reduced integral gain ($p = 0.2027$, $i = 0.01512$). Observe the response characters and overshoot. The s-domain equivalence of the PI speed controller is given in Fig. 2.2. In light of Eq. 2.38, relating the feedback gains to the natural frequency ω_n and damping ξ, discuss the results obtained from the discrete-time system.

P4.5
For the discrete-time system analyzed in P4.4, and for the parameter setting of $p = 0.1$, $i = 0.03$, and $T_S = 0.001$, find the equivalent closed-loop transfer function in the s-domain and obtain the step response. For the conversion from discrete time to continuous time, use the Matlab command *d2c*.

P4.6
Consider the pulse transfer function numerator *numd* and denominator *dend*, obtained in P4.5; determine the closed-loop poles and zeros in the z-domain. From the s-domain transfer function numerator *numc* and denominator *denc*, obtained in P4.5, determine the closed-loop poles and zeros in the s-domain. How are the z-domain poles and zeros related to their s-domain counterparts? Why are the closed-loop zeros different?

P4.7

Consider the pulse transfer function with numerator $numd(z) = z - 0.001$, and with denominator $dend(z) = z^2 + 0.5z + 0.8$. Assume that the sampling time is $T = 1$s. Using the Matlab command *d2c*, obtain the *s*-domain equivalent $W(s) = numc(s)/denc(s)$ by using *ZOH* and *matched* options. In both cases, verify the correspondence of the poles, zeros, and step responses.

5 Digital Position Control

In this chapter, position control and its role within motion-control systems is introduced and discussed. Single-axis position controllers are explained and modeled. Analytical design of the position controller structure is given, along with procedures for setting the adjustable feedback parameters. The speed and torque system limits are explained at the end of the chapter, along with the analysis of nonlinear operating modes. Nonlinear control laws, capable of securing a robust large step response, are considered.

In the succeeding sections, the structure and parameter setting of linear discrete time position controllers are discussed, analyzed and designed. Within motion-control systems, the position controller may assume different roles and forms. Its basic purpose is to provide for the corrective action that drives the mechanical load, work piece, or tool along a predefined trajectory in space. Position trajectories are created by a superior controller, such as the production-cell computer, and they depend on specific operations to be performed within a desired production cycle.

At the output, position controllers have to provide the reference values, calculated in such a way that any position error caused by the load disturbance or a reference change, is suppressed and driven to zero. In a number of motion-control applications, position controllers provide the reference of the driving torque and feed this reference to the torque actuator. In other cases, the position controller calculates the speed the system ought to assume in order that the position error is dissipated. In such cases, the output of the position controller is fed to the reference input of the speed controller. Eventually, the torque commands are calculated so as to correct the speed error, assisting the position control task in an indirect way. The physical location of position controllers varies with the application.

The host computer in traditional motion control systems, known as the CNC (Computerized Numerical Control), generates the reference profiles for a number of motors that have to effectuate a coordinated motion. In most cases, the CNC comprises individual position controllers. Position control functions are often implemented by relying on dedicated hardware units called the *axis cards*, located within the CNC. The host computer may control the individual drives by sending them the speed reference. In this case, the speed control function is performed by digital drive controllers, associated with drive power converters, and thus controlling the current torque, flux, and speed of individual motors. Control functions are shared between the CNC and the digital drive controllers. The two controllers may be separated spatially, and the overall performance may largely depend upon the quality of the signal transmission between the CNC and the drive.

In many applications, the host computer executes the position control algorithm and generates the torque reference. This reference is fed to individual drives in the form of an analog signal, or by means of a high-speed digital serial link. The signal integrity issues are less pronounced than in the case when the speed reference is transferred. Individual drive controllers perform the current, torque, and flux control.

With the position control functions distributed between the CNC and the digital drive controller, the accuracy and bandwidth of the position loop is compromised by the delay and degradation of the reference and feedback signals communicated between the control nodes. Therefore, contemporary drive controllers include single-axis positioners, capable of keeping the position of the motor at the reference. The reference trajectory can be stored within the internal memory of the drive controller, or received in real time from the host computer by means of a high-speed digital serial link.

In this chapter, single-axis discrete-time position controllers with position error at the input and torque reference at their output are discussed, analyzed, and designed.

5.1 The role and desired performance of single-axis positioners

Industrial motion-control systems utilize the electromagnetic torque or force generated by servo motors in order to effectuate controlled motion of work pieces, tools, or machine parts. Within a production cell, several motors need to move in coordination in order to provide for the desired motion. Each production cycle consists of several motion sequences. Fast motion leads to a reduced cycle time and increased productivity. On the other hand, precision in following the position reference profiles affects production quality. Fast tracking of the reference trajectory has detrimental effects on position accuracy. Therefore, productivity and production quality impose conflicting requirements on the position controller. For that reason, there is a perpetual requirement to increase the closed-loop bandwidth of position controllers, providing, in such a way, the desired accuracy at elevated speeds.

In most cases, the response character of position-controlled systems has to be aperiodic, approaching the reference without an overshoot. An overshoot may result in collision of moving parts, eventually damaging or breaking the tools and/or the work piece. A strictly aperiodic response results in driving torque transients free of oscillations and sign changes during the step response. Consequently, the backlash and flexibility in transmission elements are less emphasized, and the cumulative tracking error is smaller. For that reason, the parameter-setting procedure discussed in this chapter focuses on providing a strictly aperiodic response with a maximum workable speed. The advantages related to strict aperiodicity have been summarized in the previous chapter.

A discrete-time position controller acquires position feedback at each sampling instant, derives the tracking error, and calculates the reference for the driving torque, formulated so as to drive the position error down to zero in the shortest time possible. The torque reference is fed to the torque actuator, comprising the electrical servo motor, the drive power converter, and the digital drive controller, accountable for controlling the motor current, torque, and flux.

In the majority of applications, current and torque transients are considerably faster than the response of the position. Therefore, contemporary electrical drives, used as torque actuators in position-controlled systems, provide a driving torque that tracks the torque reference quickly and accurately. With a torque response time corresponding to the bandwidth exceeding $f_{BW} = 1$ kHz, and with a target bandwidth of the position loop on the order of 100 Hz, the torque actuator is assumed to have instantaneous action. Therefore, it is modeled as a static gain block K_M in Fig. 5.1.

It is interesting to note that the gain block K_M represents the servo motor, the drive power converter, and the corresponding motor control routines executed within the digital drive controller. The latter are designed and used to control motor flux and torque. Whether the servo motor is a brushed DC motor, a vector-controlled induction motor, or a synchronous motor with permanent magnet excitation, the electromagnetic torque T^{em} is, in all cases, the product of the fast-changing motor current and slowly varying flux. Therefore, the torque rise time is determined by the performances of the current controller.

Motion-control DSP [10, 11] provides a platform for the digital current loop, with the sampling rate exceeding $f_S = 20$ kHz. The practicable closed-loop bandwidth reaches 2 kHz, resulting in a current-loop rise time of 100–200 μs. At the same time, the rise time represents the time interval required for the driving torque to reach the torque command T^*. With such dynamics, the torque actuator is faster than the mechanical subsystem by an order of magnitude. Therefore, the analysis and design in this chapter are performed assuming that the actuator responds instantly to the reference pulses $T^*_{(n)}$. The static gain K_M defines the ratio between the torque reference $T^*_{(n)}$, represented as a digital word residing in RAM, and the actual driving torque T^{em}, expressed in [Nm], generated by the servo motor and delivered to the load.

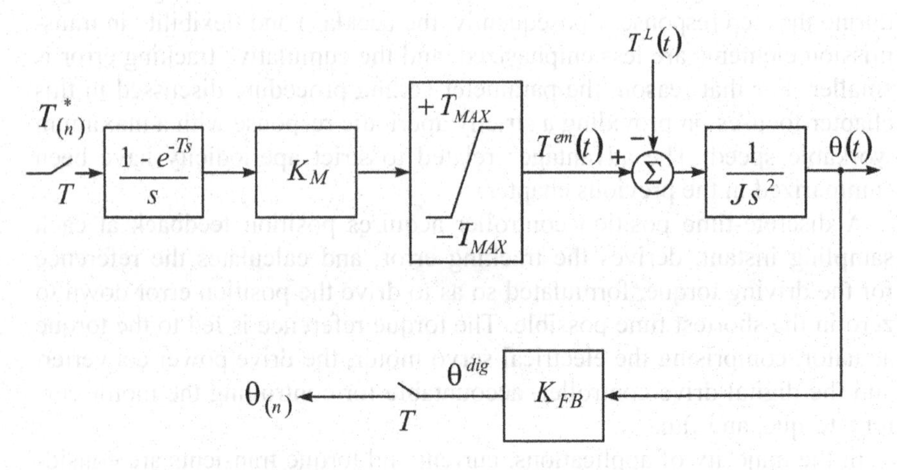

Fig. 5.1. The torque actuator within the position-controlled system has a negligible delay and a gain of K_M. The drive peak torque capability is T_{MAX}. The mechanical subsystem $1/(Js^2)$ has the position θ at the output. The parameter K_{FB} relates a digital representation of the output position to the actual position in [rad]. The discrete-time controller acquires the samples $\theta_{(n)}$ at each sampling instant $t = nT$.

The block diagram given in Fig. 5.1 represents the torque actuator and the mechanical subsystem of a position controlled system. The peak torque capability T_{MAX} of the actuator depends on the limits of the servo motor and the drive power converter. The mechanical subsystem in the figure is a double integrator, having output θ [rad].

The position feedback is obtained by means of an optical encoder or a resolver. In Fig. 5.1, the variable θ^{dig} stands for the digital word representing the output position. The value of θ^{dig} is obtained either by counting the encoder pulses or from an R/D converter [2] processing the resolver signals and thus obtaining the digital representation of θ [rad]. In Fig. 5.1, the gain $K_{FB} = \theta^{dig}/\theta$ determines the ratio between the actual position and its digital form. It depends on the number of encoder pulses per turn, or the resolution of the R/D converter used.

A discrete-time position controller with sampling period T acquires the position samples $\theta_{(n)}$ and calculates the torque reference $T^*_{(n)}$ at each sampling instant $t = nT$. The zero-order hold on the left in Fig. 5.1 provides the torque reference in such a way that the value of $T^*_{(n)}$ is maintained throughout the sampling period. At the next sampling instant $t = (n+1)T$, the torque reference is replaced with the new sample $T^*_{(n+1)}$. If we neglect delays in controlling the torque of the servo motor, the actual driving torque T^{em} will track the reference $K_M T^*$, provided that the latter remains within the system limits $\pm T_{MAX}$.

In addition to the torque constraint, the speed of the servo system must be limited as well. Within the position-controlled system, the speed may not be explicitly controlled. However, the angular speed of the revolving parts and the speed of translation of the moving parts have to be limited to the range of $\pm\omega_{MAX}$. Excess speed increases the risk of mechanical damage due to the eccentric motion of the rotating parts and the imbalance of the moving parts. The motor bearings and transmission elements have the maximum permissible speed that can limit the top speed practicable by the positioner. Therefore, the structure and control actions of position controllers have to be designed to ensure that the speed experienced during transients does not exceed the permissible range of $\pm\omega_{MAX}$.

The speed and torque system limits introduce nonlinearity in the control object. It will be shown in subsequent chapters that a stable, robust response at large disturbances requires nonlinear elements within the position controller, designed in consideration of the system limits $\pm\omega_{MAX}$ and $\pm T_{MAX}$. In this chapter, the operation of position-controlled systems is considered for small input and load disturbances and with speed and torque transients staying below the limits. Based on the analysis of position-controlled systems in the linear regime, a structure and parameter setting are proposed so as to achieve strictly aperiodic behavior with the fastest step response possible. The analysis of position-controlled systems in nonlinear mode and the appropriate extension of the controller structure are given in the next chapter.

5.2 The pulse transfer function of the control object

Consider the mechanical subsystem in Fig. 5.1 and assume that the equivalent inertia J is known and that the load torque $T^L(t)$ is available. Then, the change in the output position θ is defined by Eq. 5.1.

$$J\frac{\mathrm{d}^2\theta}{\mathrm{d}t^2} = J\frac{\mathrm{d}\omega}{\mathrm{d}t} = T^{em}(t) - T^L(t) \tag{5.1}$$

The load torque may comprise the friction and other speed-dependent components, proportional to the speed ($K_F\omega$), to the speed square ($K_V\omega^2$), or having another functional dependence on the speed. The load torque component $K_F\omega$ cannot be considered to be an external disturbance, as it depends on the state ω of the mechanical subsystem. In fact, the presence of the $K_F\omega$ component of the load torque changes the transfer function of the mechanical system from $1/(Js^2)$ (Fig. 5.1) to $1/(Js+K_F)/s$. On the other hand, the static friction is the most pronounced friction component in many applications. It can be expressed as a nonlinear function of the speed, and it shows effects when the moving parts start from a standstill. Other speed-related motion resistances may depend on specific characteristics of transmission elements and may appear only in particular machine positions and operating conditions. Although related to speed, such motion resistances are hardly predictable, and their effects are similar to those of external disturbances, that is, unrelated to the internal states of the system. Therefore, the control object model, expressed as $1/(Js^2)$in Fig. 5.1, is a viable representation, along with the assumption that all motion resistances can be treated as external disturbances and therefore can be added to the disturbance signal $T^L(t)$. Such an approach is further justified by the fact that, in a number of servo systems, the motion resistances are negligible compared with the acceleration torque $J\mathrm{d}\omega/\mathrm{d}t$. The prevalence of the acceleration torque comes from the requirement to reach the target position within a short interval. Quick motion and high acceleration require increased values of the driving torque, surpassing the rated torque and exceeding by far the friction and the parasitic motion resistances. This understanding, in turn, supports the model of the control object given in Fig. 5.1.

The torque reference $T^*(t)$ obtained at the output of the zero-order hold retains the value $T^*_{(n)}$ on the interval $[nT..(n+1))$. The sample $T^*_{(n)}$ is calculated within the position controller on the basis of the feedback sample $\theta_{(n)}$, acquired at sampling instant nT. Upon acquisition of the next feedback sample $\theta_{(n+1)} = \theta(nT+T)$, the controller evaluates the error $\Delta\theta$ and derives the subsequent sample of the torque reference $T^*_{(n+1)}$. Since the calculation time is negligible compared with the sampling interval, it is assumed henceforth that the new sample $T^*_{(n+1)}$ is made available starting from $t = (n+1)T$. In such cases, the new torque reference $K_M T^*_{(n+1)}$ is delivered to the load within the interval $[(n+1)T..(n+2))$.

With the position loop sampling periods expressed in hundreds of microseconds and with the motion-control DSP platforms [10, 11] performing numerical operations in tens of nanoseconds, the assumption of instantaneous availability of the torque reference samples holds. In cases with extreme complexity of the position control algorithms, reduced numerical capabilities of the digital controller, or significant increase of the sampling frequency, calculation delays must be modeled and accounted for.

In Fig. 5.1, it is assumed that the controlled output is the shaft position of the servo motor. In servo systems, both linear and conventional revolving motors are encountered. While the latter provides torque as the driving force and rotates, the linear motors provide the force F[N] and carry out translation along a predefined path. In both cases, the control objective is to perform the motion of the load along predefined trajectories. The load to be positioned is mostly the tool or the work piece. The servo motor is coupled to the load by means of mechanical coupling elements such as the gears, pullies, belts, and other transmission elements. In cases when the revolving servo motor is used, transmission elements may also convert the rotation into linear motion. In further considerations, it is assumed that the mechanical couplings are ideal and that the backlash, elasticity, and other imperfections in the mechanical subsystem can be neglected. In such cases, the position of the load is directly proportional to the motor shaft position. Therefore, the latter can be considered to be the system output, as shown in Fig. 5.1.

With the shaft position θ expressed in radians, the motor speed $\omega = d\theta/dt$ is obtained in [rad/s]. The change in speed within one sampling period and the corresponding increment in position can be expressed as follows:

$$\omega_{(n+1)} = \omega_{(n)} + \frac{T}{J} K_M T^*_{(n)} - \frac{1}{J} \int_{nT}^{nT+T} T^L(t)\,dt \tag{5.2}$$

$$\theta_{(n+1)} = \theta_{(n)} + \int_{nT}^{nT+T} \omega(t)\,dt \,. \tag{5.3}$$

Calculation of $\omega_{(n+1)}$ in Eq. 5.2 requires the values of the torque reference $T^*_{(n)}$ and the previous speed sample $\omega_{(n)}$. The speed also depends on the load disturbance T^L, which may exibit changes within the sampling interval observed. Further analysis and derivation of pulse transfer functions requires the conversion of differential equations, such as Eq. 5.1, into equivalent difference equations. The presence of integrals on the right-hand side in Eq. 5.2 and Eq. 5.3 hinders such a conversion.

In Eq. 5.2, the load torque integral represents the average value within the sampling period under consideration. In Eq. 5.4, it is therefore replaced with the constant $T^L_{(n)}$, representing the average value of $T^L(t)$ on the interval

$[nT..(n+1))$. Without lack of generality, the impact of the load torque $T^L(t)$ on the pulse train $\omega_{(n)}$ can be studied further by considering the train of samples $T^L_{(n)}$, comprising the average values of $T^L(t)$ within individual sampling periods (Eq. 4.18). On the other hand, the value of the integral in Eq. 5.3 can be expressed in terms of the speed samples $\omega_{(n)}$ and $\omega_{(n+1)}$ only in cases when the change of the actual speed $\omega(t)$ between the sampling instants is linear, as discussed in Section 4.2.4 (Fig. 4.6, Eq. 4.24). The speed change is linear only in cases when the load torque $T^L(t)$ does not change on the interval $[nT..(n+1))$. The support for such an assumption is given in the previous chapter, and will be restated hereafter.

Due to the load torque $T^L(t)$ affecting the speed and the output position, the choice of the sampling time T has to be made in such a way that the position controller may provide the reference torque quickly enough to compensate for load changes and to suppress the position error. With an adequate sampling period T, the train of samples $T^*_{(n)}$, refreshed at the sampling rate $f_S = 1/T$, compensates for disturbance fluctuations and keeps the output position at the desired trajectory. As discussed in the previous chapter, the load torque may comprise spectral components at frequencies beyond reasonable sampling rates. The phenomena related to such disturbances are referred to as the *unmodeled dynamics*, indicating that the associated errors cannot be suppressed by an intentional action of the position controller. The design of feedback controllers is based on the assumption that the errors associated with the high-frequency unmodeled dynamics are suppressed in a passive way, relying on the low-pass nature of the control plant. In position-controlled systems, such an assumption is justified by the control plant transfer function being a double integrator ($1/Js^2$).

For the reasons noted above and discussed in the previous chapter, the subsequent analysis assumes that load torque variations within the sampling period are negligible and that the change in the shaft speed between the sampling instants is linear. To this end, the disturbance signal $T^L(t)$ is assumed to be constant on the interval $[\,nT, (n+1)T\,)$ and equal to the average value $T^L_{(n)}$ (Eq. 4.18), obtained as

$$T^L_{(n)} = \frac{1}{T} \cdot \int_{nT}^{nT+T} T^L(t)\,\mathrm{d}t\,.$$

Under these assumptions, the speed transition from the sampling instant nT to the next sampling instant is found in the difference equation 5.4:

$$\omega_{(n+1)} = \omega_{(n)} + \frac{T}{J} K_M T^*_{(n)} - \frac{T}{J} T^L_{(n)}. \tag{5.4}$$

With a linear change in the shaft speed (Fig. 4.6, Eq. 4.24), the integral on the right in Eq. 5.3 can be expressed in terms of the shaft speed samples $\omega_{(n)}$ and $\omega_{(n+1)}$. The change in the output position can be described by the difference equation 5.5. The equation derives the next sample of the output $\theta_{(n+1)}$ from the speed and position samples $\theta_{(n)}$, $\omega_{(n)}$, and $\omega_{(n+1)}$.

$$\theta_{(n+1)} = \theta_{(n)} + \int_{nT}^{(n+1)T} \omega \, \mathrm{d}t = \theta_{(n)} + \frac{\omega_{(n+1)} + \omega_{(n)}}{2} T \tag{5.5}$$

The output position, speed, driving torque and load disturbance can be expressed in terms of their complex images $\theta(z)$, $\omega(z)$, $T^*(z)$, and $T^L(z)$, respectively, obtained by applying the z-transform to the relevant pulse trains $\theta_{(n)}$, $\omega_{(n)}$, $T^*_{(n)}$, and $T^L_{(n)}$:

$$\theta(z) = \sum_{i=0}^{\infty} \theta_{(i)} z^{-i}, \quad \omega(z) = \sum_{i=0}^{\infty} \omega_{(i)} z^{-i},$$

$$T^*(z) = \sum_{i=0}^{\infty} T^*_{(i)} z^{-i}, \quad T^L(z) = \sum_{i=0}^{\infty} T^L_{(i)} z^{-i}$$

The difference equations 5.4 and 5.5 can be converted to algebraic equations and used for deriving the pulse transfer function of the control object. The time shift property of the z-transform, expressed in Eq. 4.6, implies that multiplication by operator z creates a time shift of one sampling period:

$$\sum_{i=0}^{\infty} \theta_{(i+1)} z^{-i} = z\theta(z) - z\theta_{(0)}.$$

Introduction of the time shift property of the z-transform into difference equations 5.4 and 5.5 results in the algebraic equations 5.6 and 5.7, comprising the complex images of the position, speed, and torque:

$$\omega(z) = \frac{T}{J}\frac{1}{z-1} K_M T^*(z) - \frac{T}{J}\frac{1}{z-1} T^L(z) \tag{5.6}$$

$$\theta(z) = T\frac{z+1}{2(z-1)}\omega(z). \tag{5.7}$$

As indicated in Fig. 5.1, Eq. 5.6 holds for the linear operating mode of the system. In other words, the actual driving torque $T^{em}(t)$ corresponds to the reference $K_M T^*_{(n)}$, provided that the latter stays within the system limits $\pm T_{MAX}$, keeping the torque limiter inactive. The complex image of the output position $\theta(z)$ is given in Eq. 5.8, expressed in terms of $T^*(z)$ and $T^L(z)$.

The pulse transfer functions $W_P(z)$ and $W_{PL}(z)$ describe the signal flow from the driving torque input to the output (Fig. 5.1), and from the disturbance input to the output, respectively. In position-controlled systems, the control plant has a double real pole $z_{1/2} = 1$ and one real zero.

$$\theta(z) = \frac{K_M T^2(z+1)}{2J(z-1)^2} T^*(z) - \frac{T^2(z+1)}{2J(z-1)^2} T^L(z) \tag{5.8}$$

$$\theta(z) = W_P(z) T^*(z) - W_{PL}(z)\, T^L(z)$$

It is of interest to calculate the pulse transfer functions $W_P(z)$ and $W_{PL}(z)$ by transforming the s-domain images $W_P(s)$ and $W_{PL}(s)$ into the z-domain:

$$W_P(z) = \mathrm{Z}\left(\frac{1-e^{-sT}}{s} K_M \frac{1}{J s^2}\right) = \frac{K_M T^2(z+1)}{2J(z-1)^2}$$

$$W_{PL}(z) = \mathrm{Z}\left(\frac{1-e^{-sT}}{s} \frac{1}{J s^2}\right) = \frac{T^2(z+1)}{2J(z-1)^2} .$$

Starting from the block diagram in Fig. 5.1, comprising the zero-order hold and the control object modeled as $W_P(s) = 1/Js^2$, the pulse transfer function $W_P(z)$ can be verified by using the Matlab tools. Table 5.1 lists the set of commands that must be typed in at the Matlab command prompt.

Table 5.1. The Matlab command sequence used to convert the continuous-domain transfer function $W_P(s) = 1/Js^2$ into its discrete-time equivalent $W_P(z)$.

```
>> num = [1]                  % Defines the numerator of 1/Js^2
>> den = [1  0  0]            % Denominator is 1 s^2  + 0 s  + 0
>> sysc = tf(num,den)         % Creates continuous-domain system
>> sysd = c2d(sysc,1,'zoh')   % Conversion into discrete time, T=1, ZOH;
>>                            % sysd is the discrete-time equivalent
>> tf(sysd)                   % Deriving the pulse transfer function
>> Transfer function:         %
>>     0.5 z + 0.5            %
>>  ----------------          % Matlab replies with the pulse
>>   z^2 - 2 z + 1            % transfer function  WP(z)
```

The functions obtained in Eq. 5.8 are applied further to determine the structure and parameters of the discrete-time position controller.

5.3 The structure of position controllers

With the design procedure applied in Chapter 4, the discrete-time position controller and the feedback gains can be determined by considering the control object pulse transfer function $W_P(z)$, deriving the transfer function of the position controller $W_{PC}(z)$, calculating the closed-loop system transfer function $W_{SS}(z)$ from the previous two, discussing the characteristic polynomial and its zeros, and determining the feedback gains that result in the desired character of the step response and the desired closed-loop bandwidth. Firstly, it is necessary to determine the necessary control actions and the structure of the position controller.

5.3.1 Derivative action in position controllers

Consider the control object transfer function $1/Js^2$, comprising a double pole at the origin. It is interesting to note that derivative control action may be required in order to achieve stability. To more easily understand the role of the derivative action, the present discussion is focused on a simplified block diagram (Fig. 5.2), comprising the s-domain representation of the position controller.

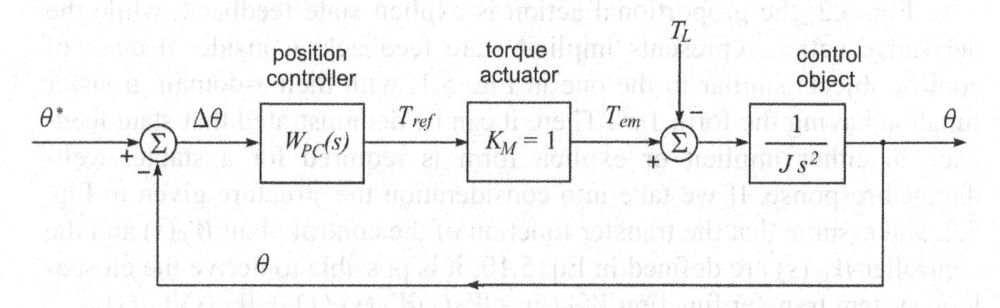

Fig. 5.2. Simplified block diagram of the s-domain position controller.

If we assume that the position controller in Fig. 5.2 has proportional and derivative action, its transfer function in the s-domain is written as $W_{PC}(s) = K_P + K_D s$. The potential need to implement integral actions is disregarded at this point and will be discussed later. With $K_M = 1$, the closed-loop system transfer function and the characteristic polynomial are obtained in Eq. 5.9, along with the undamped frequency ω_n and the damping coefficient ξ of the closed-loop poles.

The derivative action $K_D d\theta(t)/dt$, calculated from the output position, is equivalent to the control action $K_D\omega(t)$, which is proportional to the speed.

Note in Eq. 5.9 that the derivative action is indispensable in achieving a stable, well-damped closed-loop response. With $K_D = 0$, the characteristic polynomial would become $f(s) = s^2 + \omega_n^2$, resulting in the undamped closed-loop poles $s_{1/2} = \pm j\,\omega_n$.

$$W_{SS}(s) = \frac{1 + s\dfrac{K_D}{K_P}}{1 + s\dfrac{K_D}{K_P} + 1 + s^2\dfrac{J}{K_P}} \tag{5.9}$$

$$f(s) = s^2 + s\frac{K_D}{J} + \frac{K_P}{J}, \quad \omega_n^2 = \frac{K_P}{J}, \quad 2\xi\omega_n = \frac{K_D}{J}$$

If we consider the proportional and integral actions in Fig. 5.2, it is of interest to point out that both contribute to the driving torque T^* in proportion to the state variables of the control plant. In other words, both control actions can be interpreted as the state feedback. The action $K_P(\theta^* - \theta)$ is proportional to the output position, which is, at the same time, the state variable of the mechanical plant. The derivative action $K_D\,\mathrm{d}\theta(t)/\mathrm{d}t$ is proportional to the speed $\omega(t)$ of the mechanical subsystem, the second state variable of the control object.

In Fig. 5.2, the proportional action is explicit state feedback, while the derivative action represents implicit state feedback. Consider a class of control objects similar to the one in Fig. 5.1, with their s-domain transfer function having the form $1/s^n$. Then, it can be demonstrated that state feedback in either implicit or explicit form is required for a stable, well-damped response. If we take into consideration the structure given in Fig. 5.2, and assume that the transfer function of the control plant $W_P(s)$ and the controller $W_C(s)$ are defined in Eq. 5.10, it is possible to derive the closed-loop system transfer function $W_{SS}(s) = W_P(s)W_C(s) / (1 + W_P(s)W_C(s))$.

$$W_P(s) = \frac{1}{s^n}, \quad W_C(s) = K_0 + K_1 s + \ldots + K_{n-1}s^{n-1} \tag{5.10}$$

For the control object with n series-connected integrators, it is assumed that the controller $W_C(s)$ involves derivatives of the feedback signal up to the $(n-1)$th-order. According to Fig. 5.3, such a controller implements implicit state feedback, with the driving force comprising the components proportional to every state contained within $W_P(s)$. Notice that the nth-order derivative is not present, as it would represent the signal proportional to the driving force, given on the left in Fig. 5.3. The characteristic polynomial of the system is obtained from the denominator $1 + W_P(s)W_C(s)$ of the closed-loop transfer function $W_{SS}(s)$ and given in Eq. 5.11. The polynomial $f(s)$ has n zeros. The coefficients $K_0...K_{n-1}$ can be expressed in terms of the zeros $\sigma_1...\sigma_n$ (Eq. 5.12).

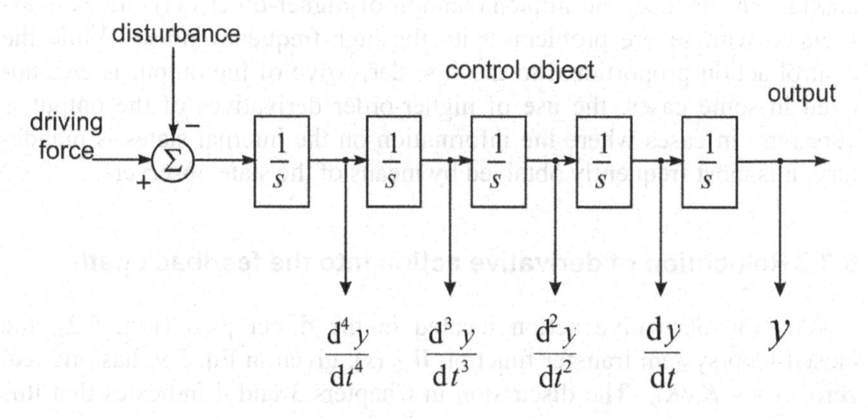

Fig. 5.3. Control action proportional to the kth-order derivative is implicit state feedback, proportional to an internal state of the control object. The state under consideration is obtained at the output of the integrator at the kth place, on the left from the output y.

$$\begin{aligned} f(s) &= K_0 + K_1 s + ... + K_{n-1} s^{n-1} + s^n \\ &= (s-\sigma_1)(s-\sigma_2)....(s-\sigma_{n-1})(s-\sigma_n) \end{aligned} \tag{5.11}$$

$$\begin{aligned} K_0 &= \sigma_1 \sigma_2 \sigma_3 ....\sigma_{n-1} \sigma_n (-1)^n \\ K_1 &= K_0 \left(\frac{1}{\sigma_1} + \frac{1}{\sigma_2} + \frac{1}{\sigma_3} + ... + \frac{1}{\sigma_n} \right) (-1)^{n-1} \\ &...... \\ K_{n-1} &= -\sigma_1 - \sigma_2 - \sigma_3 .... - \sigma_{n-1} - \sigma_n \end{aligned} \tag{5.12}$$

The roots $\sigma_1...\sigma_n$ of the equation $f(s) = 0$ can be real or conjugate complex numbers. The stability conditions require that all the roots have negative real components. If any of the feedback coefficients $K_1...K_{n-1}$ given in Eq. 5.12 becomes zero, the stability conditions cannot be met. Given $K_{n-1} = -\sigma_1-\sigma_2...-\sigma_n = 0$, either all the roots have their real components equal to zero or some of them have positive real components. In both cases, a stable, well-damped response cannot be achieved. Similar conclusions can be drawn for other coefficients. Hence, for a control object as shown in Fig. 5.3, all the feedback coefficients in Eq. 5.10 must have a nonzero value. Hence, implicit state feedback is required for the purpose of achieving stability.

The sample controller $W_S(s)$ is given in Eq. 5.10 for the purpose of discussion. In practice, the implementation of higher-order derivatives is associated with severe problems with the high-frequency noise. While the control action proportional to the first derivative of the output is encountered in some cases, the use of higher-order derivatives of the output is very rare. In cases where the information on the internal states is mandatory, it is most frequently obtained by means of the state observers.

5.3.2 Relocation of derivative action into the feedback path

With the derivative action located in the direct path (Fig. 5.2), the closed-loop system transfer function $W_{SS}(s)$, given in Eq. 5.9, has one real zero, $z_1 = -K_P/K_D$. The discussion in Chapters 3 and 4 indicates that this zero may contribute to an overshoot in the step response, even in cases when all the closed-loop poles are real, leading to a strictly aperiodic response. According to developments given in Sections 2.2.4 and 4.6, relocation of the proportional action of the speed controller into the feedback path eliminates the overshoot, provided that the closed-loop poles are real. A similar change in the structure of the position controller can remove the real zero of the closed-loop system transfer function. In Fig. 5.4, the derivative action of the position controller is relocated into the feedback path. The corresponding closed-loop transfer function $W_{SS}(s) = \theta(s)/\theta^*(s)$ is given in Eq. 5.13. It has two closed-loop poles and no finite zeros. Hence, in cases when the closed-loop poles are real and the transient response is strictly aperiodic, the step response of the output position will not overshoot the target.

$$W_{SS}(s) = \frac{\theta(s)}{\theta^*(s)} = \frac{K_P}{K_P + K_D s + J s^2} = \frac{1}{1 + s\dfrac{K_D}{K_P} + 1 + s^2\dfrac{J}{K_P}} \tag{5.13}$$

Note in Fig. 5.4 that the signal y_1 has the role of an internal speed reference. Namely, in cases when the torque $T_{em} = T_{ref}$ required to run the system is negligible, the derivative action and the signal y_1 are summed as $T_{ref} = y_1 - K_D\, d\theta/dt = y_1 - K_D\omega = 0$, and the speed of the system is close to $\omega = y_1/K_D$.

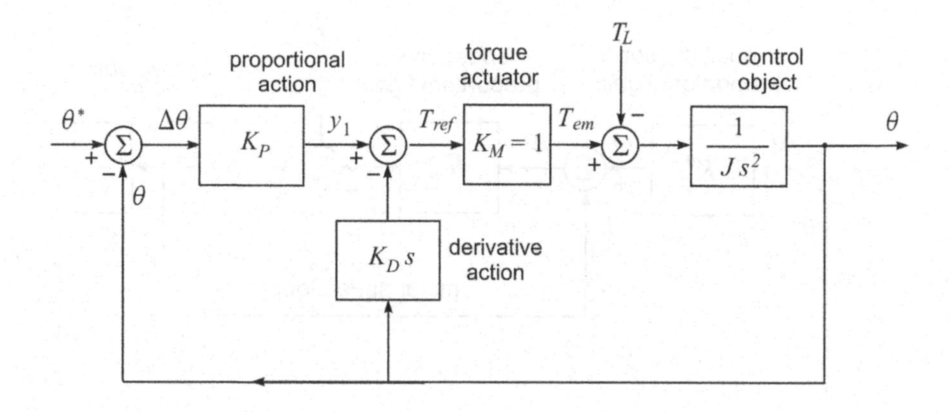

Fig. 5.4. Proportional-derivative position controller with the derivative action relocated into the feedback path.

5.3.3 The position controller with a minor speed loop

In a number of cases, position-controlled systems comprise a minor speed loop, the speed reference of which is supplied from a position controller. Specifically, the position tracking error $\Delta\theta = \theta^* - \theta$ is detected within the position controller and the appropriate speed reference is generated and supplied to the minor loop comprising the speed controller. The purpose of the discussion in this section is to derive the closed-loop transfer function for the system comprising a position controller with proportional action K_P^θ and with a minor speed loop having proportional gain K_P^ω. In addition, the equivalence is demonstrated between the derivative gain K_D^θ of the position controller and the gain K_P^ω.

In Fig. 5.5, it is assumed that the speed feedback ω is available for the inner speed loop. The position feedback θ is taken from the output of the system. The minor loop receives the speed feedback ω^*, proportional to the position error $\Delta\theta$. The driving torque is obtained by multiplying the speed error $\Delta\omega$ and the gain K_P^ω. If the minor loop with the speed controller is considered as an isolated subsystem, its closed-loop system transfer function is derived as

$$W_{SS}^{\omega}(s) = \frac{\omega(s)}{\omega^*(s)} = \frac{K_P^\omega}{Js^2 + K_P^\omega}. \tag{5.14}$$

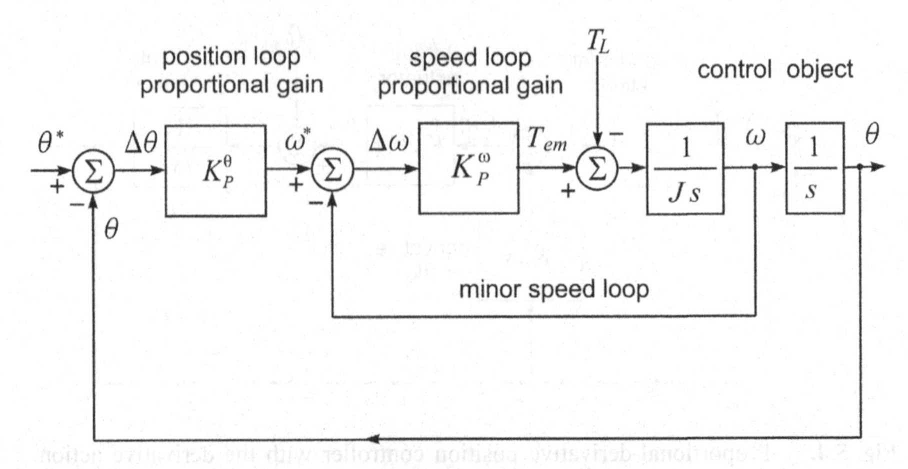

Fig. 5.5. Proportional position controller with an inner speed loop.

To introduce Eq. 5.14, the block diagram of the position controlled system under consideration reduces to the form given in Fig. 5.6. The resulting closed-loop transfer function is given in Eq. 5.15. The function is essentially the same as the closed-loop system transfer function with the position controller having proportional and derivative actions (Fig. 5.4), as given in Eq. 5.13. From a comparison of the two equations, it is evident that the proportional gain K_P^ω of the speed controller corresponds to the derivative gain of the PD position controller. The subsequent analysis is focused on position controllers of the form given in Fig. 5.4, without explicit separation of the minor-loop speed controller, and with all relevant control actions calculated from the position feedback.

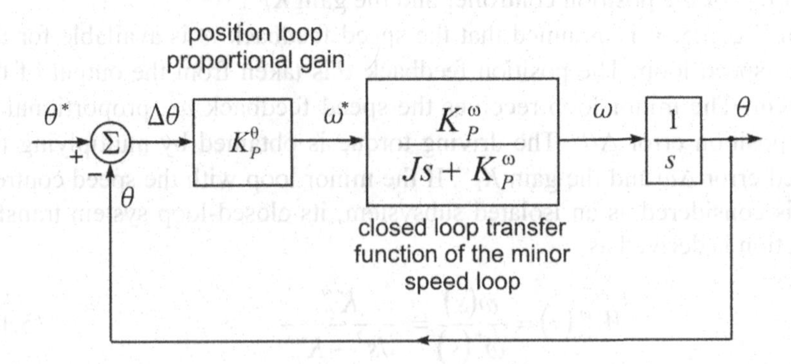

Fig. 5.6. Minor speed loop replaced by its closed-loop system transfer function.

$$W_{SS}^{\theta}(s) = \frac{\theta(s)}{\theta^*(s)} = \frac{K_P^{\omega} K_P^{\theta}}{Js^2 + K_P^{\omega} s + K_P^{\omega} K_P^{\theta}} \tag{5.15}$$

5.3.4 Stiffness of the position-controlled system

If we consider a simplified block diagram of the position controller with proportional and derivative action, as given in Fig. 5.4, it is apparent that the presence of a constant load torque $T_L(t) = T_{LOAD}$ results in a steady-state error in the output position. In the steady-state condition, the output position would not change. Therefore, the derivative action in Fig. 5.4 is equal to zero. Under these circumstances, the load torque is found to be $T_{em} = K_P(\theta^* - \theta) = K_P\,\Delta\theta$. In order to keep the system in the steady state, the driving torque must balance the load. Therefore, the steady-state error is calculated as $\Delta\theta = T_{LOAD}/K_P$. The steady-state ratio $T_{LOAD}/\Delta\theta$ is known as the *stiffness coefficient* of the servo system. In the case of the PD controller given in Fig. 5.4, the stiffness is equal to the reciprocal value of the proportional gain. An increase in K_P would reduce the steady-state position error. On the other hand, the range of applicable gains is limited, due to stability and noise problems. Therefore, the steady-state error $\Delta\theta = T_{LOAD}/K_P$ cannot be suppressed by increasing the K_P parameter.

The tracking error $\Delta\theta$ produced by the load torque disturbance can be determined from the closed-loop transfer function $W_{LS}(s)$. For the system in Fig. 5.2, with the position controller $W_{PC}(s)$, the torque actuator described as $K_M = 1$, and the control object $1/Js^2$, the response to the load disturbance T_L is described by the transfer function $W_{LS}(s)$, given in Eq. 5.16. With the reference position $\theta^*(t) = 0$, the complex image of the output is obtained as $\theta(s) = W_{LS}(s)T_L(s)$. With the proportional–derivative (PD) position controller, the function $W_{LS}(s)$ is given in Eq. 5.17. In cases with the load torque $T_L(t) = T_{LOAD}\,h(t)$, the complex image is obtained as $T_L(s) = T_{LOAD}/s$. In this case and with $\theta^*(t) = 0$, the output position in the steady state is given in Eq. 5.18, representing, at the same time, the steady-state tracking error.

$$W_{LS}(s) = \left.\frac{\theta(s)}{T_L(s)}\right|_{\theta^*=0} = -\frac{1}{Js^2 + W_{PC}(s)} \tag{5.16}$$

$$W_{PC}(s)=K_P+K_D s \quad \Rightarrow \quad W_{LS}(s)=-\frac{1}{Js^2+K_D s+K_P} \tag{5.17}$$

$$\theta(\infty)=\lim_{s\to 0}\left(s\theta(s)\right)=\lim_{s\to 0}\left(s\frac{T_{LOAD}}{s}W_{LS}(s)\right)=-\frac{T_{LOAD}}{K_P} \tag{5.18}$$

According to the analysis given in Section 2.2.2, the load step disturbance $T_L(s) = T_{LOAD}/s$ can be suppressed, provided that the disturbance dynamics $1/s$ appears in the controller $W_{PC}(s)$. This concept is referred to as the *Internal Model Principle* (IMP, see Section 2.3.2). According to IMP, a position controller with integral action K_I/s is capable of suppressing load disturbances having the form of $T_L(t) = T_{LOAD}/h(t)$. The transfer function of the PID controller is given in Eq. 5.19, along with the resulting function $W_{LS}(s)$. With $\theta^*(t) = 0$ and with the load torque assuming the Heaviside step form, the steady-state output is calculated in Eq. 5.20. The presence of the integral control action eliminates the tracking error in the steady state.

$$W_{PC}(s)=K_P+K_D s+\frac{K_I}{s}\Rightarrow W_{LS}(s)=\frac{-s}{Js^3+K_D s^2+K_P s+K_I} \tag{5.19}$$

$$\theta(\infty)=\lim_{s\to 0}\left(s\frac{T_{LOAD}}{s}\frac{-s}{Js^3+K_D s^2+K_P s+K_I}\right)=0 \tag{5.20}$$

In position-controlled servo systems, both PD and PID position controllers are used. The absence of the integral action reduces the order of the system, simplifies the parameter setting, and allows for a larger closed-loop bandwidth. In cases when the load torque is predictable, the PD controller tracking error $\Delta\theta = T_{LOAD}/K_P$ can be removed by the feedforward corrective action. Specifically, the load torque estimate is to be added to the position controller output T_{ref}, and the resulting signal is to be fed to the torque actuator. In a number of applications, the load torque is not predictable, or the accuracy of its estimate is not sufficient. In such cases, the integral control action is compulsory.

5.4 The discrete-time PD position controller

In this section, the discrete-time PD position controller is considered, and the closed-loop pulse transfer function $W_{SS}(z)$ is derived. The block diagram of the position-controlled system with a PD controller is given in Fig. 5.7. The position feedback θ^{FB}, obtained from the shaft sensor, is proportional to the output position θ. The ratio between the two is defined by the parameter K_{FB}. The feedback signal is sampled at the rate $f_S = 1/T$ and converted into the pulse train $\theta^{FB}{}_{DIG}$, comprising the samples $K_{FB}\theta_{(n)}$ of the output position. The error discriminator at the extreme left in Fig. 5.7 compares the reference $K_{FB}\,\theta^*$ and the feedback, obtaining the tracking error $K_{FB}\,\Delta\theta$. Notice in the figure that the proportional action of the controller resides in the direct path and multiplies the tracking error. The derivative action is relocated into the feedback path in order to suppress the closed-loop zeros. The proportional and derivative actions are denoted by y_1 and y_2, respectively. At the sampling instant $t = nT$, the signals y_1 and y_2 are obtained as

$$y_{1(n)} = K_{FB}\,K_P\,\Delta\theta_{(n)}$$

$$y_{2(n)} = K_{FB}\,K_D\left(\theta_{(n)} - \theta_{(n-1)}\right),$$

where the parameters K_P and K_D represent the proportional and derivative gain. At the given sampling instant, the sample of the driving torque $T^*_{(n)}$ is obtained as the difference $y_{1(n)} - y_{2(n)}$. The train of samples T^*_{DIG} is fed to the zero-order hold (Fig. 5.7), obtaining, in this way, the continuous-time torque reference $T_{ref}(t)$. Concerning the ZOH operation, refer to Fig. 4.5A, Fig. 4.5B, and the related discussion in Section 4.2.1.

In Fig. 5.7, the continuous-time torque reference $T_{ref}(t)$ is fed to the torque actuator. To simplify further developments, the actuator is assumed to have the static gain of $K_M = 1$. Where needed, different values of K_M can be taken into account by rescaling the feedback gains K_D and K_P. In Fig. 5.7, the actuator is assumed to have a torque limit of T_{MAX}. Limited torque capability is modeled by introducing a limiter, ensuring that the driving torque T_{em} remains within $\pm\,T_{MAX}$ boundaries. Further on the right, the driving torque and the load torque disturbance T_L are fed to the control object, modeled as $1/Js^2$.

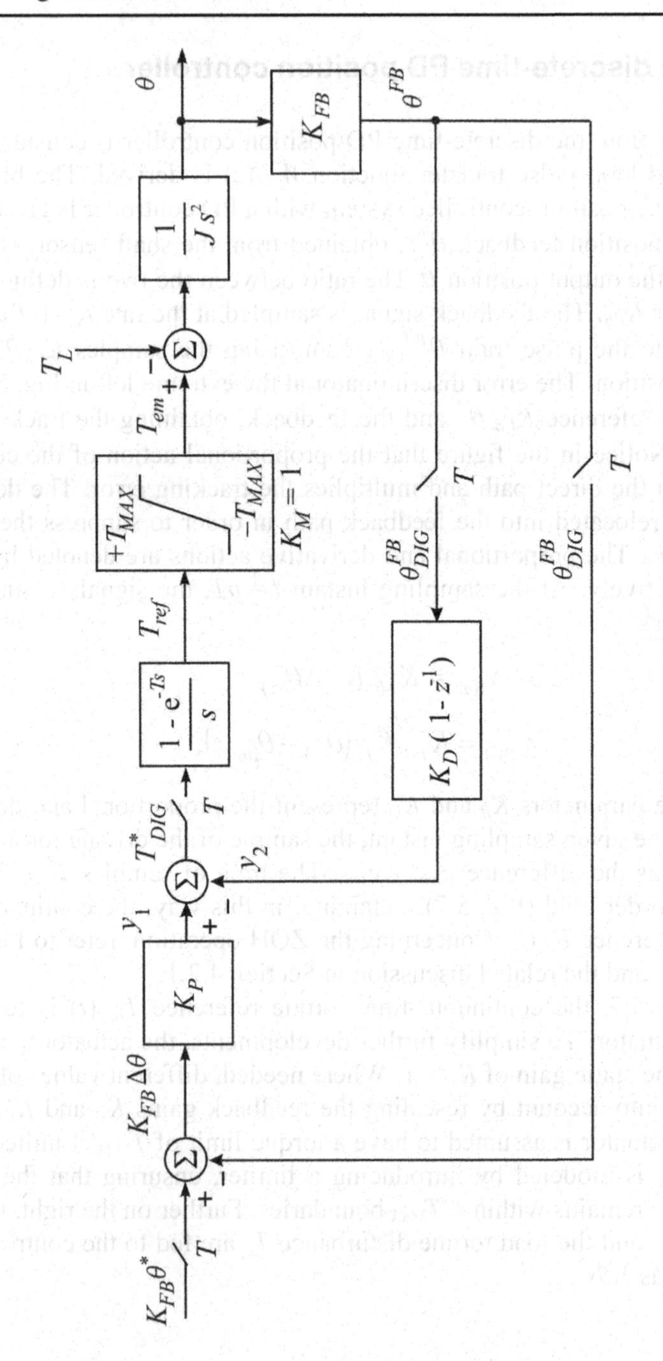

Fig. 5.7. Position-controlled system with discrete-time position controller. Derivative control action is relocated into the feedback path.

The derivative control action can be implemented in the direct path, along with the proportional action. In such a case, both actions process the tracking error $\Delta\theta$. With the proportional action defined as $K_P\Delta\theta$, the derivative action would be proportional to the error increment detected in successive sampling periods. At the instant $t = nT$ and with both actions in the direct path, the derivative action is calculated as $K_{FB} K_D (\Delta\theta_{(n)} - \Delta\theta_{(n-1)})$. This solution is discussed in Section 5.3.2. If we compare the closed-loop system transfer function in Eq. 5.9, obtained with both control actions in the direct path, and the function given in Eq. 5.13, obtained with the derivative action relocated into the feedback path, we find that the former has closed-loop zeros contributing to overshoots in the step response. The characteristic polynomial and the closed-loop poles are not affected by the relocation of the derivative gain. It is of interest to note that the derivative action in the direct path emphasizes the high-frequency content of the input signal. The structure under consideration (Fig. 5.7) has the derivative control action relocated into the feedback path in order to remove the closed-loop zeros, suppress the overshoot in the step response, and ensure a lower amplitude of the high-frequency components in the driving torque.

It is also interesting to notice in Fig. 5.7 that the mechanical subsystem with the derivative feedback y_2 constitutes a speed-controlled subsystem, with the signal y_1 assuming the role of the speed reference. The signal y_2 is proportional to the increment in the output position encountered in successive sampling instants. Hence, such a signal is proportional to the shaft speed $\omega = \mathrm{d}\theta/\mathrm{d}t$. If we assume that the speed ω experiences only a minor change within the sampling period T, the feedback signal y_2 can be approximated as

$$y_2 \approx K_{FB} \, K_D \, \omega T \, .$$

This outcome restates the results given in Section 5.3.4. and shows that the derivative position feedback and the proportional speed feedback are equivalent. If we assume that the torque T_{em} required to set the control object into motion is relatively low, then the difference between the signal y_1 and the feedback y_2 can be neglected. With $y_1 \approx y_2$, the shaft speed is given as

$$\omega \approx \frac{y_1}{K_{FB} \, K_D T} \, .$$

At this point, the subsystem comprising the control object, the torque actuator, and the derivative feedback operates as a speed-controlled subsystem with speed reference y_1 and speed feedback $y_2 = K_{FB} K_D \, \omega T$. The internal speed reference $y_1 = K_P K_{FB} \Delta\theta$ is proportional to the tracking error $\Delta\theta$. In the structure given in Fig. 5.7, the proportional control action can be considered as an outer loop, calculating the speed reference y_1 in proportion to the tracking error, while the derivative action assumes the role of a minor loop, providing proportional feedback $K_{FB} K_D \omega T$ to the local speed controller. This viewpoint will prove useful in designing nonlinear control laws, as discussed in the subsequent sections. In order to preserve the aperiodic character of the step response, even in cases with large disturbances, wherein the driving torque and the shaft speed reach the limits of the system, the speed must be subdued to a nonlinear limit. Implementation of this limit is facilitated by the presence of a local speed controller.

Consider the system in Fig. 5.7 in linear operating mode. The torque reference sample $T^*_{(n)}$, calculated at the sampling instant $t = nT$, is found in Eq. 5.21. The z-transform $T^*(z)$ of the driving torque is given in Eq. 5.22. This equation defines the transfer function of the discrete-time PD position controller.

$$T^*_{(n)} = K_P\, K_{FB} K_M \left(\theta^*_{(n)} - \theta_{(n)}\right) + K_D\, K_{FB} K_M \left(\theta_{(n-1)} - \theta_{(n)}\right) \tag{5.21}$$

$$T^*(z) = K_P\, K_{FB} K_M \left(\theta^*(z) - \theta(z)\right) - K_D\, K_{FB} K_M \theta(z)\left(1 - z^{-1}\right) \tag{5.22}$$

The pulse transfer functions $W_P(z)$ and $W_{PL}(z)$ are given in Eq. 5.8. The functions describe the signal flow from the driving torque input to the system output ($W_P(z)$), and from the disturbance input to the output ($W_{PL}(z)$).

In Eq. 5.22, the first component of the driving torque $T^*(z)$ is proportional to the tracking error, while the second depends on the output position increment. Therefore, the z-transform of the torque can be expressed as $T^*(z) = W_{KP}\, \Delta\theta(z) - W_{KD}(z)\theta(z)$, where $W_{KP} = K_P\, K_{FB}\, K_M$ and $W_{KD}(z) = K_D\, K_{FB}\, K_M\, (1 - z^{-1})$. From these expressions and from Eq. 5.8 and Eq. 5.21, the position-controlled system with a discrete-time PD controller can be represented by a simplified block diagram, given in Fig. 5.8. This representation is valid for the system operating in linear mode, with the driving torque staying away from the torque limit $\pm T_{MAX}$. At the output of the system, $\theta(z)$ represents the z-transform of the pulse train $\theta_{(n)}$, comprising the samples of the shaft position. The complex image $T^L(z)$ represents the samples $T^L_{(n)}$. Each sample $T^L_{(n)}$ corresponds to the average value of the load torque within the given sampling interval T (Eq. 4.18). For the system in Fig. 5.8, it is useful to calculate the closed-loop transfer function $W_{SS}(z) = \theta(z)/\theta^*(z)$, as well as the transfer function $W_{LS}(z) = \theta(z)/T^L(z)$, describing the output response to the load disturbance.

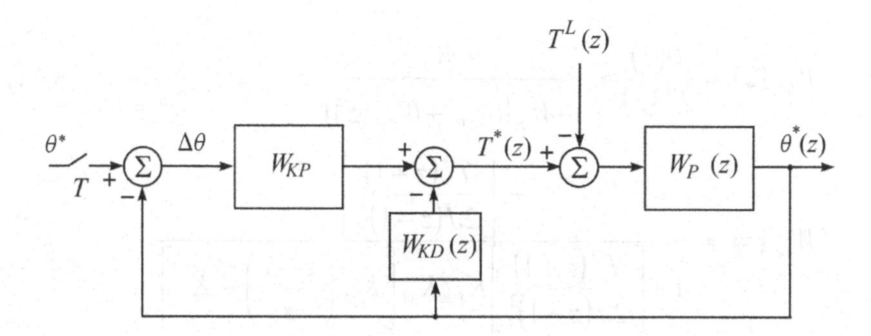

Fig. 5.8. Simplified block diagram of a position-controlled system with a discrete-time PD controller.

The closed-loop system transfer function $W_{SS}(z)$ is given in Eq. 5.23, while the load disturbance transfer function $W_{LS}(z) = \theta(z)/T^L(z)$ is derived in Eq. 5.24. In these equations, K_P and K_D are the feedback gains, K_M is the gain of the torque amplifier included in Eq. 5.8, and K_{FB} is the shaft sensor gain. The functions $W_{KD}(z)$ and W_{KP} are defined as $K_D\, K_{FB}\, K_M\,(1 - z^{-1})$ and $K_P\, K_{FB}\, K_M$, respectively, while $W_P(z)$ is the pulse transfer function of the control object, given in Eq. 5.8.

$$W_{SS}(z) = \frac{\theta(z)}{\theta^*(z)} = \frac{W_{KP}\, W_P}{1 + W_P\left(W_{KP} + W_{KD}(z)\right)} \tag{5.23}$$

$$W_{SS}(z) = \frac{\left[\dfrac{T^2(z+1)}{2J(z-1)^2}\right] K_P\, K_{FB}\, K_M}{1 + \left[\dfrac{T^2(z+1)}{2J(z-1)^2}\right] K_{FB} K_M \left[K_D\left(\dfrac{z-1}{z}\right) + K_P\right]}$$

The functions can be simplified by introducing normalized feedback gains for the proportional and derivative action. The normalized gains can be defined as $p = K_P\, K_{FB}\, K_M \cdot (T^2/2J)$ and $d = K_D\, K_{FB}\, K_M \cdot (T^2/2J)$. The introduction of normalized gains facilitates the parameter-setting procedure. For the case of the discrete-time speed controller, discussed in the previous chapter, it is shown that the optimized values of normalized feedback gains are invariable (Eq. 4.59) and do not depend on the system parameters J, K_M, and K_{FB}, nor do they change with the sampling period T. Similar findings will be proved in this chapter for the position controller. Once the optimized parameter setting (p_{OPT}, d_{OPT}) is determined, the absolute gains K_P and K_D can be calculated by multiplying the normalized gains by the scaling coefficient $2J/(K_{FB}\, K_M T^2)$.

$$W_{LS}(z)=\frac{\theta(z)}{T^L(z)}=\frac{-W_P}{1+W_P\left(W_{KP}+W_{KD}(z)\right)} \tag{5.24}$$

$$W_{LS}(z)=\frac{-\left[\dfrac{T^2(z+1)}{2J(z-1)^2}\right]}{1+\left[\dfrac{T^2(z+1)}{2J(z-1)^2}\right]K_{FB}K_M\left[K_D\left(\dfrac{z-1}{z}\right)+K_P\right]}$$

With normalized gains introduced into Eq. 5.23 and Eq. 5.24, the transfer functions $W_{SS}(z)$ and $W_{LS}(z)$ are found to be:

$$W_{SS}(z)=\frac{z(z+1)p}{z(z-1)^2+\left[d(z-1)+pz\right](z+1)} \tag{5.25}$$

$$=\frac{pz^2+pz}{z^3-(2-p-d)z^2+(1+p)z-d}$$

$$W_{LS}(z)=-\frac{1}{K_P K_{FB}K_M}\frac{pz^2+pz}{z^3-(2-p-d)z^2+(1+p)z-d} \tag{5.26}$$

$$=-\frac{T^2}{2J}\frac{z^2+z}{z^3-(2-p-d)z^2+(1+p)z-d}$$

$$=-\frac{T^2}{2J}\frac{z^2+z}{f_{PD}(z)}\ .$$

The characteristic polynomial $f_{PD}(z)=z^3-(2-p-d)z^2+(1+p)z-d$ can be written as $f_{PD}(z)=(z-\sigma_1)(z-\sigma_2)(z-\sigma_3)$. It is present in the denominators of both transfer functions. The polynomial zeros σ_1, σ_2, and σ_3 are, at the same time, the closed-loop poles, and they determine the character of the closed-loop step response and the bandwidth of the system.

With the assumption that the closed-loop poles are stable, and given the input reference $\theta^*(t)=\Theta^*\,\mathrm{h}(t)$ and the load disturbance $T^L(t)=T_{LOAD}\,\mathrm{h}(t)$, it is possible to apply the transfer functions $W_{SS}(z)$ and $W_{LS}(z)$ and obtain the steady-state value of the output position. For the input and load disturbances having the form of a Heaviside step, their z-transforms are $\theta^*(z)=\Theta^*/(1-z^{-1})$ and $T^L(z)=T_{LOAD}/(1-z^{-1})$, respectively. From Eq. 5.25 and Eq. 5.26, the complex image of the output position is obtained:

$$\theta(z) = W_{LS}(z)\, T^L(z) + W_{SS}(z)\,\theta^*(z).$$

According to the final value theorem given in Eq. 4.13, the steady-state value of the output position is found in the following manner:

$$\theta(\infty) = \lim_{z \to 1} \left[\left(1 - z^{-1}\right)\theta(z)\right] = \Theta^* - \frac{T^2}{2J}\frac{1}{p}T_{LOAD}\,. \tag{5.27}$$

In the steady state, the error $\Delta\theta(\infty) = \Theta^* - \theta(\infty)$ is proportional to the load torque, as redicted in Eq. 5.18 and derived from a simplified *s*-domain model of the system. Note in Eq. 5.27 that the error is directly proportional to the square of the sampling period T and inversely proportional to the normalized feedback gain p. The increase in p cannot drive the output error $\Delta\theta(\infty) = \Theta^* - \theta(\infty)$ down to zero, since the range of applicable gains is restricted by the stability and noise condition. In the subsequent analysis, it will be shown that the optimized values of the normalized gains p_{OPT} and d_{OPT}, applied to different control objects, retain their values. Therefore, the gain p_{OPT} cannot be arbitrarily changed in an attempt to reduce the output error.

Being proportional to T^2, the error would decrease as the sampling rate $f_S = 1/T$ increases. Doubling the sampling rate would reduce the output error by a factor of four. However, there are limits to the sampling rate, and some of them are discussed in Section 4.3. Namely, in each digital controlled system, certain high-frequency dynamic phenomena and noise components cannot be suppressed by deliberate control action. Instead, such signals are attenuated by the low-pass nature of the control object and are treated as the unmodeled dynamics. Therefore, the analog prefiltering and the sampling process need to ensure that such signals do not impair the proper acquisition of the low-frequency feedback. In most cases, the sampling rate is selected in a way that keeps the unmodeled dynamics above the Shannon frequency $f_S/2 = 1/2T$. An unreasonable increase in the sampling frequency would bring the high-frequency signals related to the unmodeled dynamics into the system. This would increase the high-frequency content of the driving torque and might render the control object model unusable. In turn, the step response character would change, leading to instability and requiring a redesign of the controller.

In practical position-controlled systems, the typical sources of high-frequency noise are Pulse-Width Modulation (PWM), quantization effects due to finite resolution of feedback sensors, and parasitic effects such as the cogging torque and slot harmonics.

In most cases, servo motors are supplied from static power converters with power semiconductor switches controlled through the PWM. The commutation frequency is limited due to commutation losses and rarely exceeds 20 kHz. Pulse-shaped supply voltage and limited commutation frequency result in a parasitic component of the motor current, and hence, the torque pulsations associated with the PWM frequency. The triangular-shaped component of the motor current at the frequency of commutation is known as the *ripple*. The current and torque ripple are the noise signals classified as the unmodeled dynamics. They should be attenuated by the analog prefilters and suppressed by the sampling process. An increase in the sampling rate would extend the bandwidth of the sampling circuit, introduce the PWM noise into the system and emphasize the effects of quantization. To avoid the effects of the PWM noise in position-controlled systems, it is common practice to select the sampling period T as an integer multiple of the PWM period ($T = n\, T_{PWM}$, with n ranging from 3 to 10).

Finite resolution of position sensors introduces quantization noise, which is emphasized by an increase in the sampling rate. Incremental encoders are frequently used as feedback devices. They provide two binary signals, known as the encoder phases A and B, each one providing N pulses per turn, with $2^8 < N < 2^{16}$. The processing of encoder signals within the encoder peripheral unit (QEP) of typical motion control DSP devices [10, 11] provides the shaft position with a resolution of $\Delta\theta = \pi/(2N)$. In order to gain insight into quantization effects, it is of interest to consider the signal y_2 in Fig. 5.7, proportional to the shaft speed, in the case when $N = 256$ and with $T = 1$ ms. When the motor runs at a speed of 1000 rpm, the shaft position changes by $(4NT\cdot 1000/60)\ \Delta\theta = 17\ \Delta\theta = 17\ \pi/(2N)$ within each sampling interval. The speed is detected with a resolution of 1000/17 rpm. In other words, one least significant bit (LSB) of the speed estimate corresponds to 58.8 rpm. Such a coarse speed reading would be further worsened by reducing the sampling time.

The impact of the sampling time on the noise content of the driving torque is investigated by means of computer simulations. A position-controlled system with a discrete-time PD controller is modeled in Simulink. The model is given in Fig. 5.9. The control actions are arranged according to the block diagram given in Fig. 5.8. Following the ZOH block, the torque limiter restricts the driving torque to $\pm T_{MAX}$ range. The control object is modeled as $W_P = 1/Js^2$. Limited resolution of the shaft sensor is modeled in the block designated *Quantizer*. Simulation traces of the output position and the driving torque are collected in arrays entitled *torque* and *position*.

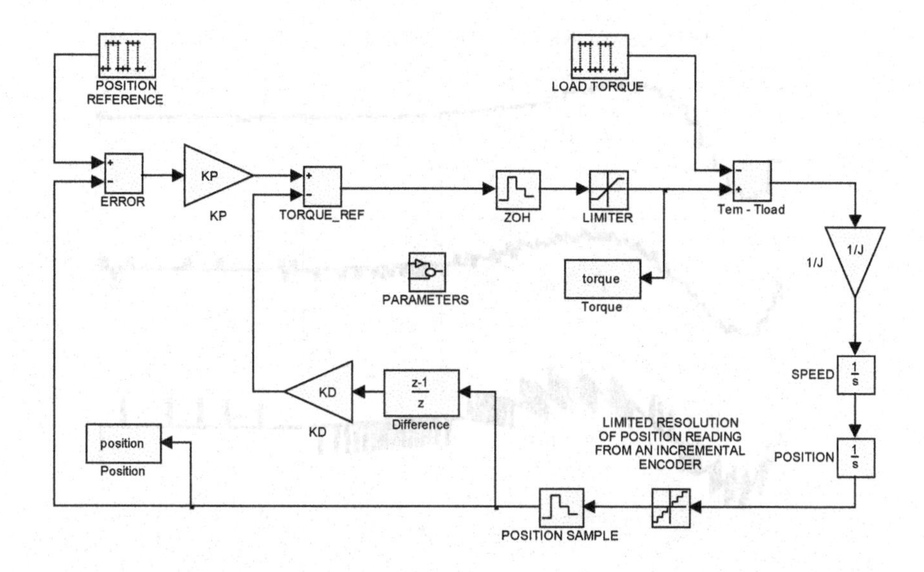

Fig. 5.9. Simulink model of a position-controlled system with a discrete-time PD controller. Derivative action is relocated into the feedback path, while the torque actuator provides the driving torque limited to $\pm T_{MAX}$. The block designated *Quantizer* provides the possibility of setting the resolution of the shaft sensor.

Within the model, it is possible to set the inertia J of the load, the feedback gains, the torque limit, the sampling time, the position reference, and the load disturbance. For simplicity, it is assumed that $K_{FB} = 1$ and $K_M = 1$.

Simulation traces of the driving torque, given in Fig. 5.10, are obtained for three distinct values of the sampling time (1 ms, 3 ms and 10 ms). The waveforms presented in Fig. 5.10 reflect the changes in the driving torque at the step change in the reference. In all three simulation runs, the feedback gains K_P and K_D are adjusted to preserve the response character at various sampling rates. It is assumed that an incremental encoder with $N =$ 2500 pulses per turn is used as the shaft sensor. The high-frequency content of the driving torque is produced by the quantization effects. As the sampling time drops down to $T = 1$ ms, the torque pulsations become unacceptable.

It is concluded that the stiffness of the position-controlled system in Fig. 5.7 cannot be increased by reducing the sampling period. The largest acceptable sampling rate depends on the quantization effects and the noise contents. The application of higher sampling rates $f_S = 1/T$ requires shaft encoders with a larger number of pulses per turn.

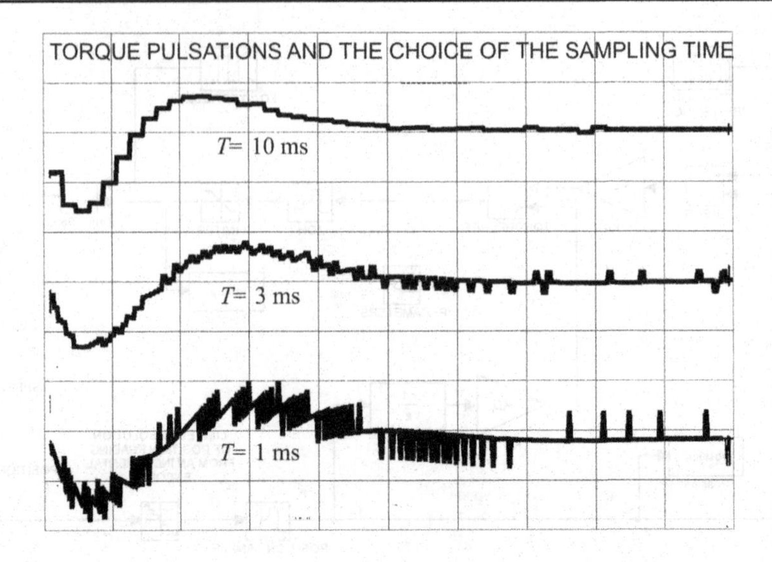

Fig. 5.10. Simulation traces of the driving torque obtained with a discrete-time PD position controller, given in Fig. 5.8. Finite resolution of the shaft sensor is considered, and the sampling time T varies from 1 ms to 10 ms.

5.5 Optimized parameter setting

The purpose of the PD position controller is to detect any discrepancy between the reference trajectory θ^* and the output position, and to provide the torque that drives the system towards the reference and suppresses the tracking error $\Delta\theta$. In the position controller with the proportional and derivative action, the feedback gains K_D and K_P multiply the error in order to provide the driving torque. According to Eq. 5.27, the higher the gains, the larger the stiffness of the system. On the other hand, the range of applicable gains is limited by stability conditions and by the requirement to achieve a well-damped step response with no overshoot in the output position. At the same time, it is desirable to avoid oscillations of the driving torque during transients. In a number of practical servo systems, the torque oscillations and frequent sign changes give rise to detrimental effects related to backlash and other imperfections in mechanical transmission elements. Therefore, the conjugate complex poles with the associated damped oscillations are to be avoided. Similar consideration is given in the previous chapter, where the parameter setting is discussed for the discrete-time speed controller. With real and positive closed-loop poles residing within the unit circle of the z-plane, the step response will have no overshoot. At the same time, the torque transients will be aperiodic as well. Such behavior is strictly aperiodic.

In this section, the parameter setting of the PD position controller is discussed, with the aim of achieving the fastest strictly aperiodic response. The zeros σ_1, σ_2, and σ_3 of the characteristic polynomial $f_{PD}(z) = z^3 - (2-p-d)z^2 + (1+p)z - d = (z-\sigma_1)(z-\sigma_2)(z-\sigma_3)$ are to be real, positive numbers residing within the interval (0 .. 1). The optimized setting of the feedback gains results in the fastest step response under the aforementioned assumptions. In order to derive the optimized gains, it is necessary to derive a criterion function expressing the response quality in numerical form. Further analysis and the parameter-setting procedure are similar to those in Section 4.7.2 and Section 4.7.3. Therefore, the developments are shortened and some in-depth explanations omitted.

In Fig. 5.11, a strictly aperiodic step response of the shaft position $\theta(t)$ is given, along with the position reference $\theta^*(t)$, which exhibits a step at the instant $t = 0$. The output error samples $\Delta\theta_{(n)} = \theta^* - \theta(nT) = \Delta\theta(nT)$ are strictly positive, as the output position does not overshoot the setpoint θ^*. The shaded area in Fig. 5.11 can serve as an indicator of the response speed. A faster response results in a smaller surface of the shaded area. Given the discrete nature of the controller, the surface S is defined by the sum Q defined in Eq. 5.28. The value of Q in this equation represents the sum of all the error samples starting from the instant when the position reference steps up. This value depends on the feedback gains and will serve as the criterion function. The choice of adjustable feedback parameters should drive Q down to the smallest possible values for the given constraints. The optimized gain setting should result in a minimum value of Q, yet preserving the strictly aperiodic character of the step response. If we consider the closed-loop system transfer function given in Eq. 5.25 and have the z-transform of the reference step $\theta^*(z)$ given as $\Theta^*/(1-z^{-1})$, the error samples $\Delta\theta_{(n)}$ are represented by their complex image $\Delta\theta(z)$, given in Eq. 5.29.

$$Q = \sum_{k=0}^{\infty} \Delta\theta(kT) \tag{5.28}$$

Consider the definition of the z-transform (Eq. 4.3). The complex image $\Delta\theta(z)$ can be obtained from the error samples $\Delta\theta_{(n)}$ according to Eq. 5.30. With the criterion function being the sum of the error samples (Eq. 5.28), the value of Q equals the complex image $\Delta\theta(z)$ calculated for $z = 1$ (Eq. 5.31).

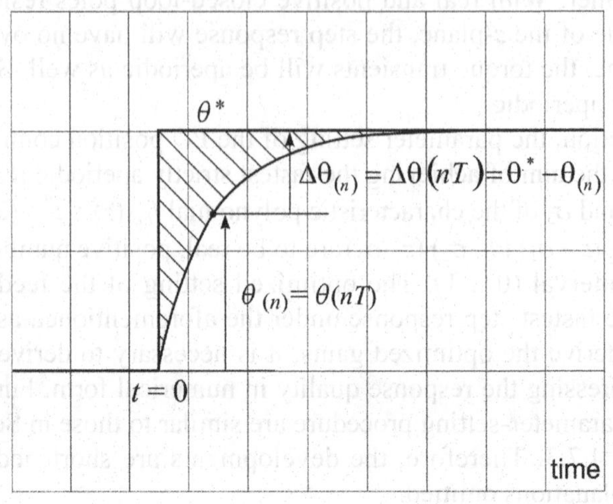

Fig. 5.11. Strictly aperiodic step response of the output position. The shaded surface corresponds to the speed error integral. The smaller the shaded area, the faster the step response.

$$\Delta\theta(z) = (1 - W_{SS}(z))\theta^*(z)$$

$$= \frac{z(z-1)^2 + d(z-1)(z+1)}{z^3 - (2-d-p)z^2 + (p+1)z - d} \frac{\Theta^*}{1-z^{-1}} \tag{5.29}$$

$$= \frac{z^2(z-1) + d(z+1)z}{z^3 - (2-d-p)z^2 + (p+1)z - d} \Theta^*$$

$$\Delta\theta(z) = \sum_{k=0}^{\infty} \Delta\theta_{(k)} z^{-k} \tag{5.30}$$

$$\Delta\theta(1) = \Delta\theta(z)\big|_{z=1} = \sum_{k=0}^{\infty} \Delta\theta_{(k)} = Q \tag{5.31}$$

If we apply results in Eq. 5.29 and Eq. 5.31, the criterion function Q can be expressed in terms of normalized feedback gains:

$$Q = \Delta\theta(1) = \frac{d(1+1)}{1-2+d+p+p+1-d} \Theta^* = \frac{d}{p} \Theta^*. \tag{5.32}$$

As in Eq. 4.50, the optimized gains are those that minimize the ratio d/p while still preserving a strictly aperiodic response. The optimized setting includes the largest applicable proportional gain under the constraints, thus increasing the stiffness and minimizing the tracking error in the presence of load disturbances (Eq. 5.27). It is of interest to derive the optimized values for both normalized gains p and d.

For a strictly aperiodic response, the closed-loop poles σ_1, σ_2, and σ_3 have to assume real, positive values on the interval (0 .. 1). The poles are, at the same time, the zeros of the characteristic polynomial:

$$f_{PD}(z) = z^3 - (2-p-d)\, z^2 + (1+p)\, z - d = (z-\sigma_1)\,(z-\sigma_2)\,(z-\sigma_3).$$

From the above expression, the values σ_1, σ_2, and σ_3 can be expressed in terms of normalized feedback gains p and i:

$$\sigma_1\sigma_2\sigma_3 = d$$
$$\sigma_1\sigma_2 + \sigma_2\sigma_3 + \sigma_3\sigma_1 = 1 + p$$
$$\sigma_1 + \sigma_2 + \sigma_3 = 2 - p - d\,.$$

The sum of the previous equations results in Eq. 5.33. This equation imposes a constraint on the pole placement: namely, the three closed-loop poles cannot be arbitrarily chosen. With the first two poles selected at will, the third will be calculated from Eq. 5.33.

$$\left(\sigma_1 + \sigma_2 + \sigma_3\right) + \left(\sigma_1\sigma_2 + \sigma_2\sigma_3 + \sigma_3\sigma_1\right) + \left(\sigma_1\sigma_2\sigma_3\right) = 3 \qquad (5.33)$$

The calculation of parameters p and d that meet the above conditions and minimize the ratio d/p is equivalent to the procedure given in Section 4.7.3, deriving the optimized gains p and i for a speed controller. The criterion function can be expressed in terms of the closed-loop poles:

$$Q_1 = \frac{1}{Q} = \frac{p}{d} = \frac{\sigma_1\sigma_2 + \sigma_2\sigma_3 + \sigma_3\sigma_1 - 1}{\sigma_1\sigma_2\sigma_3}.$$

If we introduce reciprocal values of the closed-loop poles $x = 1/\sigma_1$, $y = 1/\sigma_2$, and $v = 1/\sigma_3$ and express $v(x, y)$ in terms of $1/\sigma_1$ and $1/\sigma_2$ (Eq. 4.53), the criterion function can be written as $Q = 1/Q_1(x, y)$, where the latter has real arguments $x>1$ and $y>1$:

$$Q_1 = \frac{1}{Q} = \frac{p}{d} = x + y + \left(1 - xy\right)\frac{xy + x + y + 1}{3xy - x - y - 1}\,.$$

The feedback gains p_{OPT} and d_{OPT}, resulting in the maximum $Q_1(x, y)$ and the minimum Q, are found by equating the first derivatives to zero:

$$f_1(x,y) = \frac{\partial Q_1(x,y)}{\partial x} = 0, \quad f_2(x,y) = \frac{\partial Q_1(x,y)}{\partial y} = 0 .$$

By introducing $y = y(x)$ from Eq. 4.55 into the above equation, the optimized values for x, y, and z are obtained as $x_{OPT} = y_{OPT} = z_{OPT} = 1.7024$. Consequently, all the closed-loop poles σ_1, σ_2, and σ_3 are equal (Eq. 5.34), and the characteristic polynomial $f(z)$ assumes the form given in Eq. 5.35.

$$\sigma_1 = \sigma_2 = \sigma_3 = \sigma = 0.587 \tag{5.34}$$

$$f(z) = (z-\sigma_1)(z-\sigma_2)(z-\sigma_3) = (z-\sigma)^3 = (z-0.587)^3 \tag{5.35}$$

The optimized values of normalized gains p_{OPT} and d_{OPT} are calculated from σ_1, σ_2, and σ_3 (Eq. 5.36). Note that the values obtained correspond to the optimized setting for the speed controller proportional and integral gain, obtained in Eq. 4.57.

$$\begin{aligned} d_{OPT} &= \sigma^3 = 0.2027 \\ p_{OPT} &= 3\sigma^2 - 1 = 0.03512 \end{aligned} \tag{5.36}$$

While the optimized setting of normalized gains holds for any position-controlled system, the absolute values K_P and K_D depend on the control object inertia J [kg m^2], the parameter K_{FB} of the shaft sensor, the gain K_M of the torque actuator, and the sampling period T. The absolute values of the feedback gains can be calculated in Eq. 5.37. Note in Eq. 4.58, where the optimized gains are given for the speed controller, that these gains are proportional to the sampling rate $f_S = 1/T$. In the case of a position controller, the optimized gains change with the square of the sampling rate.

$$K_{DOPT} = 0.2027 \frac{2J}{K_{FB} K_M T^2}, \quad K_{POPT} = 0.03512 \frac{2J}{K_{FB} K_M T^2} \tag{5.37}$$

5.6 Computer simulation of the system with a PD controller

A Simulink model of the position-controlled system with a discrete-time PD controller is given in Fig. 5.9. The structure includes the torque limiter, which restricts the driving torque to $\pm T_{MAX}$ range. The model takes into account the finite resolution of the shaft sensor and applies the derivative gain relocated into the feedback path.

The dynamic characteristics of the position controller designed in the previous sections are investigated by running the Simulink model and obtaining the output position and the driving torque transients for the input reference and load torque disturbances. The system parameters are assumed to be $J = 0.11$ kg m^2, $T = 0.001$s, $K_{FB} = 1$, and $K_M = 1$. The feedback gains are tuned according to Eq. 5.37.

The transient response of the driving torque and position for the step change in position reference is given in Fig. 5.12. The amplitude of the reference step is made small, and the driving torque does not reach the limit $+T_{MAX}$. The output position reaches the reference in 11–12 sampling periods. Both torque and position exhibit aperiodic behavior. The output position reaches the target without an overshoot.

The response obtained for a larger input disturbance is given in Fig. 5.13. The reference exhibits a step change and returns to the initial value. The output position has an S-shaped response and reaches the set value without an overshoot. The input step is determined in such a way that the driving torque reaches the limit and remains at $+T_{MAX}$ for 6–7 sampling periods. As the output position moves towards the target and passes approximately one half of the pathway, the torque leaves the limit and the system returns to linear operating mode. Note at this point that the speed reaches the maximum, the torque changes sign, and the system enters into a deceleration phase. In another 11–12 T, the torque, speed, and position error reduce to zero. To avoid the overshoot, it is necessary that the speed decreases to zero at the time when the output position reaches the target. The results obtained in Fig. 5.13 prove that the step response has no overshoot and that all the traces are strictly aperiodic.

The responses obtained in the previous figures require 11–12 sampling intervals for the transient phenomena to settle to their steady-state values. Compared with the discrete-time speed controller (Fig. 4.19), where the speed and torque settling time was 7–8 T, this finding presents an increase in the duration of the transient phenomena of approximately 50%. A longer response time and comparatively smaller bandwidth of position controllers can be attributed to a more complex control object. In the case of a position controller, the control object $W_P(s) = 1/Js^2$ is of the second order, comprising two state variables (speed ω and position θ). The speed-controlled system has a control object $W_P(s) = 1/Js$, with only one state variable. Therefore, under equivalent conditions, the speed controller response is quicker and its closed-loop bandwidth higher.

In Fig. 5.14, the load torque disturbance is shifted back and forth and the transient responses of the position, speed, and driving torque are simulated.

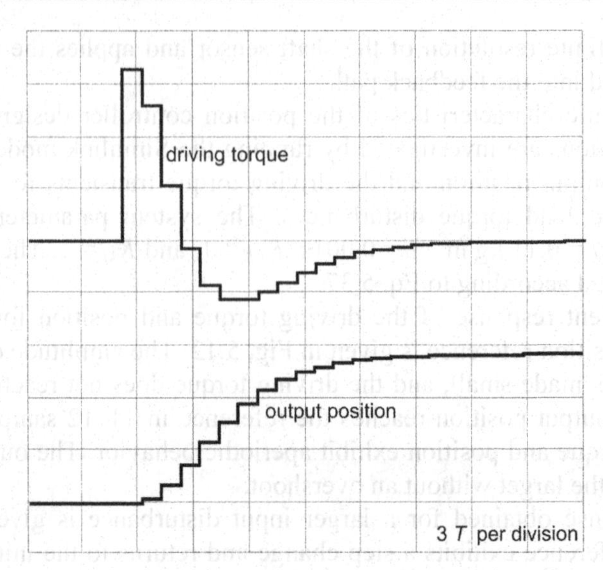

Fig. 5.12. Transient response of the driving torque (upper trace) and position (lower trace) for a small step change in the reference position. The amplitude of the input step is small, and the driving torque does not reach the system limit. The feedback gains of the PD controller are tuned to provide an aperiodic step response of the output position.

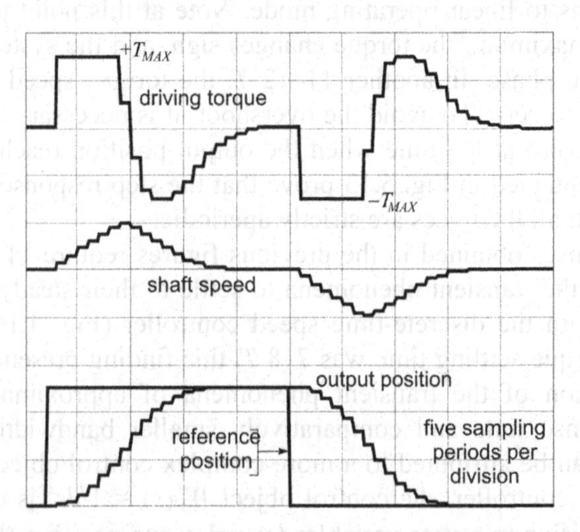

Fig. 5.13. Transient response of the driving torque (upper trace), shaft speed (middle), and position (lower trace) for the step change in the reference position. The reference position exhibits a pulse change and returns to the initial value. The driving torque reaches the limit $\pm T_{MAX}$. Other settings and conditions are the same as in Fig. 5.12.

The load step produces a speed drop and a consequential position error (Eq. 5.27). The position controller provides a driving torque to balance the load. At the instant when $T_{em} = T_L$, the speed stops decreasing. The subsequent interval with $T_{em} > T_L$ drives the speed back to zero. The resulting position error is given in Eq. 5.27 and determined by the stiffness of the system.

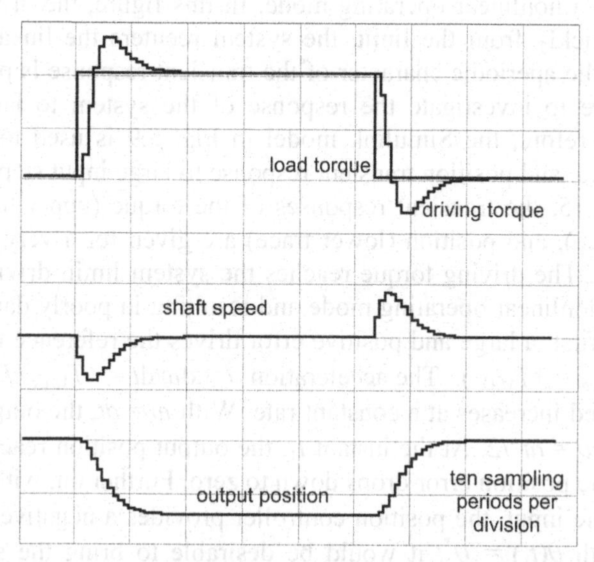

Fig. 5.14. Transient response of the driving torque (upper trace), shaft speed (middle), and position (lower trace) for the step change in the load torque. Other settings and conditions are the same as in Fig. 5.12.

5.7 Operation of the PD position controller with large disturbances

Each position-controlled system has two intrinsic limits. The driving torque T_{em} cannot go beyond the limit $\pm T_{MAX}$, defined by the servo motor characteristics and by the peak current available from the drive power converter. In addition, the shaft speed has to be restricted to the $\pm\omega_{MAX}$ and/or $\pm v_{MAX}$ boundary. The speed of translation (v_{MAX}) and/or rotation (ω_{MAX}) has to be limited in order to preserve the integrity of the moving parts and transmission elements. Excessive speed may result in unacceptable stress and wear in the mechanical subsystem components and eventually cause their failure.

In cases when the system operates in a linear regime, as shown in Fig. 5.12, the torque and speed do not reach the system limits. The position, speed, and torque transients would increase in amplitude as the input disturbance grew larger. At a certain point, the step in position reference would make the speed and /or torque reach the system limits. The torque transient response given in Fig. 5.13 reaches the limit T_{MAX}, bringing the system into a nonlinear operating mode. In this figure, the driving torque recovers quickly from the limit, the system reenters the linear operating mode, and the aperiodic character of the transient response is preserved. It is worthwile to investigate the response of the system to larger disturbances. Therefore, the Simulink model in Fig. 5.9 is used to obtain the torque, speed, and position transient response to large input steps.

In Fig. 5.15, the transient responses of the torque (upper trace), speed (middle trace), and position (lower trace) are given for a very large input disturbance. The driving torque reaches the system limit, driving the system into a nonlinear operating mode and resulting in poorly damped oscillations. At first, a large and positive error drives the reference torque T^* to the limit ($T_{em} = +T_{MAX}$). The acceleration $a = d\omega/dt = +T_{MAX}/J$ is constant, and the speed increases at a constant rate. With $\omega = at$, the output position changes as $\theta = at^2/2$. At the instant t_1, the output position reaches the setpoint and the position error drops down to zero. Further on, with a negative $\Delta\theta < 0$ at the input, the position controller provides a negative torque reference. With $\theta(t_1) = \theta^*$, it would be desirable to bring the system to a standstill and keep $\omega(t) = 0$ for $t > t_1$. However, this outcome cannot be achieved due to the speed $\omega(t_1) = at_1$ having reached a large value. The time interval T_d required to decelerate from $\omega(t_1)$ to a standstill is inversely proportional to the torque limit ($T_d = Jat_1/T_{MAX}$). Therefore, the speed remains positive after t_1, and the output position overshoots the target. Due to a large, negative position error, the driving torque reaches the negative limit and becomes $T_{em} = -T_{MAX}$.

At the instant t_2, the speed is reduced to zero, but the steady-state condition cannot be established. While decelerating in the interval (t_1 .. t_2), the system accumulates a large overshoot in the output position, resulting in a position error $\Delta\theta$ comparable to the reference step. With a negative driving torque, the system decelerates further and the shaft speed becomes negative, increasing in amplitude. At the instant t_3, the position error is once again equal to zero and $\theta(t_3) = \theta^*$. But, equilibrium cannot be restored at this instant either, due to $\omega(t_3) \neq 0$. The process continues as shown in Fig. 5.15, with the position error oscillations gradually decreasing in amplitude. The amplitude of the oscillations is inversely proportional to their frequency, indicating nonlinear behavior of the system.

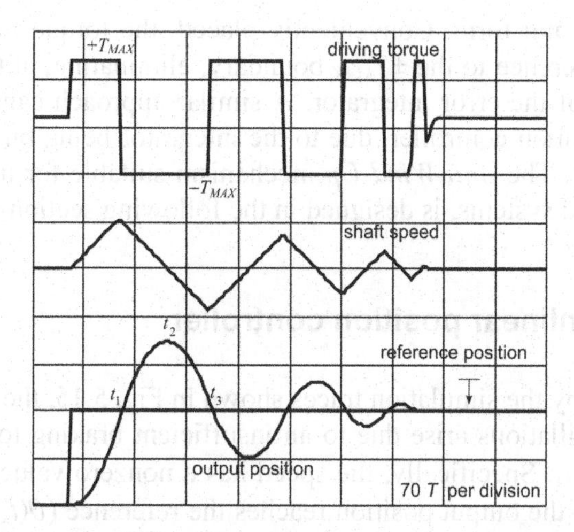

Fig. 5.15. Operation of the PD position controller with large input disturbances. The driving torque reaches the system limit, driving the system into a nonlinear operating mode and resulting in poorly damped oscillations. The amplitude of the oscillations is inversely proportional to their frequency.

It is interesting to compare the traces in Fig. 5.15 to the traces in Fig. 4.23, representing the large step response of the speed controlled system. In both cases, the closed-loop step response of the system in linear operating mode is aperiodic and without an overshoot. The presence of the torque limiter in conjunction with a large reference step results in nonlinear, poorly damped oscillations. The *wind-up* effect has been discussed in the previous chapter and is illustrated in Fig. 4.24. The difficulty in reaching the steady-state condition arises from the fact that the error integrator assumes a nonzero value at instants when $\Delta\omega = 0$ (Fig. 4.24). In a like manner, the responses in Fig. 5.15 reach the steady state through a sequence of nonlinear oscillations, with the shaft speed $\omega(t_x) \neq 0$ at instants when $\theta(t_x) = \theta^*$ and $\Delta\theta = 0$.

In the case of a speed controller, the effects are attributed to the interaction between the torque limiter and the error integrator, which winds up while the torque is in the limit. The wind-up phenomenon is present in the responses in Fig. 5.15. The integrator that winds up is buried within the control object, so, the output position is obtained by integrating the speed. While the torque is in the limit, the output position winds up and overshoots the reference.

In the previous chapter, the speed controller was modified in order to suppress the wind-up effects. In Fig. 4.25, the speed controller is implemented in an incremental form. Conveniently placed, the torque limiter confines the torque reference to the $\pm T_{MAX}$ boundary, eliminating, at the same time, the wind-up of the error integrator. A similar approach cannot be applied to the PD position controller, due to the integrator being buried within the control object. The *Anti-Wind-Up* mechanism suitable for use in the position-controlled systems, is designed in the following section.

5.8 The nonlinear position controller

As indicated by the simulation traces shown in Fig. 5.15, the overshoot and nonlinear oscillations arise due to an insufficient braking torque available at the instant t_1. Specifically, the speed has a nonzero value ($\omega(t_x) \neq 0$) at instants when the output position reaches the reference ($\theta(t_x) = \theta^*$). A limited deceleration rate $|d\omega/dt| = T_{MAX}/J$ cannot secure an abrupt reduction in speed. Therefore, the output position passes through the target with a nonzero speed, producing the overshoot and consequential nonlinear oscillations, as shown in Fig. 5.15. In order to avoid the overshoot, it is necessary to approach the target position in such a way that the speed of motion reduces to zero as the output position reaches the target. Hence, the PD position controller must be modified in such a way that the position error and the speed come to zero at the same time.

When the system moves towards the target, the absolute value of the position error represents the remaining path. Assuming that the system brakes with the maximum torque available ($-T_{MAX}$), the speed will reduce as a linear function of time. Since the output position is the integral of the speed, it is possible to express the maximum permissible speed ω_M for the remaining path $\Delta\theta$. The system with a limited driving torque T_{MAX}, running at a speed $|\omega| < \omega_M(\Delta\theta)$, will be capable of reaching zero speed at the same instant as when the output position reaches the target. In order to guarantee such behavior, it is necessary to implement the speed limit $\omega_M(\Delta\theta) = fp(\Delta\theta)$. The subsequent considerations are focused on deriving the function $fp(\Delta\theta)$.

5.8.1 The speed-limit dependence on the remaining path

Consider the system running at a speed ω towards the target position, with the remaining path $\Delta\theta$ and with the deceleration rate $-d\omega/dt = -T_{MAX}/J$ limited by the system limit $\pm T_{MAX}$. The kinetic energy of such a system is given in Eq. 5.38. During the deceleration phase, the servo motor delivers the torque $-T_{MAX}$, opposing the direction of motion. In such a case, the servo motor works as a generator. It converts the kinetic energy of the system into electrical energy. If we neglect the power losses, the electrical power obtained from the motor terminals equals $P_G = \omega\, T_{MAX}$. The kinetic energy of the mechanical subsystem reduces at the same rate, thus reducing the speed of motion. In a practical servo system, the braking power P_G obtained from decelerating axes is either restored to the primary source or converted into heat and dissipated on a dedicated Dynamic Braking Resistor (DBR) [17].

Kinetic energy is taken by the servo motor through the braking process with a constant torque $T_{em} = -T_{MAX}$. Therefore, it can be calculated by multiplying the torque and the path, as indicated in Eq. 5.39.

$$W_{KIN}(\omega) = \frac{1}{2} J\omega^2 \tag{5.38}$$

$$W_{DBR}(\Delta\theta) = T_{MAX}\, \Delta\theta \tag{5.39}$$

In order to avoid the overshoot, the system needs to arrive at the target position, while reducing the speed to zero at the same time. Hence, both $\Delta\theta$ and ω have to become zero at the same instant. Therefore, at any instant t_x during the deceleration process, characterized by the actual speed $\omega(t_x)$ and the remaining path $\Delta\theta\, t_x)$, the kinetic energy $W_{KIN}(\omega(t_x))$ must not exceed the value of $W_{DBR}(\Delta\theta(t_x))$. The maximum permissible speed ω_M of the system approaching the target position can be calculated from $W_{KIN} = W_{DBR}$. The speed is given in Eq. 5.40 as a function of the remaining path, peak torque, and inertia. This equation relates the absolute value $|\omega_M|$ to the absolute value of the remaining path.

$$|\omega_M| = fp(|\Delta\theta|) = \sqrt{\frac{2T_{MAX}|\Delta\theta|}{J}} \tag{5.40}$$

5.8.2 Enhancing the PD controller

The structure of the PD position controller, given in Fig. 5.7, does not limit the speed. Therefore, speed changes will increase with the amplitude of the input. For a sufficiently large input step (Fig. 5.15), the speed will exceed the limit ω_M. In order to implement a path-dependent speed limit, the structure in Fig. 5.7 must be changed. It is of interest to notice that the signal y_2 in Fig. 5.7 corresponds to the speed feedback:

$$y_2 \approx K_{FB} K_D \omega T .$$

Therefore, given $y_1 \approx y_2$ in the same figure, the speed of motion is proportional to y_1. In Fig. 5.7, the control object with the derivative action in its minor loop constitutes a local speed controller, wherein the signal y_1 assumes the role of an internal speed reference:

$$\omega \approx \frac{y_1}{K_{FB} K_D T} .$$

Hence, the speed of the system can be limited by intervening within the position controller structure and imposing the limit $\omega_M(|\Delta\theta|)(K_{FB} K_D T)$ on the signal y_1. Whenever $|y_1|$ exceeds the limit $K_{FB} K_D T\, \omega_M(|\Delta\theta|)$, the absolute value of the signal is to be restricted while preserving its sign. In other cases, for lower input disturbances, the signal y_1 is not affected by the functional speed limit.

The required functionality of the speed limit $\omega_M(|\Delta\theta|)$ is illustrated in Fig. 5.16. With the position error $\Delta\theta$ on the horizontal axis and the speed on the vertical axis, the diagram represents a phase plane. The permissible operating points $(\Delta\theta, \omega)$ are comprised in the shaded area. Note in the figure that the straight line *fl* has the slope $K_P/(K_D T)$.Consider the structure in Fig. 5.7 with the internal signal $y_1 = K_{FB} K_P \Delta\theta \approx K_{FB} K_D T \omega$ taking the role of the speed reference. In the phase plane of Fig. 5.16, the line *fl* provides the speed $K_P \Delta\theta/(K_D T)$ for the given error $\Delta\theta$. The operation of the position controller in Fig. 5.7 can be envisaged in such a way that the proportional action generates the internal speed reference $y_1 = K_{FB} K_P \Delta\theta$, while the position derivative feedback y_2 provides for the speed feedback and runs the system at a speed $K_P \Delta\theta/(K_D T)$, proportional to the tracking error $\Delta\theta$. It is concluded that the area above *fl* in Fig. 5.16 cannot be reached, since the proportional action generates the signal y_1 strictly on the line *fl*. Not all the points along the line are permissible, due to the need to limit the speed according to Eq. 5.40.

The square root curve *fp* in Fig. 5.16 corresponds to the speed limit to be observed in deceleration (Eq. 5.40) in order to secure a sufficient braking distance and to stop the drive at the target position. The intersection point of *fp* and *fl* is denoted by A in this figure. When the system operates with relatively small disturbances, requiring torque values below the limit, the operating point will slide along the line *fl* between the origin and the point A: that is, the square root speed limit *fp* will not be activated. As the amplitude of the reference step increases, the torque transients will reach the system limit, and the value of y_1, calculated as $K_{FB}K_P\,\Delta\theta$, will go beyond A. At this point, the system cannot be permitted to climb along the line *fl*. Instead, the internal speed reference must be limited. In the phase plane 5.16, the operating point $(\Delta\theta, \omega)$ has to be restricted to the section AB of the square root curve *fp*.

In addition to the path-dependent speed limit $\omega_M(\Delta\theta) = fp(\Delta\theta)$, the speed of motion has to be subjected to $|\omega| \le \omega_{MAX}$, wherein the limit ω_{MAX} corresponds to the largest speed that preserves the integrity of the mechanical subsystem. This limit is represented by the horizontal line *fm* in Fig. 5.16. Beyond intersection B in the figure, the signal y_1 must be limited in amplitude to $K_{FB}K_D\,T\omega_{MAX}$, in order to restrict the speed of motion to the system limit ω_{MAX}.

The desired operation of the position controller can be described as follows. For position errors $|\Delta\theta| < \Delta\theta(A)$, the system operates in linear mode, the signal flow depicted in Fig. 5.7 is not altered, and the signal y_1 retains the value of $K_{FB}K_P\,\Delta\theta$. With position errors $\Delta\theta(A) < |\Delta\theta| < \Delta\theta(B)$, the amplitude of y_1 is limited to $K_{FB}\,K_DT\,fp(|\Delta\theta|)$ and its sign is sustained. In cases when $\Delta\theta(B) < |\Delta\theta|$, the amplitude of y_1 is constant and equal to $K_{FB}K_D\,T\omega_{MAX}$.

The functional speed limit designed in Fig. 5.16 is summarized in Eq. 5.41. The block with proportional gain in Fig. 5.7, providing $y_1 = K_{FB}K_P\,\Delta\theta$, has to be replaced by another block, that calculates $y_1 = fx(\Delta\theta)$ according to Eq. 5.41. The input to the block remains the position error $\Delta\theta$. With the internal speed reference $fx(\Delta\theta)$, the operating point in the phase plane shown in Fig. 5.16 slides along the line *fl* for small disturbances until it reaches intersection A with the square root limit *fp*. This limit is followed until the maximum speed ω_{MAX} is reached at intersection B.

$$fx(\Delta\theta) = \min\left\{\frac{K_P\,|\Delta\theta|}{K_D\,T}, \omega_{MAX}, \sqrt{\frac{2T_{MAX}|\Delta\theta|}{J}}\right\} K_{FB}K_D T\ \mathrm{sgn}(\Delta\theta) \qquad (5.41)$$

It is interesting to notice that in cases with $|\Delta\theta| < \Delta\theta(A)$, the operation of the system is not affected by replacing the proportional gain block with $fx(\Delta\theta)$. In other words, with position reference steps smaller than $\Delta\theta(A)$, the function $fx(\Delta\theta)$ results in $y_1 = K_{FB}K_P\,\Delta\theta$, and the system corresponds in full to the previous block diagram given in Fig. 5.7. The system where K_P gain is replaced by $fx(\Delta\theta)$ is given in Fig. 5.17.

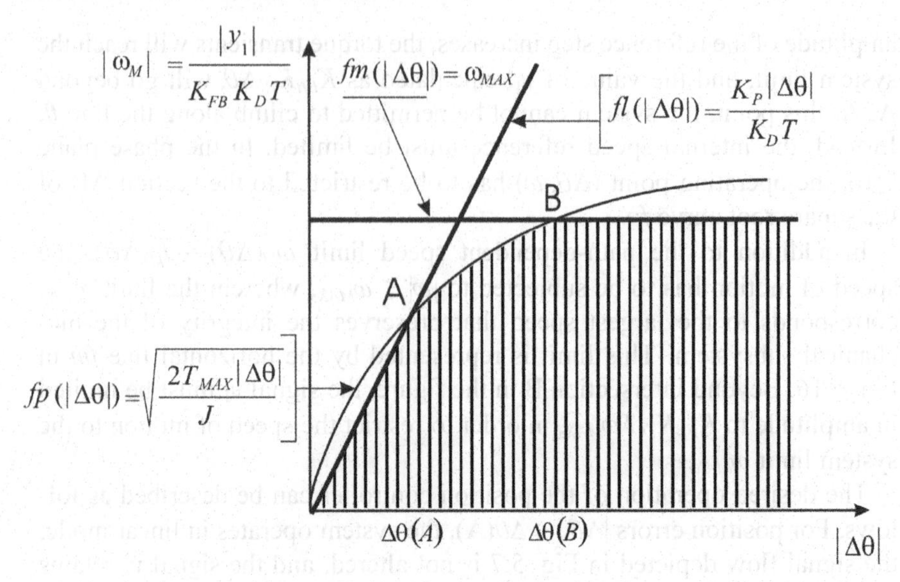

Fig. 5.16. Functionality of the path-dependent speed limit. The square root function *fp* represents the maximum speed in the deceleration (Eq. 5.40) as a function of the braking distance $\Delta\theta$. The horizontal line $fm = \omega_{MAX}$ corresponds to the maximum speed for the given mechanical subsystem. The straight line *fl* has the slope $K_P/(K_DT)$. The permissible operating points are contained within the shaded area.

It is important to consider the case when the system runs towards the target at the maximum permissible speed ω_{MAX}. In such a case, the position error $|\Delta\theta|$ exceeds $\Delta\theta$(B). The signal y_1 in Fig. 5.17 equals $K_{FB}K_D T\omega_{MAX}$. In the phase plane shown in Fig. 5.16, the operating point $(\Delta\theta, \omega)$ lies on *fm*, staying to the right of intersection B. The driving torque T_{em} (Fig. 5.17) is calculated as $K_M(y_1-y_2)$. While the system approaches intersection B and the error reduces towards $\Delta\theta$(B), the speed is constant and equal to ω_{MAX}. The torque T_{em} corresponds to motion resistances at such a speed and does not necessarily have to be at the limit ($|T_{em}| < T_{MAX}$). Having passed intersection B, the error becomes $\Delta\theta(A) < |\Delta\theta| < \Delta\theta(B)$, and the amplitude of the internal speed reference y_1 is calculated as $K_{FB}\, K_D T\, fp(|\Delta\theta|)$. The system enters a deceleration phase where the servo drive brakes the accelerated masses, applying the torque $T_{em} = -T_{MAX}\cdot\text{sign}(\omega) = -T_{MAX}\ \text{sign}(\Delta\theta)$. The operating point $(\Delta\theta, \omega)$ slides along the square-root-shaped *fp* from B towards A. After reaching intersection A, the system enters a linear operating mode. In this mode, none of the system limits is active and further dynamic behavior depends in full on the closed-loop transfer function $W_{SS}(z)$ and its poles and zeros. Therefore, the system finally reaches the target position and enters the steady state in an aperiodic manner, with an exponential decay of relevant variables and states.

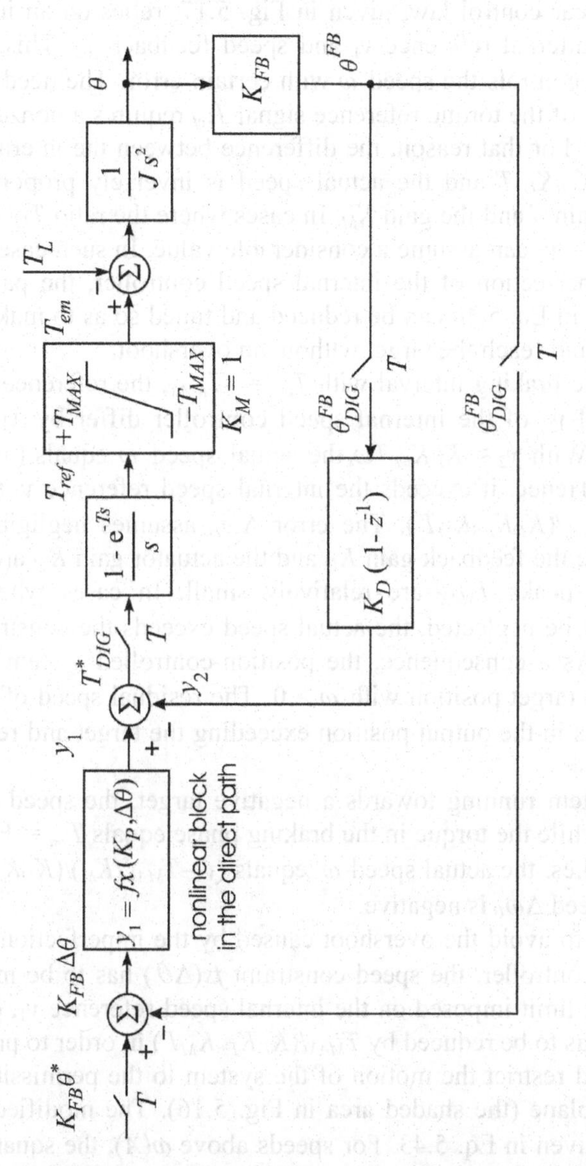

Fig. 5.17. A proportional–derivative discrete-time position controller where the proportional block is replaced with a nonlinear function $fx(\Delta\theta)$, designed in Eq. 5.41. Signal y_1 represents the internal speed reference. The function $fx(\Delta\theta)$ limits the speed to the value calculated from the braking distance $\Delta\theta$.

5.8.3 The error of the minor-loop speed controller

The nonlinear control law, given in Fig. 5.17, relies on an internal speed loop with internal reference y_1 and speed feedback y_2. This proportional speed loop controls the speed ω with certain error. The need to provide a finite value of the torque reference signal T_{ref} requires a nonzero value of $-K_M(y_1 - y_2)$. For that reason, the difference between the internal speed reference $y_1/K_{FB}K_D T$ and the actual speed is inversely proportional to the feedback gain d and the gain K_M. In cases where the ratio T_{MAX}/K_M is high, the error $y_1 - y_2$ can assume a considerable value. In such cases, to account for the imperfection of the internal speed controller, the path-dependent speed limit in Eq. 5.40 can be reduced and tuned so as to make the system decelerate and reach the target without an overshoot.

Given the braking interval with $T_{em} = -T_{MAX}$, the reference y_1 and feedback signal y_2 of the internal speed controller differ by $\Delta y = y_1 - y_2 = -T_{MAX}/K_M$. With $y_2 \approx K_D K_{FB} T\omega$, the actual speed ω equals $(y_1 + T_{MAX}/K_M)/(K_D K_{FB} T)$. Hence, it exceeds the internal speed reference $y_1/(K_D K_{FB} T)$ by $\Delta\omega_B = T_{MAX}/(K_D K_{FB} K_M T)$. The error $\Delta\omega_B$ assumes negligible values in cases where the feedback gain K_D and the actuator gain K_M are high, while the torque peaks T_{MAX} are relatively small. In cases where the error $\Delta\omega_B$ cannot be neglected, the actual speed exceeds the constraint given in Eq. 5.40. As a consequence, the position-controlled system in Fig. 5.17 reaches the target position with $\omega > 0$. The residual speed of $\omega(\Delta\theta = 0) \approx \Delta\omega_B$ results in the output position exceeding the target and resulting in an overshoot.

In a system running towards a negative target, the speed of motion is negative, while the torque in the braking phase equals $T_{em} = +T_{MAX}$. As the system brakes, the actual speed ω equals $(y_1 - T_{MAX}/K_M)/(K_D K_{FB} T)$, and the residual speed $\Delta\omega_B$ is negative.

In order to avoid the overshoot caused by the imperfection of the internal speed controller, the speed constraint $fx(\Delta\theta)$ has to be modified. The square root limit imposed on the internal speed reference y_1, derived from Eq. 5.40, has to be reduced by $T_{MAX}/(K_D K_{FB} K_M T)$ in order to provide timely braking and restrict the motion of the system to the permissible region of the phase plane (the shaded area in Fig. 5.16). The modified speed limit $fx'(\Delta\theta)$ is given in Eq. 5.43. For speeds above $\omega(A)$, the square root speed limit fp in Eq. 5.40 is reduced by $T_{MAX}/(K_D K_{FB} K_M T)$. The lower limit of fp^* in Eq. 5.42 is set to $\omega(A) = 2K_D T_{MAX} T/(JK_P)$. This speed corresponds to the intersection of curves fl and fp in Fig. 5.16. For $|\omega| \le \omega(A)$, the system operates in linear mode, where the nonlinear speed limit fx^* is inactive. The modified speed limit $fx^*(\Delta\theta)$ is used in the subsequent Simulink models and applied while performing the experimental verification.

$$fp^*(\Delta\theta) = \min\left\{\omega(A), \left(K_S\sqrt{\frac{2T_{MAX}|\Delta\theta|}{J}} - \frac{T_{MAX}}{K_D K_{FB} K_M T}\right)\right\} \tag{5.42}$$

$$fx^*(\Delta\theta) = \min\left\{\frac{K_P|\Delta\theta|}{K_D T}, \omega_{MAX}, fp^*(\Delta\theta)\right\} K_{FB} K_D T \operatorname{sgn}(\Delta\theta) \tag{5.43}$$

Reduction of the square root limit by $T_{MAX}/(K_D K_{FB} K_M T)$ resolves the problem of the steady-state error within the internal speed controller. Due to a finite K_D gain, the feedback signal y_2 tracks the internal speed reference y_1 with a certain delay. Restriction of the internal speed reference to the shaded area of Fig. 5.16 may still leave the actual speed outside the permissible region. The excess speed is proportional to transient speed errors of the internal speed controller. Even a small excursion beyond the limit $fp(\Delta\theta)$ results in a delayed braking and an overshoot. In order to provide timely braking, the square root speed limit can be scaled down, compensating in such a way for dynamic speed errors. The nominal value of the scaling coefficient in Eq. 5.42 is $K_S = 1$. In the subsequent simulation runs, the scaling is set to $K_S = 0.98$.

5.9 Computer simulation of the system with a nonlinear PD controller

The structure in Fig. 5.17 has been verified by means of computer simulations. The Simulink model, given in Fig. 5.9 is taken as the starting point. The model is modified by removing the gain block K_P and inserting in its place the function $fx^*(\Delta\theta)$, given in Eq. 5.43. The nonlinear block Y1 in Fig. 5.18 receives the position error $\Delta\theta$ and torque T_{em} and provides the internal speed reference y_1, limited according to Eq. 5.43. The subsystem Y1 is detailed in Fig. 5.19. The square root limit is reduced by subtracting the absolute value of the driving torque, obtained from the block *Abs1*. The block named *LIMITER* gives a lower bound to the signal in accordance with Eq. 5.42. Simulation runs are performed with the system parameters $K_M = 1$, $T_{MAX} = 10$ Nm, $K_{FB} = 1$, $T = 1$ms, $J = 0.01$ kgm^2, and $\omega_{MAX} = 100$ rad/s. The feedback gains are set according to the design rule in Eq. 5.37.

The response to a large step in the position reference is given in Fig. 5.20. The traces represent the driving torque (top trace), speed (middle), and position (lower trace). The system first accelerates using the maximum torque T_{MAX}, then runs at the maximum speed ω_{MAX}, and eventually brakes, using all the available braking torque $-T_{MAX}$, just in time to stop at the target without an overshoot. It is interesting to note that the response given in the figure reaches the target in the shortest time possible, given the system limits T_{MAX} and ω_{MAX}.

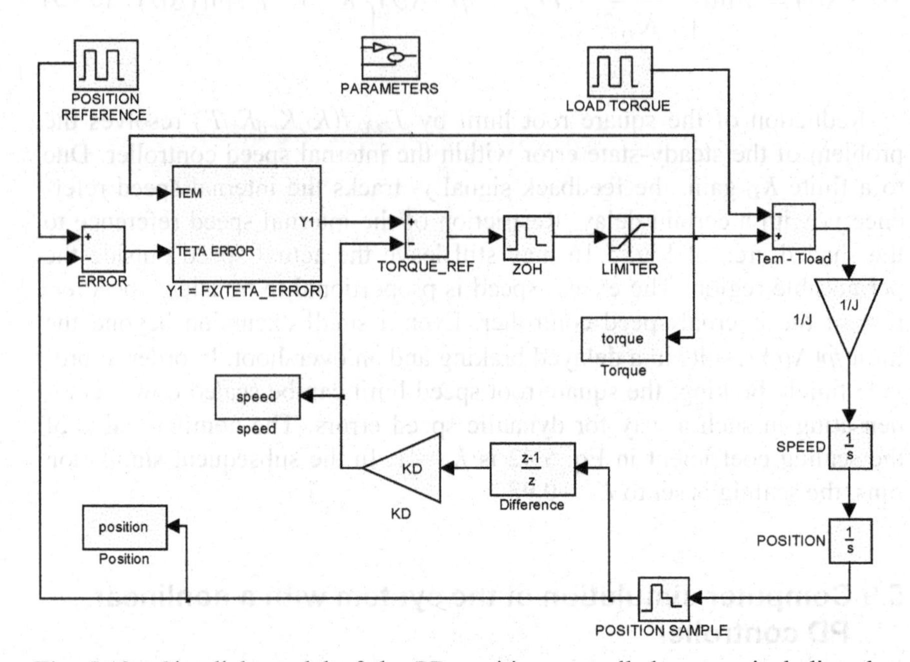

Fig. 5.18. Simulink model of the PD position-controlled system including the nonlinear speed limit $fx^*(\Delta\theta)$, given in Eq. 5.43. The nonlinear block Y1 = FX replaces the proportional gain K_P (Fig. 5.9). The block receives the position error $\Delta\theta$ and torque T_{em} and implements the formula in Eq. 5.43.

The transient response caused by a large position step takes approximately 600 sampling periods to complete. During the braking process, the nonlinear speed limit $fx^*(\Delta\theta)$ is applied. At the instant t_x, the error $\Delta\theta$ in the output position is still positive. In such a case, a linear position controller, as given in Fig. 5.7, would proceed by generating a positive torque reference. This would result in the output position passing through the target at the maximum speed, creating a large overshoot. Due to the nonlinearity $y_1(\Delta\theta) = fx^*(|\Delta\theta|)$ given in Eq. 5.43 and built into the system, deceleration commences at t_x, and the drive reduces the speed before reaching the target. At this instant, the actual speed (i.e., ω_{MAX}) exceeds the value permitted in Eq. 5.40. Therefore, y_1 is limited and the difference $y_1 - y_2$ becomes negative. Hence, the torque reference T_{ref} becomes negative, starting the deceleration phase.

Note in Fig. 5.20 that the braking torque differs in shape from the torque pulse encountered in the acceleration phase. At the instant t_y, the braking torque exhibits a small notch and then decays to zero. The effect is related to the design of the functional limit $fx^*(\Delta\theta)$, given in Eq. 5.43 and Fig. 5.16. During the braking interval, the operating point $(\Delta\theta, \omega)$ slides along fp until intersection A is reached and then it passes on the straight line fl. The first derivative of $fx(\Delta\theta)$ exhibits an abrupt change at $\Delta\theta(A)$. This change and the imperfection of the internal speed controller cause the torque pulse at braking to deviate from the rectangular shape, observed during acceleration in Fig. 5.20. The function $fx(\Delta\theta)$ designed in Fig. 5.16 can be enhanced by introducing a smooth transition between fm, fp, and fl at intersections A and B. Further considerations and experimental verification of the nonlinear position control law are based on the functional limit $fx^*(\Delta\theta)$ as specified by Eq. 5.43.

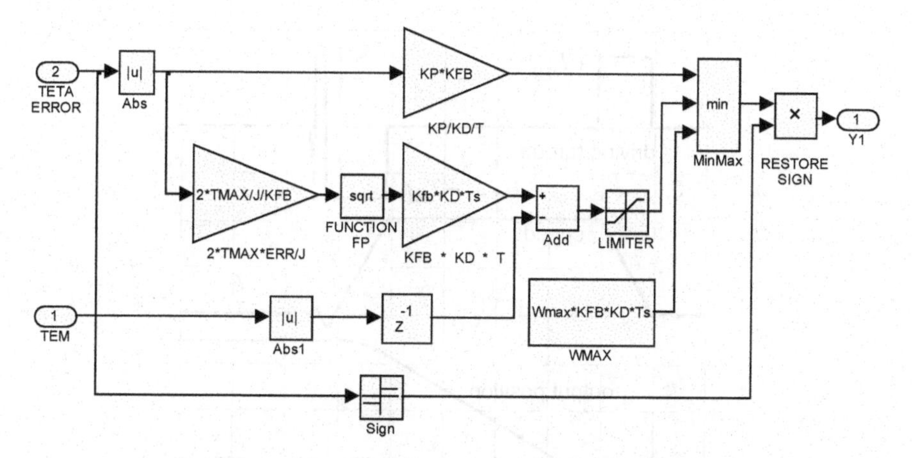

Fig. 5.19. Simulink subsystem implementing the nonlinear function $fx^*(\Delta\theta)$, as defined in Eq. 5.43. This subsystem replaces the gain block K_P in the model given in Fig. 5.9. The square root limit is reduced by subtracting the absolute value of the driving torque (*Abs1*). The block designated *LIMITER* implements the restriction in Eq. 5.42. The output of the subsystem (right) is the internal speed reference y_1.

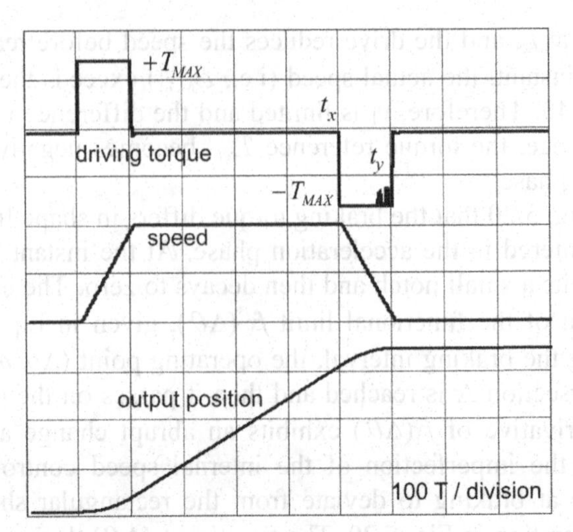

Fig. 5.20. Large step response of the discrete-time PD position controller with path-dependent speed limit. The proportional gain block is replaced with the function $fx^*(\Delta\theta)$, given in Eq. 5.43. The time required to reach the reference position depends on the system limits ω_{MAX} and T_{MAX}.

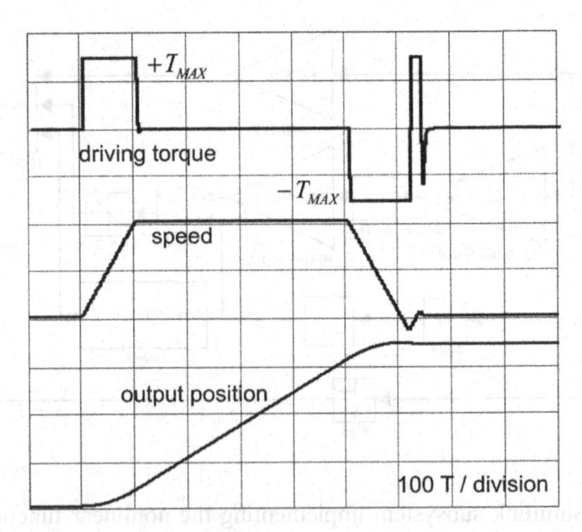

Fig. 5.21. The simulation run, given in Fig. 5.20, is repeated without the speed limit correction, defined in Eq. 5.42 and Eq. 5.43. The internal speed reference y_1 is calculated as $fx(\Delta\theta)$, given in Eq. 5.41. Delays and static error of the internal speed controller result in the speed going beyond the limit $fp(\Delta\theta)$. An excess speed results in an overshoot. The overshoot in the output position can be inferred from the speed and torque waveforms.

It is useful to examine the impact of the internal speed controller imperfection on the large step response. Therefore, the simulations are repeated, omitting the corrective actions designed in Eq. 5.42 and Eq. 5.43. Due to delays and static error of the speed controller, the actual running speed exceeds the square root limit $fp(\Delta\theta)$, resulting in an overshoot. Simulation traces are given in Fig. 5.21. The overshoot in the output position is hardly visible, but it is suggested by the speed and torque waveforms.

5.10 Experimental evaluation of performances

The position controller designed in the previous section is experimentally verified. A set of experimental runs is performed with the linear PD controller, given in Fig. 5.7. The response to large input disturbances is tested with the control structure in Fig. 5.17, including the nonlinearity $fx(\Delta\theta)$, defined by Eq. 5.41. In both cases, the adjustable feedback parameters are set according to Eq. 5.37, in order to obtain the fastest strictly aperiodic response.

The experimental setup is similar to the one described in Chapter 4 and detailed in [16]. It consists of a three-phase induction motor with $P_{nom} = 1$ HP and $n_{nom} = 1410$ rpm. The motor is supplied from a Current Regulated Pulse Width Modulated (CRPWM) inverter, equipped with an incremental shaft encoder having $N = 1250$ pulses per revolution, and controlled through the Indirect Field Oriented Control algorithm. The system comprising the induction motor, the CRPWM inverter, and the IFOC controller acts as a torque actuator. The driving torque obtained from such a controller tracks the torque reference with a rise time of 250 μs. This delay is considered negligible compared with the sampling period of the position loop, set to $T = 10$ ms. The motor is coupled to an inertial load.

The equivalent inertia of the system comprising the motor, the load, and the inertial disk amounts to $J = 0.032$ kgm^2. The peak torque capability of the motor-CRPWM torque actuator is $T_{MAX} = 13.6$ Nm. The maximum permissible speed ω_{MAX} is set to the motor rated speed $\omega_{MAX} = \omega_{nom} = 145$ rad/s. The control structures in Fig. 5.7 and Fig. 5.17 are implemented in ANSI C and listed in [16]. The traces given in the subsequent figures are obtained by writing the numbers corresponding to the relevant torque and position samples on dedicated D/A converters and feeding their analog outputs to the oscilloscope.

In Fig. 5.22, the torque and position traces are given for a relatively small position step of 0.2 rad. During the transient, the torque stays away from the limit T_{MAX} and the system operates in linear mode. The output position reaches the setpoint in 10–11 T. This corresponds to the result obtained by simulation and shown in Fig. 5.12.

With an increase in the input step, the amplitude of the driving torque increases as well. In Fig. 5.23, the torque and position traces are given for a reference step of 0.314 rad. In the acceleration phase, the driving torque reaches the limit $T_{MAX} = 13.6$ Nm, and the rate of the speed increase is limited. Therefore, the response time is prolonged compared with the previous case (Fig. 5.22). When the torque leaves the limit, the system returns to linear mode and all the transients decay without oscillations.

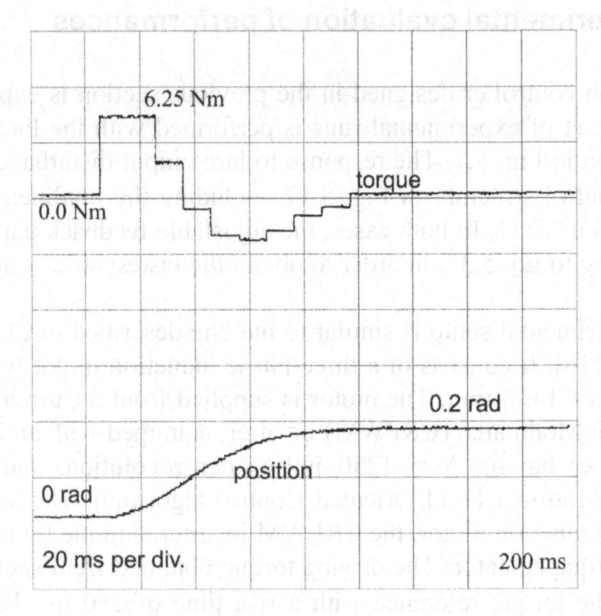

Fig. 5.22. Experimental traces of the driving torque (top) and the output position (bottom) obtained for the reference step of 0.2 rad, with the controller given in Fig. 5.7. The torque involved does not reach the limit T_{MAX}, and the system operates in linear mode. The output position settles in 10–11 sampling periods.

The responses in Fig. 5.24 are obtained for a position step of 0.628 rad, corresponding to 10% of one mechanical turn. The driving torque remains at the limit for more than 20 ms. Therefore, the overall response time is further prolonged, compared with the responses in Fig. 5.22 and Fig. 5.23. Note in the deceleration phase that the torque waveform stays away from the limit $-T_{MAX}$. Hence, the braking torque is not limited. Therefore, the system is capable of reducing the speed before the target position is reached and the response has no overshoot. In its final phase, the response corresponds to the strictly aperiodic closed loop poles.

Further tests proceed with the linear position controller, given in Fig. 5.7. In Fig. 5.25, the torque and position traces are given for a relatively large reference step of 75 rad ($\theta^*(t) = \Theta_{REF}\, h(t) = 75\, h(t)$). Within the first 800 ms, the system accelerates using all of the available torque (T_{MAX}). At the instant when the target position is reached, the shaft speed is very large and the accelerated inertia cannot be stopped. Therefore, a large overshoot in the output position is created, followed by nonlinear oscillations, corresponding to the simulation traces given previously in Fig. 5.15. In order to prevent this from happening, the speed of the system, as it approaches the target, has to be limited according to the residual path $\Delta\theta$, as defined in Eq. 5.40.

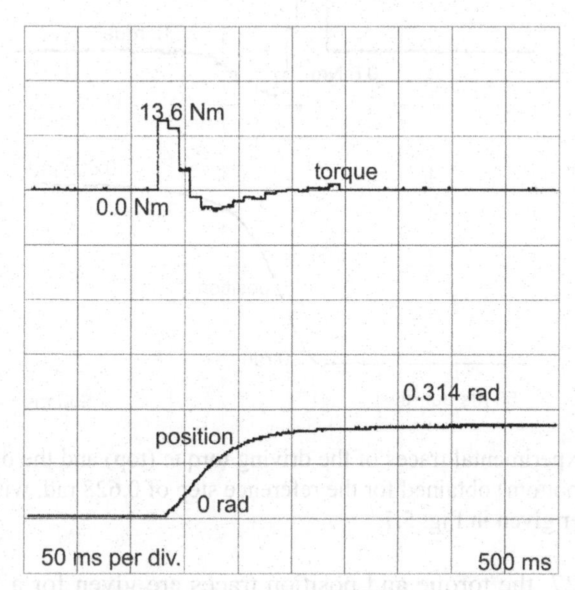

Fig. 5.23. Experimental traces of the driving torque (top) and the output position (bottom) obtained for the reference step of 0.314 rad, with the controller given in Fig. 5.7. During the acceleration phase, the torque reaches the system limit T_{MAX}. Due to limited acceleration, the response time is prolonged compared with Fig. 5.22. In the final stage, the system returns to linear mode and the error exhibits an aperiodic decay.

A further increase in the position reference results in the experimental traces shown in Fig. 5.26. In the absence of a path-dependent speed limit, the linear position controller produces an overshoot even larger than the one in the previous case. The torque and position exhibit poorly damped oscillatory behavior with a relatively large amplitude. Notice in the figure that the frequency of damped oscillations increases as their amplitudes decay. This effect is due to the activation of T_{MAX} and ω_{MAX} limits, causing the system to operate in the nonlinear mode.

Experiments presented in Figs. 5.27 and 5.28 are performed with the enhanced position controller given in Fig. 5.17, comprising the path-dependent speed limit (Eq. 5.41). The function $fx(\Delta\theta)$ replaces the proportional gain block and ensures that the braking starts on time, so as to bring the speed to zero at the same time that the output position reaches the target.

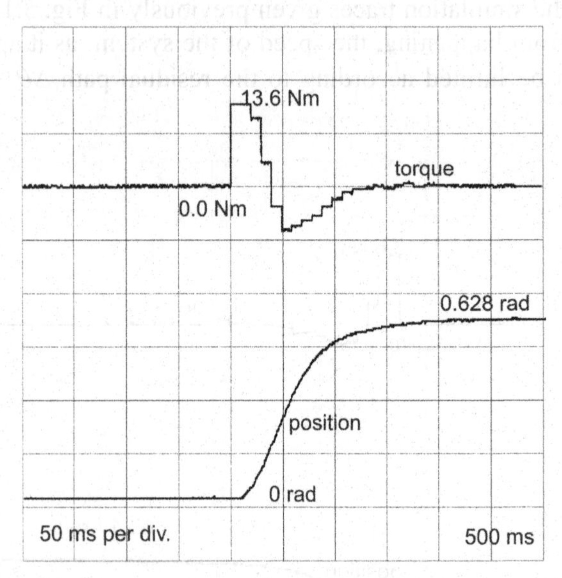

Fig. 5.24. Experimental traces of the driving torque (top) and the output position (bottom) obtained for the reference step of 0.628 rad, with the controller given in Fig. 5.7.

In Fig. 5.27, the torque and position traces are given for a large step of the input. The reference position steps forward by 75 rad, and then returns to the initial state. Within each sampling period, the nonlinear block $fx(\Delta\theta)$ calculates the permissible speed according to Eq. 5.40 and Eq. 5.41. With such a path-dependent speed limit, the braking commences after 600 ms, prior to the output position reaching the target. The system enters the target with the speed reduced to zero. During the acceleration and braking intervals, the torque reaches the system limit T_{MAX}. On the other hand, the speed limit ω_{MAX} is not reached. The interval with the constant speed $\omega(t) = \omega_{MAX}$, obtained by the simulation in Fig. 5.20, is not encountered in this experiment. For that reason, the breaking torque pulse follows the acceleration, and the speed transient assumes an S shape. During deceleration, the torque departs from $-T_{MAX}$ at certain instants. Due to finite friction and motion resistances $T_L(\omega)$, not modeled in Fig. 5.20, the braking torque required from the servo drive in order to provide the deceleration rate of $-\mathrm{d}\omega/\mathrm{d}t = -T_{MAX}/J$ is smaller than T_{MAX} by $T_L(\omega)$ ($T_{em} = -T_{MAX} + T_L(\omega)$). In addition, abrupt changes in the first derivative of the function $fx(\Delta\theta)$, at the intersections A and B in Fig. 5.16, may contribute to the torque changes during the braking interval.

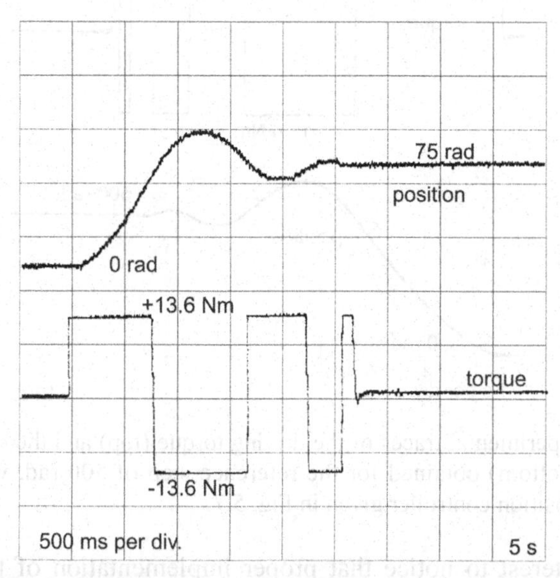

Fig. 5.25. Experimental traces of the driving torque (bottom) and the output position (top) obtained for the reference step of 75 rad, with the position controller given in Fig. 5.7. Insufficient torque in braking results in an overshoot and consequential oscillations in the torque and position responses.

With the position step of 500 rad, the experimental traces of the driving torque and the output position are given in Fig. 5.28. The nonlinear position controller in Fig. 5.17 provides an aperiodic step response without an overshoot. During the transient, both system limits are reached. In acceleration and braking, the torque stays at the limit T_{MAX}. In between, the running speed is limited to ω_{MAX}. During that interval, the acceleration $a = \mathrm{d}\omega/\mathrm{d}t$ is equal to zero. The driving torque is relatively low, and it corresponds to motion resistances at the given speed. The response in the figure represents the fastest transition from the initial to the final position, considering the limits T_{MAX} and ω_{MAX}. The nonlinear control action $fx(\Delta\theta)$ provides for a timely start of the braking interval. Within the torque response, the braking pulse is slightly shorter than the acceleration pulse, due to friction and other motion resistances assisting the torque actuator in decelerating the system during the braking interval.

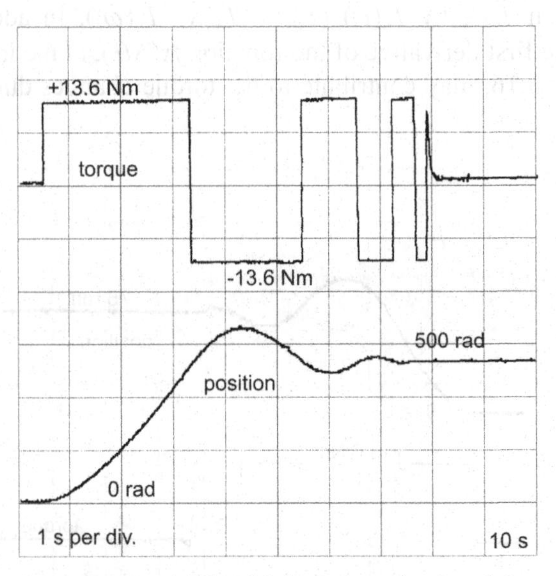

Fig. 5.26. Experimental traces of the driving torque (top) and the output position (bottom) obtained for the reference step of 500 rad, with the linear position controller given in Fig. 5.7.

It is of interest to notice that proper implementation of the nonlinear control law in Eq. 5.41 and the path-dependent speed limit (Eq. 5.40) requires inertia J of the control object. Specifically, for a system running at a speed ω, the braking distance $\Delta\theta$ (i.e., the path required to stop the system) is proportional to $J\omega^2/2/T_{MAX}$, where T_{MAX} stands for the braking torque (Eq. 5.38, Eq. 5.39). Therefore, the path-related speed limit $fx(\Delta\theta)$, given in Eq. 5.41, depends on the control object inertia J. The availability of the parameter J depends on the specific application. In cases where the mechanical subsystem of a position-controlled servo system comprises massive transmission elements and where the inertia of the servo motor prevails, the weight of the work piece has an insignificant impact on the equivalent inertia. However, in cases where an industrial manipulator with light transmission elements and a low-inertia servo motor moves heavy parts or work pieces over distances, the equivalent inertia of the control object is considerably affected by the manipulator having grabbed a heavy work piece. Note at this point that precise knowledge about the system inertia is required not only for the implementation of the nonlinear control law $fx(\Delta\theta)$ but also for the proper setting of the feedback gain, as defined in Eq. 5.37.

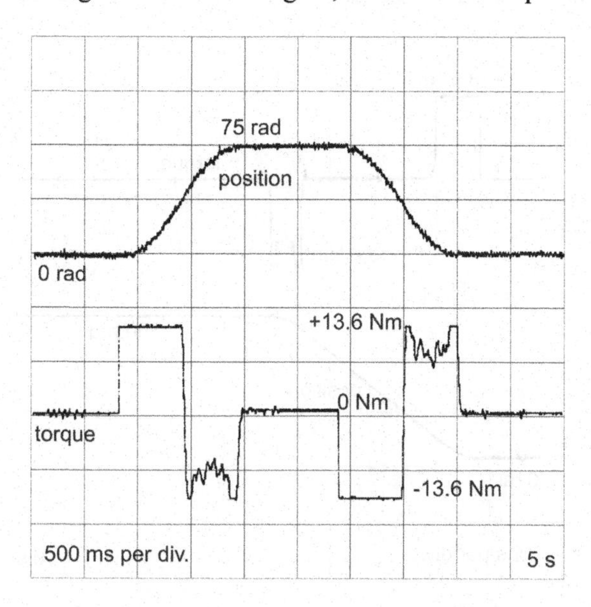

Fig. 5.27. Experimental traces of the driving torque (bottom) and the output position (top) obtained for the reference step of 75 rad, with the position controller including the path-dependent speed limit, given in Fig. 5.17. The torque operates at the limit T_{MAX}, while the speed does not reach the system limit ω_{MAX}. The function $fx(\Delta\theta)$ (Eq. 5.41) replaces the proportional gain. It ensures that the braking starts on time, so as to bring the speed to zero at the same time that the output position reaches the target.

In some cases, the changes in J can be predicted for known loads and weights of the work pieces. Applications where large, unpredictable fluctuations of the equivalent inertia are expected call for the online identification of the parameter J and the corresponding adaptation of adjustable feedback parameters and control laws. In cases when the actual inertia J is smaller than the parameter J_x, used in $fx(\Delta\theta)$ calculations, the value of $fx(\Delta\theta)$ will suggest a lower permissible speed. This outcome is due to the fx depending on the ratio T_{MAX}/J_x, which is smaller than the available deceleration T_{MAX}/J. Therefore, the situation with $J < J_x$ will not result in an overshoot. As a consequence, the braking intervals will commence earlier, reducing the average speed and extending the response time. The problem arises when $J > J_x$, since the formula f_x requires the deceleration of T_{MAX}/J_x, while the available deceleration T_{MAX}/J is lower. This situation results in an unpunctual and delayed braking, causing the system to pass through the target position and create an overshoot.

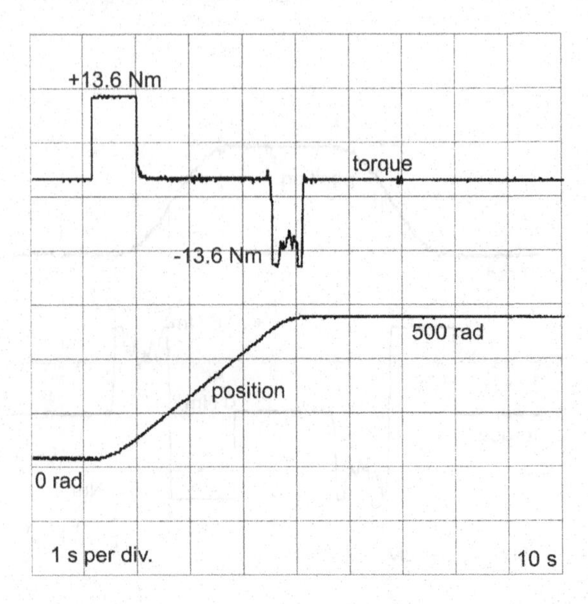

Fig. 5.28. Experimental traces of the driving torque (top) and the output position (bottom) obtained for the reference step of 500 rad, with the position controller given in Fig. 5.17. During acceleration and braking, the torque stays at the limit T_{MAX}. In between, the speed is limited to ω_{MAX}. The nonlinear control action $fx(\Delta\theta)$ ensures that the braking starts on time, so as to bring the speed to zero at the same time that the output position reaches the target.

The operation of the position controller with a nonlinear block $fx(\Delta\theta)$ that has a detuned value of J is presented in Fig. 5.29. It is assumed that the value of J_x used in $fx(\Delta\theta)$ calculations is lower than the actual inertia by 25%. Therefore, the braking sequence does not start on time, and the system cannot properly brake and stop at the target position. For that reason, the output position overshoots the target by roughly 7%. The torque response in the figure begins with an acceleration pulse that has the amplitude of T_{MAX}, proceeds with a delayed braking pulse, and finally ends with a brief pulse of the opposite sign, related to the speed change during the overshoot in the output position.

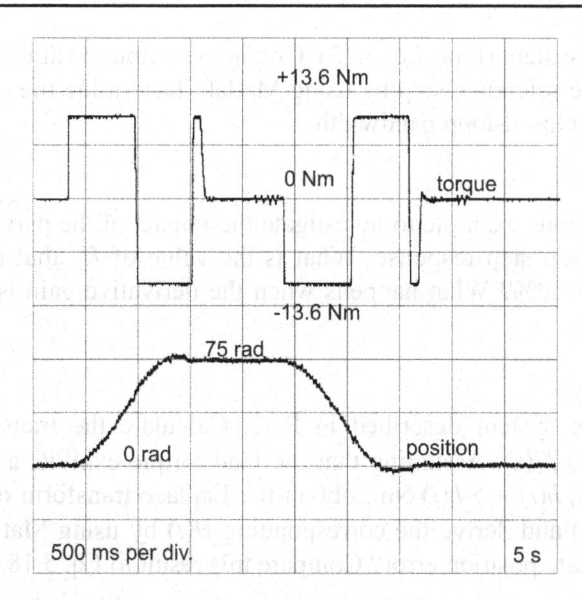

Fig. 5.29. Experimental traces of the driving torque (top) and the output position (bottom) obtained for the reference step of 75 rad, with the position controller given in Fig. 5.17. The nonlinear control action $fx(\Delta\theta)$ is calculated with an erroneous value of the inertia J_x, smaller than the actual inertia J of the control object by 25%. An incorrect speed limit $fp(\Delta\theta)$ (Fig. 5.15) results in an overshoot in the output position of roughly 7% N.

Problems

P5.1
The mechanical subsystem of a position-controlled system is described by $J = 0.01$ kgm^2 and $B = 0.01$ Nm/(rad/s). The torque actuator gain is $K_M = 1$. Assuming the sampling time of $T = 10$ ms, obtain the pulse transfer function $W_P(z)$ of the control object. Note: An appropriate sequence of Matlab commands is included in Section 5.2.

P5.2
Consider the position-controlled system with the PD controller implemented in the s-domain, as given in Fig. 5.4. The torque actuator gain is $K_M = 1$, the friction B has a negligible value, and the derivative gain is relocated into the feedback path. Given $J = 1$ kgm^2, $K_P = 1$ Nm/rad, and $K_D = 2$ Nm/(rad/s), calculate the s-domain transfer function $W_{SS}(s)$ of the

closed-loop system (hint: Eq. 5.13). Obtain the output position transient response to the reference step by using Matlab. Determine the rise time and estimate the closed-loop bandwidth.

P5.3
Use the previous example to investigate the impact of the parameter K_D on the closed-loop step response. What is the value of K_D that results in an overshoot of 50%? What happens when the derivative gain is completely removed?

P5.4
Consider the system described in P5.2. Calculate the transfer function $W_{LS}(s) = \theta(s)/T_L(s)$. Assuming that the load torque exhibits a step change $T_L(t) = T_{LOAD}\, h(t) = 5\, h(t)$ Nm, obtain the Laplace transform of the output position $\theta(s)$ and derive the corresponding $\theta(t)$ by using Matlab. What is the steady-state position error? Compare this result to Eq. 5.18.

P5.5
Consider the position controlled system in Fig. 5.7. The discrete-time position controller has the proportional gain in the direct path and the derivative gain in the feedback path. The signals internal to the controller, such as $\Delta\theta$, y_1, T_{ref}, and θ^{FB}, are digital words residing within the DSP controller RAM memory. These signal have no units. What are the units of K_{FB}, K_M, K_P, and K_D? Consider the case where $J = 0.01$ kgm^2, $T = 1$ms, with $K_{FB} = 1$ [] and $K_M = 1$ []. Determine the optimized gains resulting in the fastest strictly aperiodic response.

P5.6
Use Matlab and the closed-loop transfer function of the position controlled system with the PD controller given in Eq. 5.25 to obtain the response of the output position to the step change in the reference position. Use the *dstep* command to obtain the output position and the *diff* command to obtain the torque waveform. Assume that normalized gains are set to $p = 0.03512$ and $d = 0.2027$. What is the rise time expressed in terms of the sampling period?

P5.7
The position-controlled system in Fig. 5.7 has a torque actuator with a peak torque of $T_{MAX} = 10$ Nm and an inertia of $J = 0.01$ kgm^2. The system approaches the target position at a speed of ω_1. At a given instant, the remaining path $\Delta\theta$ is equal to 100 rad. Considering the fact that the braking torque is limited, what is the maximum value of ω_1? How is it related to $\Delta\theta$?

6 The Position Controller with Integral Action

A position controller with integral action suppresses the output error caused by a constant load. It also provides the system with the capability of tracking the constant slope profile without an error. In this chapter, the controller structure and parameter setting are considered for linear operating mode. The ability of the system to track the reference trajectory is discussed. The impact of the controller structure and the trajectory properties is evaluated. For the operation with large input disturbances, where the torque and speed limits of the system are reached, a nonlinear modification of the discrete-time position controller is proposed, providing a robust, aperiodic, and time-optimal response to large disturbances.

The design of a discrete-time position controller with proportional and derivative actions has been discussed in the previous chapter. In the absence of the load torque disturbance, and with the position reference $\theta^*(t) = \Theta h(t)$, a PD position controller is capable of eliminating the output error ($\Delta\theta = 0$). When the load torque is present (Eq. 5.27), the output error of a PD controller is proportional to the load and inversely proportional to the proportional gain. Due to a finite range of applicable gains, the stiffness of the system with a discrete-time PD controller is limited, and, in most cases, unacceptable.

The analysis given in Chapter 2 and the Internal Model Principle summarized in Section 2.3.2 propose that the output error produced by the load disturbance W_1 can be suppressed by extending the controller with the control action $W_2 = W_1$, wherein W_1 is the complex image of the load disturbance in either the z- or s-domain, while W_2 is the transfer function of the added control action. Hence, in cases when the load disturbance is constant ($W_1(s) = T^L/s$), the controller has to be extended with the integral action ($W_2(s) = G_I/s$). With discrete-time implementation, the integral control action is obtained as the sum of the error samples within the interval $[0, nT]$, multiplied by the integral gain K_I. A block diagram of a linear, discrete-time PID position controller is given in Fig. 6.1.

The torque actuator, mechanical subsystem, and shaft sensor are modeled in the same way as in the previous chapter (Fig. 5.1). The torque actuator is assumed to have a negligible delay and a static gain of K_M. The maximum driving torque available is T_{MAX}. The control object has the transfer function $1/Js^2$ and the position θ at the output. The position feedback is obtained from the sensor attached to the shaft of the servo motor. In some cases, the sensor detects the position of the tool or the payload, performing a linear motion. The shaft sensor and the feedback acquisition circuits are characterized by the parameter K_{FB}, relating the numerical representation of the output position θ^{FB} to the actual position, expressed in [rad]. The sampling circuit, represented by the switch designated by T in Fig. 6.1, acquires the samples $\theta_{(n)}$ at each sampling instant $t = nT$, providing, in such a way, the train of pulses $\theta^{FB}{}_{DIG}$.

The proportional, integral, and derivative actions in Fig. 6.1 are arranged to remove the closed-loop zeros and suppress the overshoot associated with such zeros. At the same time, relocation of proportional and derivative actions into the feedback path and their implementation in an incremental form facilitate the subsequent implementation of the nonlinear control law, suited for the operation with large disturbances, where the torque and speed enter the system limits T_{MAX} and ω_{MAX}.

The integral action is located in the direct path, and it processes the tracking error $\Delta\theta$. At the output of the gain block K_I, the increment in the integral action is obtained as $K_{FB}K_I\Delta\theta$. The increment in the proportional action is obtained as the difference between the two successive samples of the feedback, $y_{3(n)} = K_{FB}K_P(\theta_{(n)} - \theta_{(n-1)})$. The signal y_3 is subtracted from $K_{FB}K_I\Delta\theta$, obtaining the increment Δy_1 of the signal y_1. The increments Δy_1 are summed within the digital integrator, denoted *INT* in Fig. 6.1. The derivative feedback is obtained as $y_{2(n)} = K_{FB}K_P(\theta_{(n)} - \theta_{(n-1)})$. The torque reference samples $T^*_{(n)}$ are calculated as $y_{1(n)} - y_{2(n)}$ and fed to the zero-order hold. The actual torque T_{em} is assumed to be proportional to the zero-order hold output T_{ref} ($T_{em}(t) = K_M T_{ref}(t)$). It is worthwile to note that the signal y_1 represents an internal speed reference; that is, for the reasons explained in the previous chapter, the derivative feedback y_2 is in proportion to the shaft speed, and it constitutes a local speed loop with the reference y_1 and the feedback signal y_2.

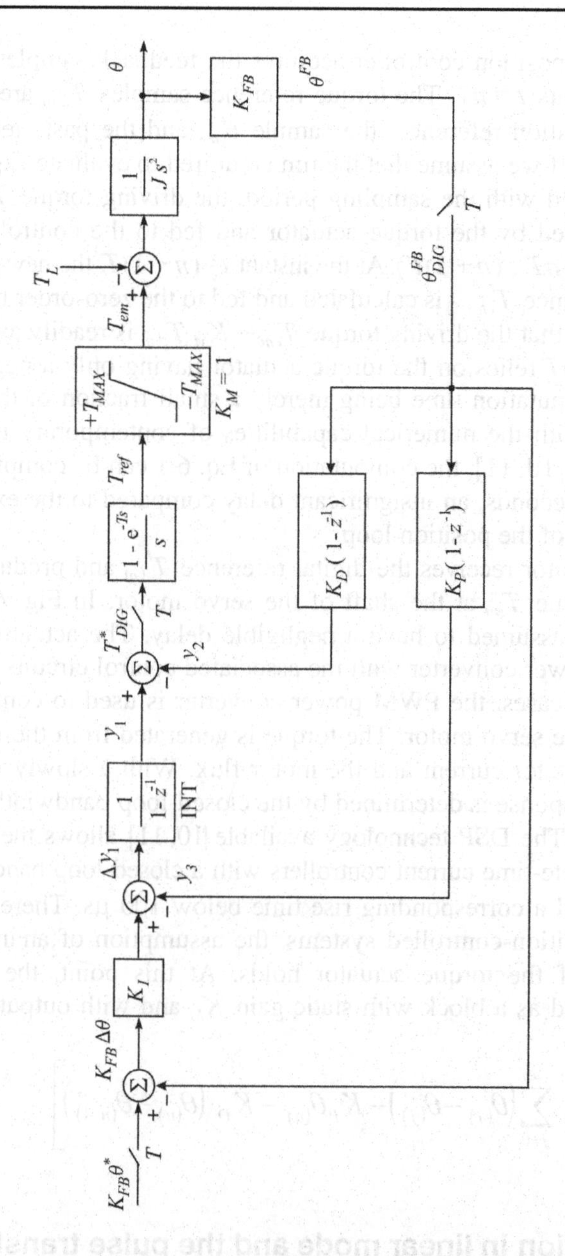

Fig. 6.1. Linear, discrete-time position controller with integral, proportional and derivative actions. The integral action is in the direct path and processes the tracking error $\Delta\theta$. The proportional and derivative actions are placed in the feedback path.

A discrete-time position controller acquires the feedback samples $\theta_{(n)}$ at the sampling instants $t = nT$. The torque reference samples $T^*_{(n)}$ are calculated from the position reference, the sample $\theta_{(n)}$, and the past feedback samples (Eq. 6.1). If we assume that the time required to evaluate Eq. 6.1 is negligible compared with the sampling period, the driving torque $T_{em}(t) = K_M T^*_{(n)}$ is generated by the torque actuator and fed to the control object during the interval $[nT .. (n+1)T)$. At the instant $t=(n+1)T$, the new sample of the torque reference $T^*_{(n+1)}$ is calculated and fed to the zero-order hold.

The assumption that the driving torque $T_{em} = K_M T^*_{(n)}$ is readily available at the instant $t = nT$ relies on the torque actuator having only a negligible delay and the computation time being merely a small fraction of the sampling period T. With the numerical capabilities of contemporary motion-control processors [10, 11], the computation of Eq. 6.1 can be completed in a couple of microseconds, an insignificant delay compared to the expected sampling period T of the position loop.

The torque actuator receives the digital reference $T^*_{(n)}$ and produces the actual driving torque T_{em} at the shaft of the servo motor. In Fig. 6.1, the torque actuator is assumed to have a negligible delay. The actuator comprises the drive power converter with the associated control circuits and algorithms. In most cases, the PWM power converter is used to control the stator current of the servo motor. The torque is generated from the interaction between the stator current and the motor flux. With a slowly varying flux, the torque response is determined by the closed-loop bandwidth of the current controller. The DSP technology available [10, 11] allows the implementation of discrete-time current controllers with a closed-loop band-width of several kHz and a corresponding rise time below 100 μs. Therefore, in most practical position-controlled systems, the assumption of an instantaneous response of the torque actuator holds. At this point, the torque actuator is modeled as a block with static gain K_M and with output torque limited to $\pm T_{MAX}$.

$$T^*_{(n)} = K_{FB}\left[K_I \sum_{j=0}^{j=n}\left(\theta^*_{(j)} - \theta_{(j)}\right) - K_P\theta_{(n)} - K_D\left(\theta_{(n)} - \theta_{(n-1)}\right)\right] \qquad (6.1)$$

6.1 The operation in linear mode and the pulse transfer functions

The driving torque and speed of a servo system are limited to T_{MAX} and ω_{MAX}, respectively. In response to a large input disturbance, the torque and/or speed may reach their limits and bring the system into a nonlinear operating mode. With the control variable at the limit, the driving torque is not affected by the feedback and the loop is broken. At this point, further changes in the output position cannot be controlled. In order to provide controlled behavior in such cases, the position controller has to be extended with a nonlinear control law, similar to the one given in Fig. 5.17 and Eq. 5.41.

When the disturbances are relatively small, the variables of the system in Fig. 6.1 stay within the system limits. The system is linear, and the relation between the output position and the input and load disturbances can be described by means of the pulse transfer function. The response character is determined by the poles and zeros of the closed-loop system transfer function.

The control object is assumed to have an inertia J and a negligible friction. The pulse transfer function of the control object is discussed in detail in Section 5.2 and is briefly reviewed below. Under these assumptions, the transition of the system between successive sampling instants can be obtained by integrating Eq. 6.2 within the interval $[nT..(n+1)T]$. Successive speed samples $\omega_{(n)}$ and $\omega_{(n+1)}$ are related in Eq. 6.3.

$$J\frac{\mathrm{d}^2\theta}{\mathrm{d}t^2} = J\frac{\mathrm{d}\omega}{\mathrm{d}t} = T_{em}(t) - T_L(t) \tag{6.2}$$

$$\begin{aligned}\omega_{(n+1)} &= \omega_{(n)} + \frac{T}{J}K_M T^*_{(n)} - \frac{1}{J}\int_{nT}^{nT+T} T_L(t)\,\mathrm{d}t \\ &= \omega_{(n)} + \frac{T}{J}K_M T^*_{(n)} - \frac{T}{J}T^L_{(n)}\end{aligned} \tag{6.3}$$

For reasons discussed in Chapters 4 and 5, the disturbance signal $T^L(t)$ is assumed to be constant for the interval $[nT, (n+1)T)$. The value of $T^L_{(n)}$ in Eq. 6.3 corresponds to the average load torque during the interval. With a constant driving torque $T^*_{(n)}$ and a constant load torque $T^L_{(n)}$, the right side of Eq. 6.2 and, hence, the acceleration $\mathrm{d}\omega/\mathrm{d}t$, is constant for the interval. In such cases, the change in shaft speed between the two successive sampling instants assumes a trapezoidal form (Fig. 4.6). Therefore, the next sample of the shaft position $\theta_{(n+1)}$ can be calculated from the previous sample $\theta_{(n)}$ and from the speed samples $\omega_{(n)}$ and $\omega_{(n+1)}$:

$$\theta_{(n+1)} = \theta_{(n)} + \int_{nT}^{nT+T}\omega(t)\mathrm{d}t = \theta_{(n)} + T\frac{\omega_{(n)} + \omega_{(n+1)}}{2}. \tag{6.4}$$

Notice that the assumption $T^L(t) \approx T^L_{(n)}$ does not affect the accuracy of modeling the signal flow from the reference input to the output. Instead, it influences the modeling of the impact that the load disturbance produces upon the shaft and position samples. The closed-loop system transfer function $W_{SS}(z) = \theta(z)/\theta^*(z)$ and the values of the closed-loop poles are not affected by the assumption $T^L(t) \approx T^L_{(n)}$, whatever the spectral contents of the load. On the other hand, in cases where the load torque exhibits high-frequency changes, the speed change between the samples $\omega_{(n)}$ and $\omega_{(n+1)}$ may be other than linear, and the increment $\Delta\theta_{(n+1)} = \theta_{(n+1)} - \theta_{(n)}$ cannot be expressed as $T(\omega_{(n+1)} + \omega_{(n)})/2$. In such cases, the impact of the load torque on the speed and position samples cannot be modeled by Eq. 6.3 and Eq. 6.4. However, in most position-controlled systems, the ratio T/J is rather small. The change in shaft speed over the sampling period T is fairly low, and the difference between the average speed for the interval T and the average of the speed samples $\omega_{(n)}$ and $\omega_{(n+1)}$ is negligible, even in cases where the load torque exhibits significant changes within the interval. From difference equations 6.3 and 6.4, the complex images of the speed, torque, and position are related in Eq. 6.5. The complex image $T^L(z)$ is the z-transform of the pulse train comprising the samples of load torque average values $T^L_{(n)}$.

$$\begin{aligned}
(z-1)\omega(z) &= \frac{T}{J}\left(K_M T^*(z) - T^L(z)\right) \Rightarrow \\
\omega(z) &= \frac{T}{J(z-1)}\left(K_M T^*(z) - T^L(z)\right) \\
(z-1)\theta(z) &= T\frac{z+1}{2}\omega(z) \Rightarrow \\
\theta(z) &= \frac{T^2(z+1)}{2J(z-1)^2}\left(K_M T^*(z) - T^L(z)\right)
\end{aligned} \tag{6.5}$$

The complex image of the output position $\theta(z)$ can be expressed in terms of the driving torque and the load:

$$\theta(z) = W_P(z)T^*(z) - W_{PL}(z)\,T^L(z)$$

where the pulse transfer functions $W_P(z)$ and $W_{PL}(z)$ are obtained as

$$W_P(z) = \mathrm{Z}\left(\frac{1-e^{-sT}}{s}K_M\frac{1}{Js^2}\right) = \frac{K_M T^2(z+1)}{2J(z-1)^2} = \frac{K_M T^2(z+1)}{2Jz^2(1-z^{-1})^2} \tag{6.6}$$

$$W_{PL}(z) = \mathrm{Z}\left(\frac{1-e^{-sT}}{s}\frac{1}{Js^2}\right) = \frac{T^2(z+1)}{2J(z-1)^2}.$$

The calculation of the closed-loop system transfer function $W_{SS}(z)$ requires the function of the control object and the z-domain representation of the position controller. The difference equation 6.1 describes the calculation of the next sample of the torque reference. Therefore, the increment $\Delta T^*_{(n+1)}$ in the torque reference is obtained as

$$\begin{aligned}\Delta T^*_{(n+1)} &= T^*_{(n+1)} - T^*_{(n)} \qquad (6.7)\\ &= K_{FB}\left[K_I\left(\theta^*_{(n+1)} - \theta_{(n+1)}\right) - K_P\left(\theta_{(n+1)} - \theta_{(n)}\right) - K_D\left(\theta_{(n+1)} - 2\theta_{(n)} + \theta_{(n-1)}\right)\right].\end{aligned}$$

If we apply the time shift property of the z-transform, given in Eq. 4.7, the difference equation is converted into the following form:

$$T^*(z)\,(z-1) = K_{FB}\left\{K_I z\left(\theta^*(z) - \theta(z)\right) - \theta(z)\left[K_P(z-1) + K_D\left(z - 2 + z^{-1}\right)\right]\right\}.$$

The complex image of the torque reference is obtained as

$$\begin{aligned}T^*(z) &= K_{FB}K_I\frac{z}{z-1}\left(\theta^*(z) - \theta(z)\right) - K_{FB}K_P\theta(z) - K_{FB}K_D\frac{z-1}{z}\theta(z)\\ &= W_I(z)\left(\theta^*(z) - \theta(z)\right) - W_P(z)\theta(z) - W_D(z)\,\theta(z). \qquad (6.8)\end{aligned}$$

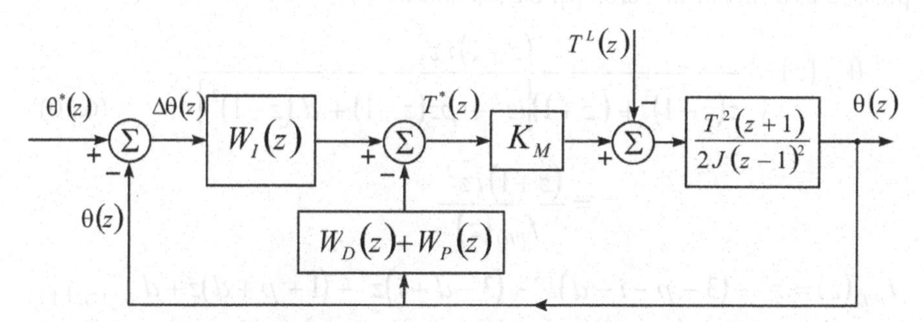

Fig. 6.2. Simplified block diagram of the position controller operating in linear mode. The pulse transfer functions $W_I(z)$, $W_P(z)$, and $W_D(z)$ represent the integral, proportional, and derivative actions, respectively.

A simplified block diagram of the system, derived from Eq. 6.6 and Eq. 6.8, is given in Fig. 6.2. From the control object transfer function $W_P(z)$ and the functions $W_I(z)$, $W_P(z)$, and $W_D(z)$, defined in Eq. 6.8, the closed-loop system transfer function $W_{SS}(z) = \theta(z)/\theta^*(z)$ is obtained in Eq. 6.9.

$$W_{SS}(z) = \frac{\theta(z)}{\theta^*(z)} = \frac{W_P W_I}{1 + W_P (W_I + W_P + W_D)}$$

$$= \frac{\left[\frac{K_M T^2 (z+1)}{2J(z-1)^2}\right] K_I K_{FB} \frac{z}{z-1}}{1 + \left[\frac{K_M T^2 (z+1)}{2J(z-1)^2}\right]\left[K_I K_{FB} \frac{z}{z-1} + K_P K_{FB} + K_D K_{FB} \frac{z-1}{z}\right]} \tag{6.9}$$

Within the transfer function, all the feedback gains K_I, K_P, and K_D are multiplied by the factor $K_M T^2 K_{FB}/2J$. Therefore, it is convenient to normalize gains *i*, *p*, and *d*, defined as $i = K_I K_{FB} K_M (T^2/2J)$, $p = K_P K_{FB} K_M (T^2/2J)$, and $d = K_D K_{FB} K_M (T^2/2J)$. The normalized gains encompass the system parameters and simplify transfer functions and the characteristic polynomial. As concluded in Chapters 4 and 5, the optimized values of normalized gains do not change with the system parameters K_M, T, K_{FB}, and J, thus simplifying the tuning and adaptation.

With the introduction of normalized gains, the closed-loop transfer function is obtained in Eq. 6.10, where $f_{PID}(z)$ designates the characteristic polynomial of the system. In Eq. 6.11, the fourth-order polynomial is expressed in terms of its zeros σ_1, σ_2, σ_3, and σ_4.

$$W_{SS}(z) = \frac{(z+1)\, i z^2}{z(z-1)^3 + (z+1)\left[iz^2 + pz(z-1) + d(z-1)^2\right]} = \frac{(z+1)\, i z^2}{f_{PID}(z)} \tag{6.10}$$

$$f_{PID}(z) = z^4 - (3 - p - i - d)z^3 + (3 - d + i)z^2 - (1 + p + d)z + d = (z - \sigma_1)(z - \sigma_2)(z - \sigma_3)(z - \sigma_4) \tag{6.11}$$

The response of the output position to changes in the load torque is described by the pulse transfer function $W_{LS}(z) = \theta(z)/T^L(z)$. This function is derived from Fig. 6.2 and obtained as

$$W_{LS}(z)=\frac{\theta(z)}{T^L(z)}=\frac{-W_{PL}}{1+W_P\left(W_I+W_P+W_D\right)}$$

$$=\frac{-\left[\dfrac{T^2(z+1)}{2J(z-1)^2}\right]}{1+\left[\dfrac{K_M T^2(z+1)}{2J(z-1)^2}\right]\left[K_I K_{FB}\dfrac{z}{z-1}+K_P K_{FB}+K_D K_{FB}\dfrac{z-1}{z}\right]} \qquad (6.12)$$

$$=\frac{T^2}{2J}\frac{-\left(z^2-1\right)z}{f_{PID}(z)}.$$

In the presence of a constant position reference $\theta^*(t) = \Theta^* h(t)$ and the step form of the load disturbance $T^L(t) = T^{LOAD}h(t)$, the steady-state value of the output position can be found from the pulse transfer functions $W_{SS}(z)$ and $W_{LS}(z)$, on the basis of the final value theorem (Eq. 4.13). With the input and load disturbances being Heaviside functions, their complex images are $\Theta^*/(1-z^{-1})$ and $T^{LOAD}/(1-z^{-1})$, respectively. Given that the closed loop poles are stable, the final value of the output is found as

$$\theta(\infty)=\lim_{z\to 1}\left\{\left(1-z^{-1}\right)\left[W_{SS}(z)\frac{\Theta^*}{1-z^{-1}}+W_{LS}(z)\frac{T^{LOAD}}{1-z^{-1}}\right]\right\}=\Theta^*. \qquad (6.13)$$

From the above expression, in the presence of constant input and load disturbances $|T^{LOAD}| \le T_{MAX}$, the steady-state value of the output position corresponds to the reference. According to Eq. 6.13, the stiffness of the PID controller is infinite. The error observed in similar conditions with the PD controller is eliminated by introducing the integral action. The error integrator in Fig. 6.1 increases the overall number of states. Therefore, the characteristic polynomial $f_{PID}(z)$ is of the fourth-order, as compared with the third-order polynomial $f_{PD}(z)$ obtained in Eq. 5.26 for the PD position controller. An increase in the order and a larger number of closed-loop poles lead to a longer rise time and smaller bandwidth, as demonstrated by the subsequent analysis in this chapter.

6.2 Parameter setting of PID position controllers

The character of the transient response to small disturbances, where the system operates in linear mode, depends on the zeros of the characteristic polynomial $f_{PID}(z)$. The presence of oscillations in the step response depends on the damping factor of the polynomial zeros. These also determine the rise time and the closed-loop bandwidth. The zeros σ_1, σ_2, σ_3, and σ_4 are determined by the normalized values of the feedback gains p, d, and i. Circumstances in which the four closed-loop poles σ_1, σ_2, σ_3, and σ_4 depend on the three parameters p, d, and i lead to a conclusion that the pole placement cannot be performed in an arbitrary way. The closed-loop poles and the feedback gains are tied by the following relations:

$$
\begin{aligned}
&\sigma_1 + \sigma_2 + \sigma_3 + \sigma_4 = 3 - p - i - d \\
&\sigma_1\sigma_2 + \sigma_1\sigma_3 + \sigma_1\sigma_4 + \sigma_2\sigma_3 + \sigma_2\sigma_4 + \sigma_3\sigma_4 = 3 + i - d \\
&\sigma_1\sigma_2\sigma_3 + \sigma_1\sigma_2\sigma_4 + \sigma_1\sigma_3\sigma_4 + \sigma_2\sigma_3\sigma_4 = 1 + p + d \\
&\sigma_1\sigma_2\sigma_3\sigma_4 = d \ .
\end{aligned}
\tag{6.14}
$$

The sum of the four relations results in Eq. 6.15, presenting the constraint for the closed-loop poles:

$$
\begin{aligned}
&\sigma_1 + \sigma_2 + \sigma_3 + \sigma_4 + \sigma_1\sigma_2 + \sigma_1\sigma_3 + \sigma_1\sigma_4 + \sigma_2\sigma_3 + \sigma_2\sigma_4 + \sigma_3\sigma_4 \\
&\quad + \sigma_1\sigma_2\sigma_3 + \sigma_1\sigma_2\sigma_4 + \sigma_1\sigma_3\sigma_4 + \sigma_2\sigma_3\sigma_4 + \sigma_1\sigma_2\sigma_3\sigma_4 = 7 \ .
\end{aligned}
\tag{6.15}
$$

With the position controller structure in Fig. 6.1, an arbitrary setting of the closed-loop poles is not feasible. When three of them are known, the fourth can be determined from Eq. 6.15. It is well known [1] that an unconstrained pole placement is feasible in systems where the controller performs either explicit or implicit state feedback. In such cases, the driving force comprises control actions proportional to each individual system state. In the structure shown in Fig. 6.1, the state feedback seems to be complete due to the derivative action being proportional to the shaft speed, a state variable internal to the control object that is not measured. Moreover, with the two state variables observed within the control object (ω, θ) and with the third one being the output of the error integrator, one would expect a characteristic polynomial of the third order. The existence of the fourth, hidden state comes from the intrinsic delay of one sampling period T, observed from Eq. 6.3. The next (n +1) sample of the shaft speed depends on the previous (n) sample of the torque reference. This adds a factor $1/z$ into the pulse transfer function of the control object (Eq. 6.6), resulting in the fourth order of the characteristic polynomial. Further analysis is focused on establishing the procedure for setting the three normalized feedback gains *p, i,* and *d* in a way that results in the desired spectrum of closed-loop poles, respecting, at the same time, the constraint in Eq. 6.15.

In a position-controlled servo system, the desired step response is aperiodic, arriving at the target position without an overshoot. Transient responses of the torque, speed, and position have to be aperiodic as well, thus avoiding undesirable changes in the torque sign and the speed direction. A strictly aperiodic response requires the closed-loop poles to be real, positive numbers within the unit circle. The advantages of a strictly aperiodic response have been discussed in depth in Chapters 4 and 5.

In order to devise the optimized parameter setting, resulting in strictly aperiodic dynamic behavior with the shortest rise time, the response speed must be evaluated numerically by formulating an appropriate criterion function. The function has to be in a monotonous relation with the response speed. This function will depend on the adjustable feedback parameters, and the calculation of its extremum will result in the optimized parameter setting.

As in Eq. 5.28, the criterion function can be defined as the sum of the position error samples. With an aperiodic response, the error samples $\Delta\theta_{(n)}$ are strictly positive. The sum of the error samples during the step response is given in Eq. 6.16, and is proportional to the shaded surface in Fig. 5.11, representing a strictly aperiodic step response of the output position and the corresponding error. The smaller the shaded area (i.e., the value of Q), the faster the step response.

$$Q = \sum_{k=0}^{\infty} \Delta\theta(kT) \tag{6.16}$$

It is necessary to express the value of Q in terms of the normalized feedback gains. Given the complex image of the position reference θ^* $(z) = \Theta^*/(1-z^{-1})$ and the closed-loop system transfer function $W_{SS}(z)$, the z-transform $\Delta\theta(z)$ of the error samples is expressed in the following form:

$$\Delta\theta(z) = \sum_{k=0}^{\infty} \Delta\theta_{(k)} z^{-k} = \theta^*(z) - \theta(z)$$

$$= [1 - W_{SS}(z)]\theta^*(z) = [1 - W_{SS}(z)]\frac{\Theta^*}{1 - z^{-1}}$$

$$= \frac{\left[z^4 - (3-p-i-d)z^3 + (3-d+i)z^2 - (1+p+d)z + d\right] - (z+1)iz^2}{z^4 - (3-p-i-d)z^3 + (3-d+i)z^2 - (1+p+d)z + d}\frac{z}{z-1}\Theta^*.$$

If we introduce the characteristic polynomial $f_{PID}(z)$ from Eq. 6.11 into the previous expression, the z-transform of the position error is obtained as

$$\Delta\theta(z)=\frac{z\left[z^3-(2-p-d)z^2+(1+p)z-d\right]}{f_{PID}(z)}\Theta^* . \tag{6.17}$$

The value of Q is found by introducing $z=1$ into Eq. 6.17:

$$\Delta\theta(z)\big|_{z=1}=\Delta\theta(1)=\sum_{k=0}^{\infty}\Delta\theta_{(k)}=\frac{p}{i}\Theta^*=Q.$$

Hence, the sum of the error samples will be smaller and the response faster for the setting where the ratio p/i is lower. Therefore, the optimized parameter setting has to provide the feedback gains p, i, and d resulting in positive, real, and stable zeros of $f_{PID}(z)$, minimizing, at the same, time the ratio p/i.

If we introduce $x=1/\sigma_1$, $y=1/\sigma_2$, $v=1/\sigma_3$, and $w=1/\sigma_4$, the criterion function can be expressed in terms of $x>1$, $y>1$, $v>1$, and $w>1$. From Eq. 6.15, the variable w can be expressed in terms of the remaining three:

$$w(x,y,v)=\frac{1+x+y+v+xy+xv+yv+xyv}{7xyv-xy-xv-yv-x-y-v-1}.$$

From the previous expression and Eq. 6.14, the criterion function Q can be expressed in terms of x, y, and v:

$$Q(x,y,v)$$
$$=\frac{(x+y+v+w(x,y,v)-xyvw(x,y,v)-1)\,\Theta^*}{1+xy+yv+vw(x,y,v)+xv+yw(x,y,v)+xw(x,y,v)-3xyvw(x,y,v)}.$$

The minimum of $Q(x, y, v)$ has to be found within the stability region $x>1$, $y>1$, and $v>1$. The procedure has been explained previously in Chapter 4 (Section 4.7.3) and in Chapter 5 (Section 5.5). For simplicity, the search for a minimum of Q is replaced by a quest for the maximum of $Q_1 = 1/Q$. It is necessary to prove that the function $Q_1(x, y, v)$ does not have a maximum on the boundary of the region of interest. Then, the partial derivatives of Q_1 are to be derived. The values of x_{OPT}, y_{OPT}, and v_{OPT} are calculated from $\partial Q_1/\partial x = 0$, $\partial Q_1/\partial y = 0$, and $\partial Q_1/\partial v = 0$. Eventually, the optimized values for variables x, y, and v are found as $x_{OPT}= y_{OPT}= v_{OPT} =$ 1.4667. The fourth variable is calculated as $w(x_{OPT}, y_{OPT}, v_{OPT}) = x_{OPT} =$ 1.4667. From this result, it is concluded that the fastest strictly aperiodic step response is obtained with the closed-loop poles $\sigma_1 = \sigma_2 = \sigma_3 = \sigma_4 =$ 0.6818. The optimized form of the characteristic polynomial and the optimized values of the feedback gains are given in Eq. 6.18 and Eq. 6.19.

$$f_{PID}(z) = (z-\sigma_1)(z-\sigma_2)(z-\sigma_3)(z-\sigma_4) = (z-\sigma)^4 \tag{6.18}$$

$$\begin{aligned} \sigma_1 = \sigma_2 = \sigma_3 = \sigma_3 = \sigma = \frac{1}{1.4667} = 0.6818 \\ d_{OPT} = \sigma^4 = 0.216 \\ p_{OPT} = 4\sigma^3 - \sigma^4 - 1 = 0.0516 \\ i_{OPT} = 6\sigma^2 + \sigma^4 - 3 = 0.0052195 \end{aligned} \tag{6.19}$$

The practical steps of tuning the absolute values of the feedback gains K_P, K_I, and K_D require the control object inertia J, the sampling period T, the static gain of the torque amplifier K_M, and the shaft sensor gain K_{FB}. The gains K_P, K_I, and K_D are then found as

$$\begin{aligned} K_I = (2J\, i_{OPT})/(K_{FB}\, K_M\, T^2) \\ K_P = (2J\, p_{OPT})/(K_{FB}\, K_M\, T^2) \\ K_D = (2J\, d_{OPT})/(K_{FB}\, K_M\, T^2)\,. \end{aligned} \tag{6.20}$$

In cases where the load inertia changes during the operating cycle of the positioner, the normalized gains i, p, and d are to be kept constant, while the absolute gains K_P, K_I, and K_D are to be adjusted according to Eq. 6.20.

The optimized parameter setting in Eq. 6.20 minimizes the sum of the error samples in the step response and results in a spectrum of the closed-loop poles where all four z-domain poles are equal to $\sigma_{1/2/3/4} = 0.6818$. A rough estimate of the closed-loop bandwidth can be calculated from the equivalent s-domain pole. With a real, negative s-domain pole s_x and its frequency $f_{PF} = s_x/2/\pi$, the corresponding pole in the z-domain is found as $z_x = \exp(-2\pi\, f_{PF}T)$. With $z_x = 0.6818$, the frequency f_{PF} obtained for the sampling periods of 0.5 ms, 2 ms and 10 ms amounts to 120 Hz, 30 Hz and 6 Hz, respectively. This result gives a rough idea of the closed-loop bandwidth. A more accurate evaluation of the bandwidth frequency is given in the next section.

6.3 The step response and bandwidth of the PD and PID controller

It is interesting to compare the step response rise time and the closed-loop bandwidth of the PD position controller, discussed in the previous chapter, and of the PID controller, comprising the integral action and an infinite stiffness. Both controllers are considered in linear operating mode, with the output position defined by the closed-loop system transfer function $W_{SS}(z)$ given in Eq. 5.25 for the PD controller and the function $W_{SS}(z)$ given in Eq. 6.10 for the PID controller. It is assumed that the feedback gains are set according to Eq. 5.36 and Eq. 6.19 so as to provide the fastest strictly aperiodic response. The step responses and the relevant Bode plots are obtained from Matlab, using the command sequence in Table 6.1.

Table 6.1. The Matlab command sequence used to obtain the step response and Bode plot for the position-controlled systems with PD and PID controllers.

```
>> d = 0.2027;  p = 0.03512;               % The feedback gains for the PD
>>                                          % controller
>> num1 = [ p p 0];                         % Numerator of the transfer
>>                                          % function in Eq. 5.25
>> den1 = [1  -(2-p-d) (1+p) -d];           % Characteristic polynominal with
>>                                          % PD controller
>> step_pd = dstep(num1,den1);              % Vector step_pd keeps the step
>>                                          % response with PD
>>
>> d = 0.216; p = 0.0516; i = 0.0052195;    % PID controller gains
>> num2 = [i i  0 0 ];                      % Numerator of WSS(z) in Eq. 6.10
>> den2 = [1  -(3-p-i-d) +(3-d+i) -(1+d+p) +d];
>>                                          % Characteristic polynominal
>> step_pid = dstep(num2,den2);             % Step response with PID controller
>> stairs(step_pd); hold on; stairs (step_pid);  grid;
>>                                          % Plotting the step responses
>>                                          % Obtaining the Bode plot
>> [a1,ph1,W1] = DBODE(num1,den1,0.001)
>> [a2,ph2,W2] = DBODE(num2,den2,0.001);
>>                                          % Plotting and comparing the
>>                                          % amplitude characteristics
>> plot(W1, a1); hold on; plot(W2,a2); axis([0 500 0 1 ]); grid;
```

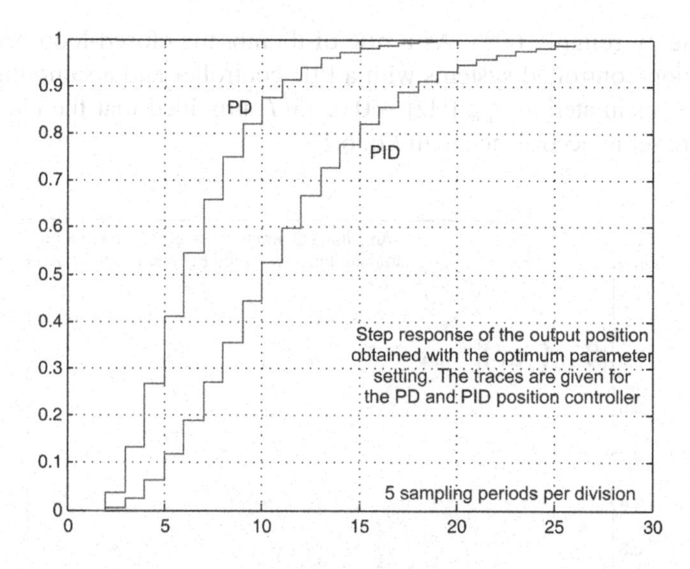

Fig. 6.3. The closed-loop step response of the PD and PID position controller. In both cases, the parameter setting results in the fastest strictly aperiodic response. The corresponding rise times are $t_{PD} = 8\,T$ and $t_{PID} = 13\,T$.

The step response of the PD controller, given in Fig. 6.3, increases from 10% to 90% of its steady-state value in eight sampling periods. With the PID controller, the rise time t_{PID} takes approximately 13 sampling periods. The closed-loop bandwith of the system can be evaluated from the rise time t_R. In systems with a well-damped closed-loop response, the bandwidth frequency f_{BW} and rise time are related by the approximate expression $t_R\, f_{BW} = 1/3$. In the case when the sampling period is $T = 1$ ms, the rise times are $t_{PD} =$ 8 ms and $t_{PID} = 13$ ms, and the corresponding bandwidth frequencies are $f_{BW}^{PD} = 1/3/0.008$ s $= 41.6$ Hz and $f_{BW}^{PID} = 25.6$ Hz.

The closed-loop bandwidth can be evaluated from the amplitude characteristic $|W_{SS}(j\omega)|$ of the closed-loop system transfer function. The amplitude characteristic is obtained by means of the Matlab function *dbode*, as explained in the command sequence given in Table 6.1. The results are plotted in Fig. 6.4, scaled to facilitate the conclusion on the bandwidth frequency. With the PD controller, the value of $|W_{SS}(j\omega)|$ drops down to -3dB (i.e., 0.707) for $\omega_{BW}^{PD} = 260$ rad/s. Hence, $f_{BW}^{PD} = 41.4$ Hz. With the PID controller, $f_{BW}^{PID} = 25.5$ Hz. The values obtained from Fig. 6.4 correspond to the estimates calculated from the rise time.

A controller with integral action has a longer rise time and a lower bandwidth frequency than a controller having only the proportional and derivative action. In Fig. 6.3, the introduction of integral action prolongs the rise time by roughly 60%. As a rule of thumb, the closed-loop bandwidth of position-controlled systems with a PID controller and a sampling period T can be estimated as f_{BW} [Hz] = 0.0255/T, provided that the closed-loop gains are set in accordance with Eq. 6.19.

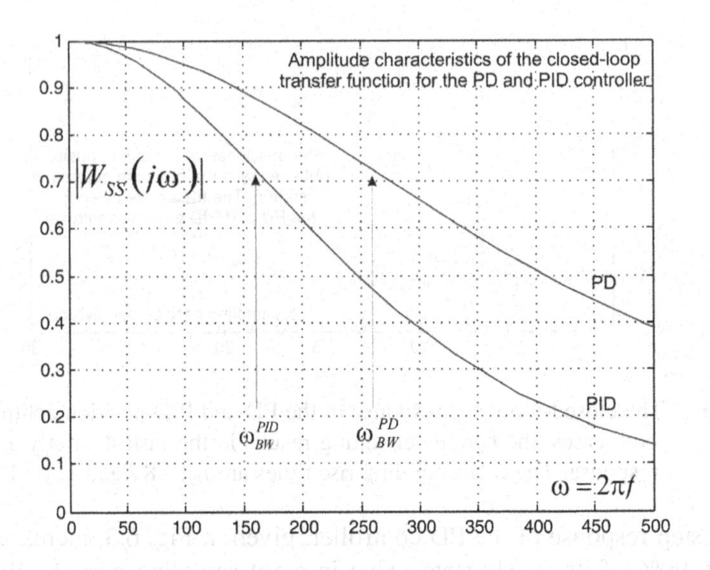

Fig. 6.4. The amplitude characteristics $|W_{SS}(j\omega)|$ of the PD and the PID position controller. It is assumed that the sampling time is $T = 1$ ms. The level of 0.707 (i.e., –3 dB) is reached for $f_{BW}^{PID} = 25.5$ Hz and $f_{BW}^{PD} = 41.4$ Hz.

6.4 Computer simulation of the input step and load step response

In order to verify the results obtained in the previous section and investigate further the dynamic performance of the discrete-time position controller with integral action, the system in Fig. 6.1 has been converted into a Simulink model. In the sample system, the inertia is set to $J = 0.11$ kgm^2 and the sampling time to $T = 0.001$ s. For simplicity, the scaling coefficients K_M and K_{FB} are set to one. The model, given in Fig. 6.5, takes into account the limited resolution of the shaft sensor. The block *Quantizer*, connected to the output of the *POSITION* integrator at the lower right of the figure, can be used to adjust the resolution of the shaft sensor. For proper use of the model, the system parameters J and T and the feedback gains K_I, K_D, and K_D must be entered at the Matlab command prompt. In the subsequent simulation runs, it is assumed that the feedback gains are set according to Eq. 6.19 and Eq. 6.20. The position controller is implemented according to the block diagrams in Fig. 6.1 and Fig. 6.2. In the model, the position reference block and the load torque block provide the input excitation and the load disturbance. The torque actuator is assumed to have an instantaneous response and a static gain of $K_M = 1$. The block denoted *LIMITER* takes into account the limited torque capability of practical torque actuators ($|T_{em}| \leq T_{MAX}$). The block entitled *position* and the block *torque* capture the samples of the shaft position and the driving torque and store them in arrays *position* and *torque* for plotting and further processing. In the *configuration parameters* menu of the Simulink, it is necessary to enter the simulation step. In order to suppress the errors in simulating the continuous part of the system (i.e., the speed and position integrators), the simulation step is to be set to a value of $T/10$ or even smaller.

In Fig. 6.6, the simulation traces of the position reference, output position, driving torque, and load torque are given, obtained from the simulation model in Fig. 6.5 for a step change in the position reference (on the left in Fig. 6.6) and the subsequent step change in the load torque. The output position reaches the target without an overshoot. The rise time is roughly 13 sampling periods, in accordance with the result obtained with Matlab function *dstep* and given in Fig. 6.3. Notice in Fig. 6.6 (left) that the driving torque exhibits an aperiodic change as well. It changes sign only once, when the system completes the acceleration phase and starts braking. On the right side of the figure, the load torque steps up. In return, the output position sags. Due to integral action, the driving torque reaches the load and goes beyond for a brief interval of time, allowing for a transient acceleration sequence, required for the position error to be dissipated and driven down to zero. In accordance with Eq. 6.13, the steady-state value of the output position is not affected by the load.

For the purpose of comparison, the traces obtained with the PD position controller are given in Fig. 6.7, using the same position reference and load disturbance. Compared with Fig. 6.6, the step response is faster. However, the load step produces an error in the output position, due to the finite stiffness obtained with the PD controller.

It is also useful to investigate the effects of a limited resolution of the shaft sensor. To this end, the experiment shown in Fig. 6.6 is repeated with the *Quantizer* block, at the lower left in Fig. 6.5, adjusted to the resolution of 25 μrad. The corresponding traces are given in Fig. 6.8, comprising small changes in the output position, observable even in the steady state.

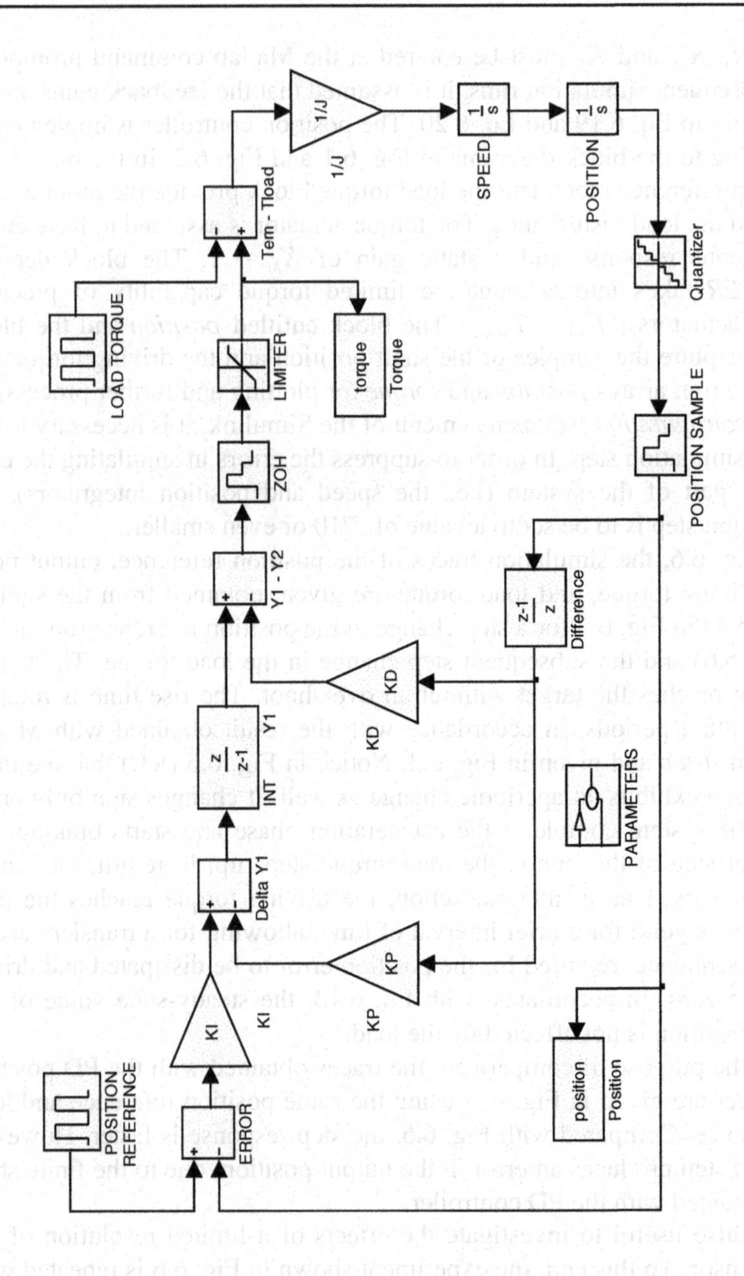

Fig. 6.5. Simulink model of linear discrete-time PID position controller.

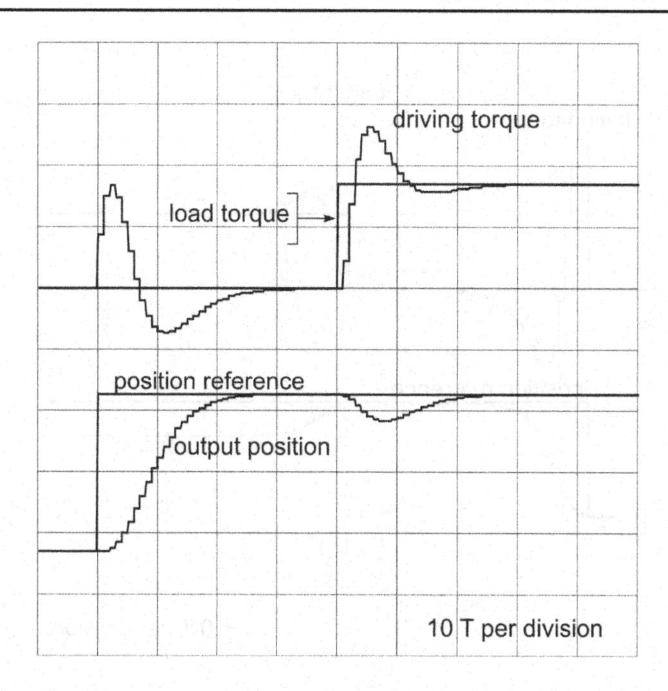

Fig. 6.6. Transient responses of the output position and the driving torque of a position-controlled system with PID controller. The step in the position reference is followed by the load step.

The driving torque comprises chattering in both the steady state and transients. In the figure, the torque pulsations, produced by the quantization effect, amount to 20–25% of the load torque. The amplitude of the torque pulsations is inversely proportional to the resolution of the shaft sensor. With an incremental encoder, the number of pulses per turn should be as high as possible.

Manufacturing of optical encoders with a large resolution may be complex. In order to achieve an n-bit resolution in the shaft position reading, the encoder glass disk must have 2^n transparent openings incised on its dark circumference. Where the number of encoder pulses per turn is insufficient, the resolution of the position reading can be enhanced by interpolation within each pulse width. A new generation of optical encoders, called *sincoders* [15], provides both the traditional digital signals with n-pulses per turn and their analog, sinusoidal counterparts. The presence of analog signals provides the means for position interpolation within one-pulse boundaries, thus increasing significantly the effective resolution in position reading.

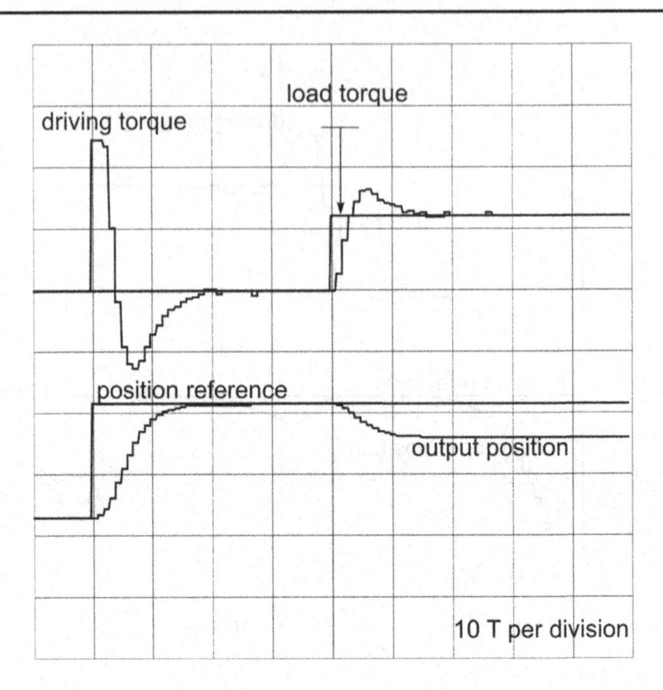

Fig. 6.7. Transient response of the output position and the driving torque of a position-controlled system with PD controller. The input and load disturbances are the same as in Fig. 6.6.

It is of interest to note in Fig. 6.8 that the system with a limited resolution of the feedback transducer does not have a steady state: variables such as the driving torque and position comprise jitters that do not cease, even after a very long period of time. The amplitude of the torque chattering reduces with an increase in resolution, but it persists in all discrete-time systems due to the amplitude quantization, intrinsic to the sampling process.

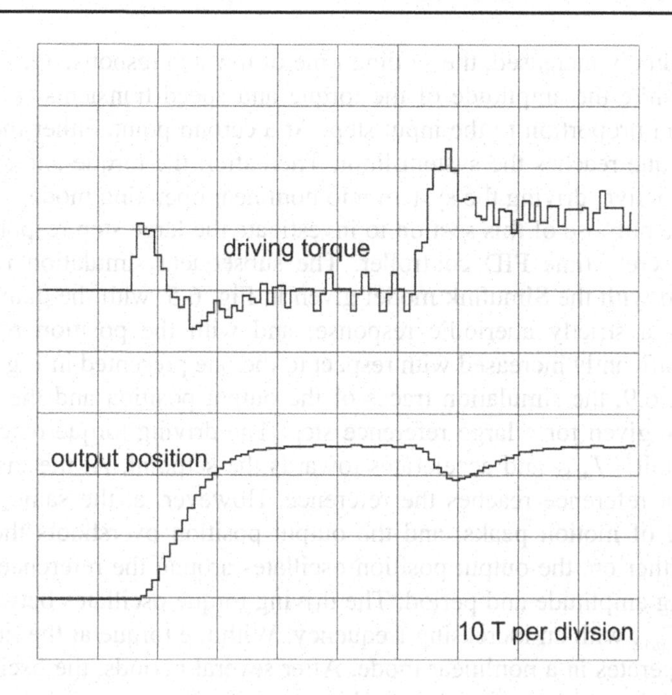

Fig. 6.8. The impact of a limited resolution on the transient response of the system with PID controller. Other signals and settings are the same as in Fig. 6.6.

6.5 Large step response with a linear PID position controller

In Fig. 6.6, the simulation traces of the driving torque and the output position are given, obtained by running the Simulink model shown in Fig. 6.5. The model comprises a linear, discrete time PID position controller implemented in the incremental form (Fig. 6.1). Following the reference step, the driving torque increases, and the system accelerates towards the target. Further on, as the output error decreases, the torque becomes negative, and the system decelerates. Within the transient response shown in Fig. 6.6, the speed and torque do not reach the system limits ω_{MAX} and T_{MAX}. Hence, the system operates in its linear mode. During the reference step transient of a linear system, the amplitude of the torque, speed, and position changes stays in proportion to the amplitude of the input disturbance: the positive and negative peaks of the driving torque in Fig. 6.6 are proportional to the reference step. With $|\omega| \leq \omega_{MAX}$ and $|T_{em}| \leq T_{MAX}$, and with the reference steps gradually increased, the settling time of the step response remains unaltered, while the amplitude of the torque and speed transients keeps increasing in proportion to the input step. At a certain point, either the speed or the torque reaches the system limit. Thereafter, the torque and/or speed limiter is active, driving the system into nonlinear operating mode.

It is the purpose of this section to investigate the large step response of a linear, discrete time PID controller. The subsequent simulation runs are performed with the Simulink model given in Fig. 6.5, with the gain setting providing a strictly aperiodic response, and with the position reference steps significantly increased with respect to the one presented in Fig. 6.6.

In Fig. 6.9, the simulation traces of the output position and the driving torque are given for a large reference step. The driving torque reaches the system limit $+T_{MAX}$ and accelerates towards the setpoint. At the instant t_X, the output reference reaches the reference. However, at the same instant, the speed of motion peaks, and the output position overshoots the reference. Further on, the output position oscillates around the reference with a decreasing amplitude and period. The driving torque oscillates between the limits $\pm T_{MAX}$ with an increasing frequency. With the torque at the limit, the system operates in a nonlinear mode. After several periods, the oscillations decay and the system enters the steady state.

The transient phenomena in Fig. 6.9 are similar to the traces obtained with the PD position controller given in Fig. 5.15 and detailed in Section 5.7. Nonlinear oscillations of the same kind, known as the *wind-up*, are encountered with the speed controller. The wind-up effect is discussed in Chapter 4 and illustrated in Fig. 4.24. In the case of a speed controller, the wind-up is attributed to the interaction between the torque limiter and the error integrator, which winds up while the driving torque is at the limit. In Fig. 4.24, the difficulty of reaching the steady state arises from the fact that the error integrator assumes a nonzero value at instants when the error $\Delta\omega$ reaches zero.

With a position controller, effects similar to wind-up take place through an interaction between the torque limiter and the integrator buried within the control object. In other words, the output position is obtained by integrating the speed. While the torque is at the limit, the output position winds up and overshoots the reference. As indicated by the traces in Fig. 6.9, the shaft speed assumes a nonzero value $\omega(t_x) \neq 0$ at instants when the output position $\theta(t_x)$ reaches the reference, resulting in $\Delta\theta = 0$. The overshoot and the consequential nonlinear oscillations can be attributed to the torque limiter. With a limited torque, the applicable deceleration rates $d\omega/dt$ are limited to T_{MAX}/J. Therefore, at instants when the output position reaches the setpoint ($\Delta\theta = 0$), the residual speed $\omega(t_x)$ cannot be driven down to zero, and the system overshoots.

An increase in the input disturbance can bring the nonlinear oscillations to instability. In Fig. 6.10, the amplitude of the reference step is increased, resulting in sustained oscillations. The effects of a further increase in the position reference step are simulated in Fig. 6.11. The amplitude of the nonlinear oscillations is increasing, while their frequency decreases. The driving torque pulses of the amplitude T_{MAX} become longer, and the position error continuously enlarges.

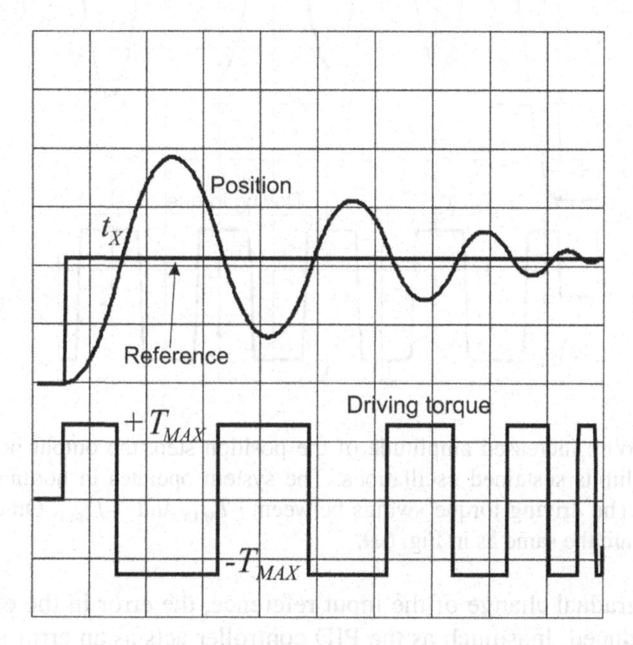

Fig. 6.9. Large step response of the discrete-time position-controlled system with the PID controller. The driving torque reaches the system limit T_{MAX} and enters nonlinear control mode. The output position oscillates around the reference. The oscillations decrease in amplitude and increase in frequency.

From the responses given in Fig. 6.9 – Fig. 6.11, it is concluded that the linear PID controller exhibits a stable, aperiodic response only in cases when the input disturbance is relatively small, resulting in the torque and speed transients staying within the system limits ω_{MAX} and T_{MAX}. Larger input steps (Fig. 6.9) bring the torque to the limit, resulting in nonlinear behavior with poorly damped oscillations. Even larger input disturbances drive the system into instability.

In a practical application, the problem can be solved by replacing the step change of the position reference by a ramp profile, increasing with a constant slope, and reaching the target position.

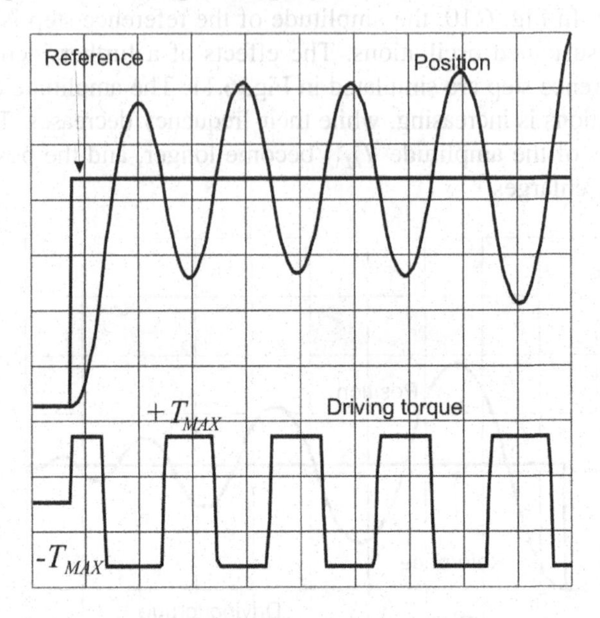

Fig. 6.10. With increased amplitude of the position step, the output position exhibits sustained oscillations. The system operates in nonlinear mode. The driving torque swings between $+T_{MAX}$ and $-T_{MAX}$. Other settings are the same as in Fig. 6.9.

With a gradual change of the input reference, the error in the output position is reduced. Inasmuch as the PID controller acts as an error amplifier, providing the torque reference at its output, restricted values of the error are unlikely to drive the torque reference to the limit T_{MAX}. Hence, with a ramp-shaped position reference having a reasonable slope $d\theta^*/dt$, the system remains in linear operating mode, preserving stability and a strictly aperiodic character of the transients. The simulation run given in Fig. 6.11 has been repeated with the position reference profile having a constant slope, and the torque and position traces are given in Fig. 6.12. The reference ramp reaches the target position equal to the reference step in Fig. 6.11. Note in Fig. 6.12 that the torque transients are much smaller than in the previous case.

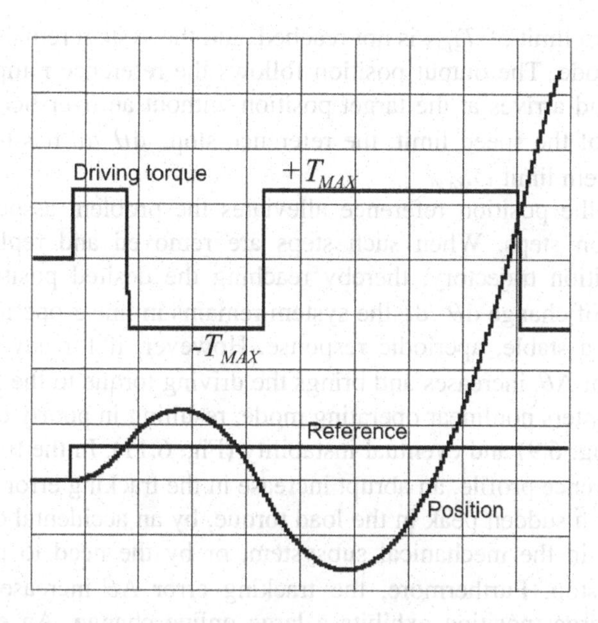

Fig. 6.11. A further increase in amplitude of the input disturbance brings the system to instability. Other settings are the same as in Fig. 6.9.

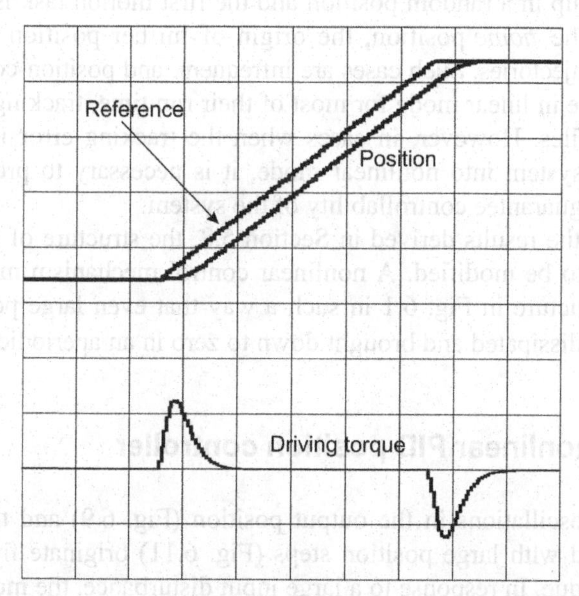

Fig. 6.12. The simulation run presented in Fig. 6.11 is repeated with the position reference step replaced by the ramp-shaped profile. The target position is equal to the reference step in the previous figure. Other settings are the same as in Fig. 6.9.

The system limit of T_{MAX} is not reached, and the system remains in linear operating mode. The output position follows the reference ramp with certain delay and arrives at the target position without an overshoot. In order to stay out of the speed limit, the reference slope $d\theta^*/dt$ has to be lower than the system limit ω_{MAX}.

Profiling the position reference alleviates the problem associated with large position steps. When such steps are removed and replaced by a smooth position trajectory, thereby reaching the desired position with a limited rate of change $d\theta^*/dt$, the system remains in linear operating mode, resulting in a stable, aperiodic response. However, if for any reason the tracking error $\Delta\theta$ increases and brings the driving torque to the limit T_{MAX}, the system enters nonlinear operating mode, resulting in poorly damped oscillations (Fig. 6.9) and eventual instability (Fig. 6.11). In the tracking of a smooth reference profile, an abrupt increase in the tracking error $\Delta\theta$ can be produced by a sudden peak in the load torque, by an accidental collision or malfunction in the mechanical subsystem, or by the need to perform an emergency stop. Furthermore, the tracking error $\Delta\theta$ increases in cases where the target position exhibits a large online change. An example of where a large position error is encountered is the startup, where the system is powered up in a random position and the first motion task is to bring the system to the *home* position, the origin of further position profiles and reference trajectories. Such cases are infrequent, and position-controlled systems operate in linear mode for most of their run time, tracking smooth reference profiles. However, in cases when the tracking error increases and brings the system into nonlinear mode, it is necessary to provide control means and guarantee controllability of the system.

As with the results derived in Section 5.8, the structure of the PID controller has to be modified. A nonlinear control mechanism must be added into the structure in Fig. 6.1 in such a way that even large position errors $\Delta\theta$ can be dissipated and brought down to zero in an aperiodic manner.

6.6 The nonlinear PID position controller

Nonlinear oscillations in the output position (Fig. 6.9) and the instability encountered with large position steps (Fig. 6.11) originate from a limited braking torque. In response to a large input disturbance, the motion speed is increased to a level that cannot be brought back to zero by the time the output position reaches the target. This instant is designated t_X in Fig. 6.9. With $\theta(t_x) = \theta^*$, $\omega(t_x) \neq 0$. Therefore, the output position overshoots, resulting in oscillations that are poorly damped (Fig. 6.9), sustained (Fig. 6.10) or unstable (Fig. 6.11), depending on the amplitude of the input disturbance. The residual speed $\omega(t_x)$ appears due to a limited braking torque and a limited deceleration ($d\omega/dt = -T_{MAX}/J$). Given the deceleration limit, the only viable approach in suppressing the consequential overshoot and oscillations is to begin the deceleration phase before the output position reaches the target. As previously found in Eq. 5.40, the speed of the system approaching the target has to be limited in proportion to the residual path $\Delta\theta$.

The functional speed limit $\omega_M = fp(|\Delta\theta|)$ is derived in Section 5.8 by equating the kinetic energy of the system, $W_{KIN}(\omega(t_x))$, to the braking energy $W_{DBR}(\Delta\theta(t_x))$, where W_{DBR} is the product of the remaining path $\Delta\theta(t_x)$ and an assumed constant braking torque T_{MAX}. The same result is obtained below in a different way. Consider the system in its braking phase and observe the instant t_y, where the speed of motion is $\omega(t_y) = \omega_y$. Then, the residual path can be expressed as $\Delta\theta_y = \theta^* - \theta(t_y)$. It is reasonable to assume that the maximum available braking torque $-T_{MAX}$ is applied in an attempt to arrest the system prior to its arrival at the target. Assuming, also, that the friction and other motion resistances are negligible compared with the peak driving torque T_{MAX}, the deceleration rate is found as $d\omega/dt = -T_{MAX}/J$. Hence, the speed changes according to

$$\omega(t) = \omega_y - \frac{T_{MAX}}{J}(t - t_y). \tag{6.21}$$

The speed decreases to zero and the system comes to arrest at the instant $t_{STOP} = t_y + J\omega_y/T_{MAX}$. Between the two instants, the output position moves from $\theta(t_y)$ to $\theta(t_{STOP})$:

$$\theta(t_{STOP}) - \theta(t_y) = \int_{t_y}^{t_{STOP}} \omega(t)\,dt = \frac{T_{MAX}}{2J}(t_{STOP} - t_y)^2. \tag{6.22}$$

In order to achieve the output response without an overshoot, the error $\Delta\theta_{STOP} = \theta^* - \theta(t_{STOP})$, encountered at the instant t_{STOP}, where $\omega(t_{STOP}) = 0$, has to be $\Delta\theta_{STOP} = 0$. Hence, the value in Eq. 6.22 has to be equal to $\Delta\theta_y$. From Eq. 6.21, $t_{STOP} - t_y = J\omega_y / T_{MAX}$. Thereby, the relation between the residual speed and the braking distance $\Delta\theta_y$ is obtained as

$$\Delta\theta_y = \frac{J\omega_y^2}{2T_{MAX}}. \tag{6.23}$$

The previous formula suggests that a position-controlled system, approaching the target at a speed of ω, does not produce an overshoot provided that the remaining path $\Delta\theta=\theta^*-\theta$ is at least $J\omega^2/2/T_{MAX}$. In Eq. 6.23, it is assumed that both the speed and the remaining path are positive. There are cases when the position-controlled system moves towards negative targets, where both the speed and the position error assume negative values. In such cases, Eq. 6.23 provides the minimum absolute value of the remaining path required to avoid an overshoot. From the previous equation, the maximum speed of approach to the target can be expressed in terms of the residual error $|\Delta\theta|$. The function $\omega_M = fp(|\Delta\theta|)$ in Eq. 6.24 is obtained from Eq. 6.23 and corresponds in full to Eq. 5.40, derived in the previous chapter, when discussing the PD controller. Eq. 6.24 gives the functional speed limit to be respected and built into the control structure in order to prevent the overshoots, nonlinear oscillations, and instability encountered with large input disturbances.

$$|\omega_M| = fp(|\Delta\theta|) = \sqrt{\frac{2T_{MAX}|\Delta\theta|}{J}} \tag{6.24}$$

In addition to the speed limit in the previous equation, the limit $|\omega| \le \omega_{MAX}$ has to be implemented as well: that is, the system limit ω_{MAX} must not be exceeded, for reasons explained in the previous chapters. On the other hand, the speed transients increase with the amplitude of the input disturbance. For large reference steps and in the cases when the system tracks the reference profiles $\theta^*(t)$ with a large rate of change, the speed of the system may exceed the limit. To prevent this from happening, the structure of the PID controller must be modified. As in the block diagram in Fig. 5.17, a nonlinear block has to be entered into the direct path of the controller, thereby ensuring that the constraint $|\omega| \le fp(|\Delta\theta|) \le \omega_{MAX}$ is met under all conditions. With the PD controller (Fig. 5.17), it is sufficient to replace the proportional gain block K_P by a nonlinear function $fx(\Delta\theta)$, given in Eq. 5.41 and explained in Fig. 5.16. Due to the presence of integral action within the PID controller, the implementation of the functional speed limit is more involved, as described in subsequent sections.

6.6.1 The maximum speed in linear operating mode

While input disturbances are small, the PID controller operates in linear mode, where the system limits ω_{MAX} and T_{MAX} are not reached. With large reference steps, the torque and/or speed reach the limits, and the system operates in nonlinear mode. The nonlinear speed limit in Eq. 6.24 must be applied in order to preserve the aperiodic nature of the response. It is of interest to distinguish between the two operating modes and to identify the characteristic values of the tracking error $\Delta\theta_{(\max)}$ and speed $\omega_{(\max)}$ delimiting the linear and nonlinear modes. With the tracking errors $|\Delta\theta| \le \Delta\theta_{(\max)}$ and with speed transients $|\omega| \le \omega_{(\max)}$, the nonlinear mechanisms, such as the limiters, are inactive. In such conditions, the linear PID controller is capable of providing an aperiodic response without an overshoot.

In the block diagram of a linear PID controller in Fig. 6.1, the signal Δy_1 represents the sum of the increments of the proportional and integral action:

$$\Delta y_{1(n+1)} = K_I K_{FB} \Delta\theta - K_P K_{FB} \left(\theta_{(n+1)} - \theta_{(n)}\right).$$

Given the speed ω, the change in the output position within the sampling interval is obtained as

$$\left(\theta_{(n+1)} - \theta_{(n)}\right) \approx \omega_{(n+1)} T.$$

The increment Δy_1 is now calculated as

$$\Delta y_{1(n+1)} \approx K_I \, K_{FB} \Delta\theta - K_P \, K_{FB} \, \omega_{(n+1)} T. \tag{6.25}$$

With the aim of determining the values $\Delta\theta_{(\max)}$ and $\omega_{(\max)}$, it is important to consider the traces given in Fig. 6.6 for the PID controller operating in linear mode. Within the step response, the speed peaks after 3–4 T. At this point, the output position ascends linearly. With the speed transient reaching the maximum, the speed changes observed at the same instant are relatively small. Since the signal y_1 represents the internal speed reference, its change Δy_1 is expected to be close to zero. If we introduce $\Delta y_1 = 0$ in Eq. 6.25, the peak value $\omega_{(\max)}$ can be calculated and given in Eq. 6.26. The error $\Delta\theta$ corresponds to the position error observed at the instant when the speed ω reaches the maximum $\omega_{(\max)}$.

$$\omega_{(\max)} \approx \frac{K_I \Delta\theta}{K_P T} \tag{6.26}$$

The proportion between the speed transient $\omega_{(\max)}$ and the position error $\Delta\theta$, given in the previous equation, is an expected behavior of a linear system. An increase in the input disturbance enlarges both the speed and the output error. At a certain point, the speed $\omega_{(\max)}$ reaches the limit $fp(|\Delta\theta|)$. A further increase in the disturbance amplitude results in speed peaks exceeding the braking limit in Eq. 6.24. With higher speeds, the braking torque $T_{em} = -T_{MAX}$ is incapable of decelerating and arresting the system prior to the output position reaching the target. Therefore, it is concluded that the condition $\omega_{(\max)} = fp(|\Delta\theta|)$ delimits both the linear and nonlinear operating modes. The error $\Delta\theta_{(\max)}$ obtained by solving $\omega_{(\max)} = fp(|\Delta\theta|)$

represents the largest error attainable in the linear operating mode. This value corresponds to the intersection of the square root function in Eq. 6.24 with the straight line of Eq. 6.26 in the phase plane, with $\Delta\theta$ and ω on the horizontal and vertical axis, respectively.

$$\frac{K_I \Delta\theta}{K_P T} = \sqrt{\frac{2T_{MAX}|\Delta\theta|}{J}} \Rightarrow \frac{2T_{MAX}|\Delta\theta|}{J} = \left(\frac{K_I}{K_P T}\right)^2 (\Delta\theta)^2$$
$$\Delta\theta_{(\max)} = \frac{2T_{MAX}}{J}\left(\frac{K_P T}{K_I}\right)^2 \tag{6.27}$$

The value of $\Delta\theta_{(\max)}$, obtained in the previous formula, represents the largest position error the system can withstand without entering nonlinear modes. To gain insight into the practical values of such an error, consider the parameter setting suggested in Eq. 6.20, with the ratio $K_P/K_I \approx 10$, and with $\Delta\theta_{(max)} \approx 200\, T_{MAX}T^2/J$. This value represents the change in the output position of the system which departs from standstill and accelerates with $T_{em} = T_{MAX}$ over 20 sampling periods. Hence, most position-controlled systems would move by $\Delta\theta_{(\max)}$ in 5–20 ms. In other words, the value of the position error that delimits the linear and nonlinear operating modes is relatively small.

From Eq. 6.26, the maximum speed at the border between linear and nonlinear modes is obtained as

$$\omega_{(\max)} = \frac{2T_{MAX}}{J}\left(\frac{K_P T}{K_I}\right). \tag{6.28}$$

With an optimized parameter setting, the speed $\omega_{(\max)}$ is obtained in 20 sampling periods, departing from the standstill and accelerating with the maximum rate available.

In a practical position-controlled system, the values of $\Delta\theta_{(\max)}$ and $\omega_{(\max)}$ are relatively small. Consider the system where the top speed of $\omega_{\text{MAX}} = 100\pi$ rad/s (3000 rpm) is reached in 250 ms. Then, the peak acceleration is $a_{\max} = T_{MAX}/J = 1256.6$ rad/s^2. With a sampling time of $T = 500$ μs, the transition between linear and nonlinear operating modes is delimited by the speed transient $\omega_{(\max)} = 12.566$ rad/s (120 rpm) and the error amplitude of $\Delta\theta_{(\max)} = 0.063$ rad. Recall that the linear PID controller in Fig. 6.1 provides a stable, aperiodic response only in cases where the speed and position transients stay below $\Delta\theta_{(\max)}$ and $\omega_{(\max)}$. Due to very low values of such transients, any attempt to use the linear PID controller is limited to exceptionally low speeds and extremely low position steps. Therefore, it is necessary to modify the controller structure and add the nonlinear speed constraint $|\omega| \le fp(|\Delta\theta|)$, given in Eq. 6.24.

In the course of installing and testing, a motion-control system is controlled manually by running individual axes (motors) in the speed control mode. In each axis, the motion is initiated in the direct or reverse direction by pressing appropriate buttons on the control panel. Such an operating mode is frequently referred to as *jog*. The speed reference issued is close to the value of $\omega_{(max)}$, given in Eq. 6.28.

6.6.2 Enhancing the PID controller with nonlinear action

A linear PID controller needs to be modified to ensure that the speed of motion during transients does not exceed the limit $|\omega| \le fp(|\Delta\theta|) \le \omega_{MAX}$. Similar modification is done with the PD controller, given in Fig. 5.17, where the proportional gain block K_P is replaced by a nonlinear function $fx(\Delta\theta)$. Given in Eq. 5.41 and explained in Fig. 5.16, the nonlinear speed limit ensures that the speed of approach to the target position does not exceed the functional limit $fx(\Delta\theta)$, calculated from the braking distance $\Delta\theta$. The value of $fx(|\Delta\theta|)$ corresponds to $fp(|\Delta\theta|)$, obtained in Eq. 6.24. With $|\omega| \le fp(|\Delta\theta|)$, the braking torque $-T_{MAX}$ is sufficient to decelerate the system before the output position reaches and overshoots the target.

Within the PID controller in Fig. 6.1, the derivative action $y_{2(n)}$, obtained at the sampling instant $t = nT$, is $K_{FB}K_D(\theta_{(n)} - \theta_{(n-1)})$. If the speed changes during the interval $[(n-1)T .. nT]$ are small, the sample $y_{2(n)}$ is proportional to the speed $\omega_{(n)}$, $y_{2(n)} \approx K_{FB}K_D\omega_{(n)}T$. Hence, the control object with a torque actuator and a minor control loop, comprising the derivative action, represents a local speed controller, with the signal y_2 as the feedback and y_1 as the speed reference. If the closed-loop gain of such a speed controller is sufficiently high, the error Δy would be small: $|\Delta y| = |y_1 - y_2| \le |y_2|$. Therefore, the speed of the system is determined by $\omega \approx y_1/(K_{FB}K_DT)$. Consequently, the functional speed limit $fp(|\Delta\theta|)$ can be applied by restricting the amplitude of y_1 to $K_{FB}K_DT\,fp(|\Delta\theta|)$.

In Fig. 6.1, y_1 is obtained at the output of the integrator *INT*, which sums the increments in proportional and integral actions (Δy_1). Implementation of the speed limit $|y_1| \le K_{FB}K_DT\,fp(|\Delta\theta|)$ has to be performed in such a way that the integrator *INT* does not wind up while the signal y_1 remains at the limit. In other words, when the speed is to be restricted to the path-dependent limit $fp(|\Delta\theta|)$, the content of the integrator has to be modified in order to prevent an uncontainable increase in its value (i.e., the *wind-up*).

In Fig. 6.13, a limiter is merged with an integrator in a way that prevents the wind-up phenomenon. The output of the integrator $y_{(n)}$ is stored in the block *MEM*, providing the past value $y_{(n-1)}$ at the output. The summing junction S_1 adds the increment $\Delta y_{(n)}$ to the past value $y_{(n-1)}$ and feeds the sum $\Delta y_{(n)} + y_{(n-1)}$ to the limiter. In cases when the variable y does not reach the limit, the signal $\Delta y_{(n)} + y_{(n-1)}$ passes through the limiter and provides the next sample. The solution given in Fig. 6.13 operates on the same principles as the discrete-time integrator D2 in Fig. 4.25, presenting the incremental implementation of the digital speed controller.

The limiter in that previous figure has to ensure that the signal y_1 does not exceed the value $K_{FB}K_DT\, fp\,(|\Delta\theta|)$. At the same time, it is necessary to make sure that the maximum permissible speed ω_{MAX} of the system is not exceeded as well. Therefore, whatever the value of $fp\,(|\Delta\theta|)$, it is necessary to ensure that $|y_1| \le K_{FB}K_DT\,\omega_{MAX}$. Both speed limits are given in Fig. 6.14. With the position error $\Delta\theta$ on the abscissa and the speed on the ordinate axis, the diagram represents a phase plane. Any state $(\omega, \Delta\theta)$ outside the permissible region results in an overshoot, caused by an insufficient braking torque. Hence, the internal speed reference $y_1/(K_{FB}K_DT)$ has to stay below the square root limit fp, respecting, at the same time, the system limit of ω_{MAX}.

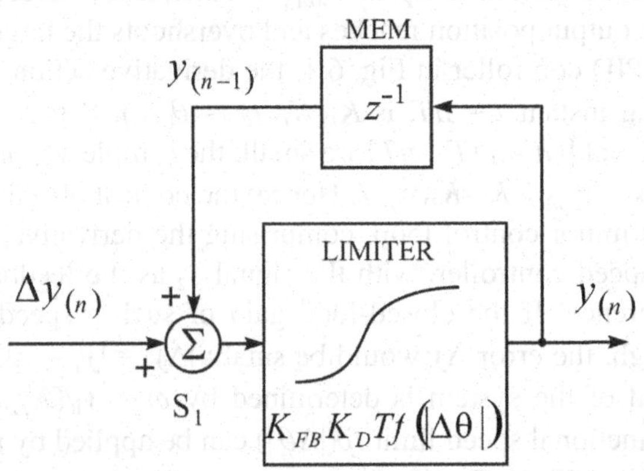

Fig. 6.13. Limiting the output of the integrator in a way that prevents the *wind-up* phenomenon.

In Fig. 6.15, a solution of a saturable discrete-time integrator is given, restricting the output $y_{1(n)}$ to the permissible region, as defined in Fig. 6.14. The block *MEM* stores the past value of the output and provides the sample $y_{1(n-1)}$. The input of the block receives the value of $y_{1(n)}$, already processed through the limiter. In such a way, the absolute value of $y_{1(n)}/(K_{FB}K_DT)$

does not exceed the system limit ω_{MAX}, nor does it exceed the functional limit $fp(|\Delta\theta|)$. At each sampling instant, the increment $\Delta y_{1(n)}$ is added to the past value $y_{1(n-1)}$. The junction S_1 in Fig. 6.15 provides the temporary sum y^*. The sign of y^* is preserved, while the amplitude is processed through the limiter, designated as *MIN* in the figure. The limiter receives three values: $x_1 = |y^*|$, $x_2 = K_{FB}K_D T fp(|\Delta\theta|)$, and $x_3 = K_{FB}K_D T\omega_{MAX}$. The output $|y_{1(n)}|$ is obtained as the smallest of x_1, x_2, and x_3. The new sample of $y_{1(n)}$ is obtained by attributing the sign of y^* to the output of the limiter. Further to the right in the figure, the derivative feedback y_2 is subtracted in S_2.

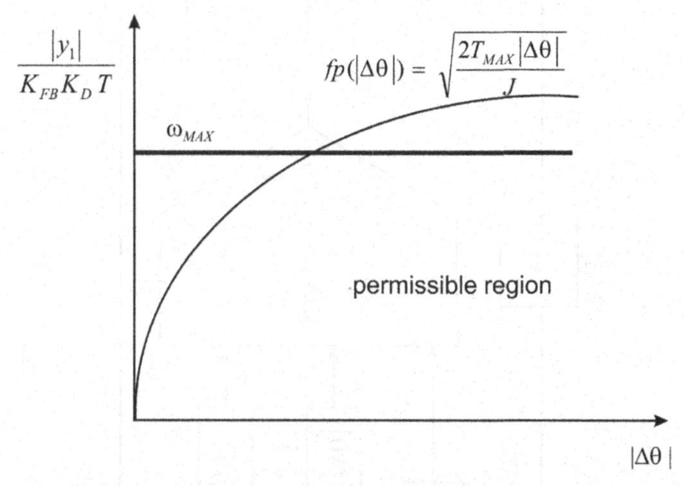

Fig. 6.14. Permissible operating region in the phase plane. In order to avoid the overshoots caused by an insufficient braking torque, the internal speed reference $y_1/(K_{FB}K_D T)$ must stay below the square root limit *fp*.

A complete block diagram of the PID position controller comprising the nonlinear speed limit is given in Fig. 6.16. Essentially, the structure is obtained by replacing the discrete-time integrator *INT* of the linear controller (Fig. 6.1) with the saturable integrator, specified in Fig. 6.15. With the signal y_1 assuming the role of the internal speed reference, the speed of the system approaching the target position will be limited according to Eq. 6.24. Therefore, a well-timed braking interval brings the system to a halt prior to exceeding the target. The nonlinear speed limit $fp(|\Delta\theta|)$ depends on the torque limit T_{MAX}. The higher the peak braking torque, the larger the admissible speed for the given braking distance $\Delta\theta$.

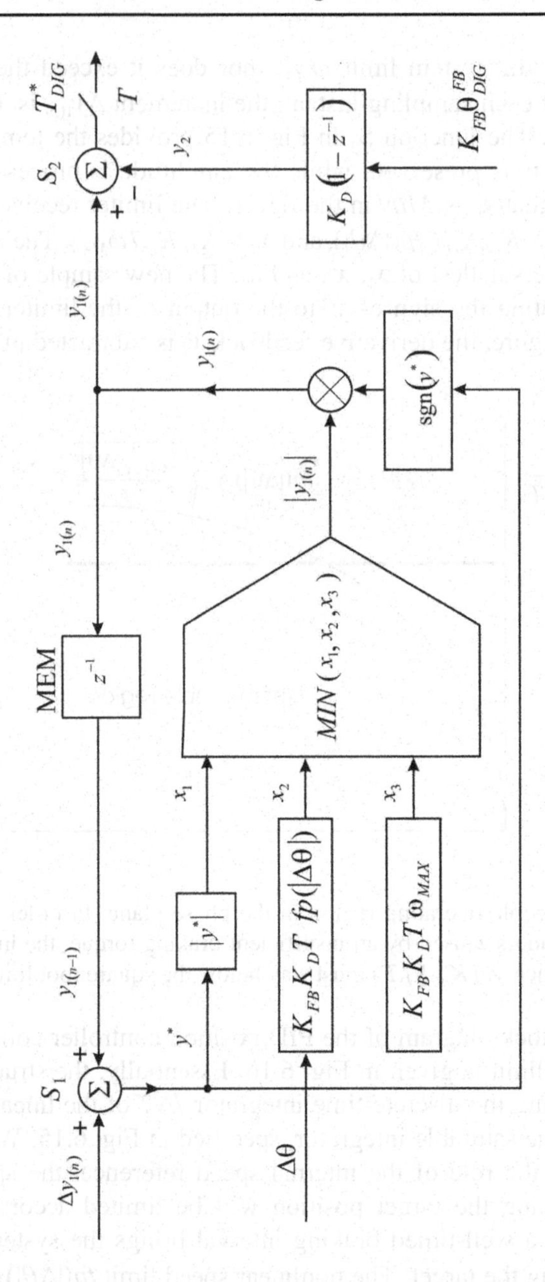

Fig. 6.15. A solution of a saturable discrete-time integrator, introduced in Fig. 6.13, comprising the nonlinear functional speed limit $fp(|\Delta\theta|)$ and restricting the motion of the system to the permissible region, as defined in Fig. 6.14.

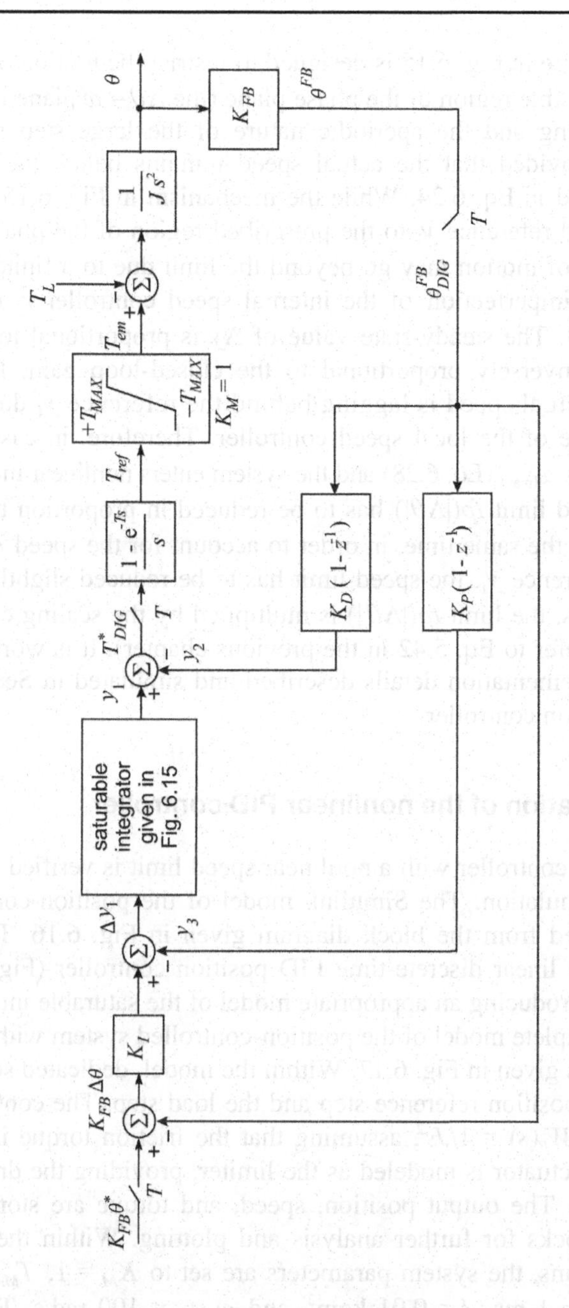

Fig. 6.16. The discrete time PID position controller with a nonlinear speed constraint. The block *INT* in Fig. 6.1 is replaced by the saturable discrete-time integrator, detailed in Fig. 6.15. With this controller, the motion of the system is restricted to the permissible region in Fig. 6.14.

The structure in Fig. 6.15 is designed to restrict the motion of the system to the permissible region of the phase plane (the $\Delta\theta - \omega$ plane in Fig. 6.14). Timely braking and the aperiodic nature of the large step response are achieved, provided that the actual speed remains below the square root limit specified in Eq. 6.24. While the mechanism in Fig. 6.15 restricts the internal speed reference y_1 to the prescribed region of the phase plane, the actual speed of motion may go beyond the limit due to a finite error $\Delta y = y_1 - y_2$. The imperfection of the internal speed controller is discussed in Section 5.8.3. The steady-state value of Δy is proportional to the driving torque and inversely proportional to the closed-loop gain. During transients, the actual speed is lagging behind the reference y_1 due to a finite response time of the local speed controller. Therefore, in cases when the speed exceeds $\omega_{(max)}$ (Eq. 6.28) and the system enters nonlinear mode, the square root speed limit $fp(|\Delta\theta|)|$ has to be reduced in proportion to the torque reference. At the same time, in order to account for the speed ω falling behind the reference y_1, the speed limit has to be reduced slightly. In further developments, the limit $fp(|\Delta\theta|)$ is multiplied by the scaling coefficient of $K_S = 0.98$ (refer to Eq. 5.42 in the previous chapter). It is worthwile to recall the implementation details described and simulated in Section 5.9 for the PD position controller.

6.6.3 Evaluation of the nonlinear PID controller

The position controller with a nonlinear speed limit is verified by means of computer simulation. The Simulink model of the position-controlled system is derived from the block diagram given in Fig. 6.16. The previous model of the linear discrete-time PID position controller (Fig. 6.5) is extended by introducing an appropriate model of the saturable integrator (Fig. 6.15). A complete model of the position-controlled system with a nonlinear speed limit is given in Fig. 6.17. Within the model, dedicated source blocks provide the position reference step and the load step. The control object is modeled as $W_P(s) = 1/Js^2$, assuming that the friction torque is negligible. The torque actuator is modeled as the limiter, providing the driving torque $|T_{em}| \le T_{MAX}$. The output position, speed, and torque are stored in corresponding blocks for further analysis and plotting. Within the subsequent simulation runs, the system parameters are set to $K_M = 1$, $T_{MAX} = 10$ Nm, $K_{FB} = 1$, $T = 1$ ms, $J = 0.01$ kgm^2, and $\omega_{MAX} = 100$ rad/s. The feedback gains are set according to the design rule in Eq. 6.20.

The subsystem with the saturable integrator receives the error signal, torque reference, and increment Δy_1, providing, at the same time, the output

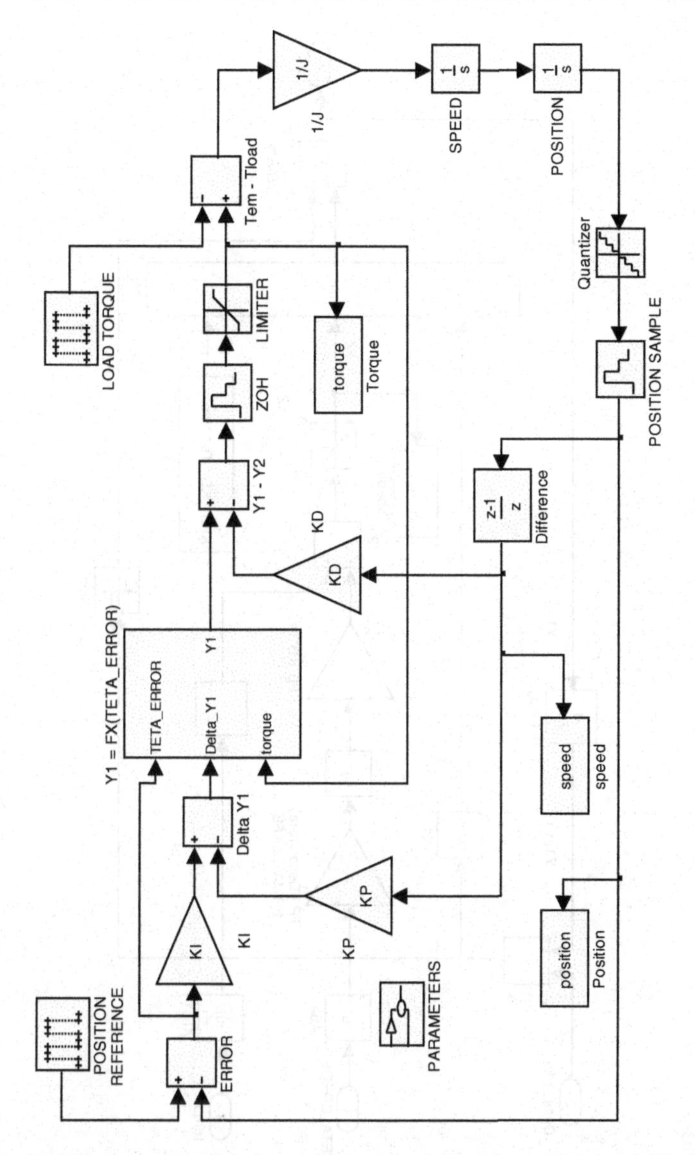

Fig. 6.17. Simulink model of the position-controlled system with a nonlinear PID position controller. The saturable integrator, providing y_1, is given in Fig. 6.18.

signal y_1. The corresponding subsystem is detailed in Fig. 6.18. The subsystem is based on the structure given in Fig. 6.15. Within the structure, the path-dependent speed limit x_2 is slightly changed due to imperfections of the internal speed controller. The limit $fp\,(\Delta\theta)$, obtained from the block *FP*

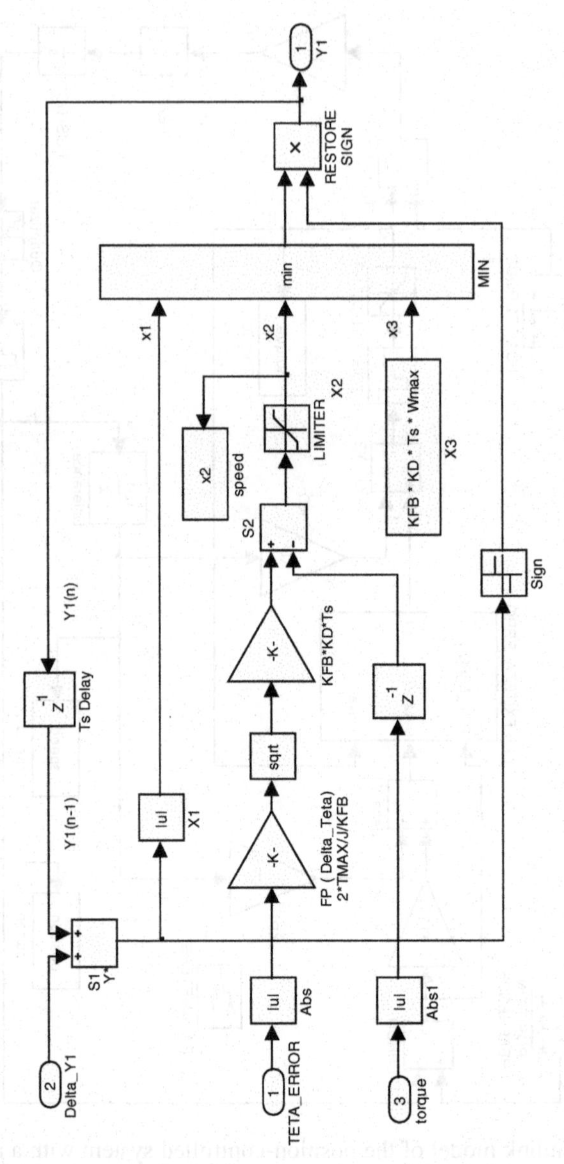

Fig. 6.18. Simulink subsystem implementing the saturable integration, designed in Fig. 6.15. The square root speed limit $fp(\Delta\theta)$ (block *FP*) is modified as suggested in Eq. 5.43. The absolute value of the driving torque (block *Abs1*) is subtracted (block *S2*) from the limit. The signal x_2 is scaled down with $K_s = 0.98$ (block *KS*) and lower bounded in accordance with $\omega_{(max)}$ in Eq. 6.28) (block *LIMITER*). The output of the subsystem (right) is the internal speed reference y_1.

in Fig. 6.18, is scaled down by the factor $K_S = 0.98$. The absolute value of the torque reference is obtained from the block *Abs1* and subtracted from x_2. The limiter suppresses the eventual negative values of x_2 and ensures that $x_2 > x_{2min} = K_D K_{FB} T \omega_{(max)}$. The speed $\omega_{(max)}$, obtained in Eq. 6.28, represents the boundary between the linear and nonlinear operating modes.

In Fig. 6.19, the simulation traces of the output position, driving torque, and speed are given. The responses are obtained for a large reference step, driving the system into nonlinear mode and reaching both the speed and torque limits. The speed of motion reaches the system limit ω_{MAX}, while the torque actuator provides the peak T_{MAX} in both acceleration and braking. The speed (middle trace) increases with a constant slope and reaches the maximum permissible value. While the speed is at the limit, the output position (bottom trace) rises linearly. The linear change in the output position ends at the instant t_x, when the driving torque assumes the value of $T_{em} = -T_{MAX}$, commencing the braking interval. The braking starts when the path-dependent speed limit $fp(\Delta\theta)$ in Eq. 6.24 reduces to ω_{MAX} and becomes equal to the running speed. At the instant t_x, the speed and remaining path are $\omega(t_x) = \omega_{MAX}$ and $\Delta\theta(t_x) = J\omega_{MAX}^2/2/T_{MAX}$, respectively. Note that a linear position controller, such as the one in Fig. 6.1, does not start braking at $t = t_x$. With $\Delta\theta(t_x) \gg 0$, it keeps providing $y_1 > y_2$ and $T_{em}(t_x) > 0$, fails to decelerate, and overshoots the target (Fig. 6.11). With a nonlinear PID controller, $y_1(t_x)$ is derived from the structure in Fig. 6.18. With $y_1(t_x) < y_2(t_x)$, the torque becomes negative and reaches $-T_{MAX}$.

Within the interval $[t_x..t_y]$ in Fig. 6.19, the braking torque is constant, the speed reduces linearly, and the slope of the output position decreases as it approaches the target. In this phase, the system slides along the square root boundary of the permissible region of the phase plane (Fig. 6.14). When the speed drops below $\omega_{(max)}$ (Eq. 6.28) and with $\Delta\theta < \Delta\theta_{(max)}$ (Eq. 6.27), the system returns to linear mode, where none of the limiters included in Fig. 6.15 is active, resulting in $|y_{1(n)}| = x_1$ and $y_{1(n)} = y^*$. Thereafter, the torque, speed and position errors decay aperiodically, and the system enters the steady state. While decelerating, the torque does not exhibit further sign changes, reducing, in this way, the effects of imperfections in transmission elements, such as the backlash. The torque, speed, and position transients obtained in Fig. 6.19 represent the fastest possible transition from the initial to the final position, considering the speed and torque constraints ω_{MAX} and T_{MAX}.

While the torque pulse encountered during acceleration is rectangular, the braking pulse slightly differs. Prior to reaching the target position, the braking torque briefly departs from $-T_{MAX}$ and exhibits two relatively small glitches (t_y in Fig. 6.19). The effect is related to the implementation of the saturable integrator with the nonlinear speed limit (Fig. 6.18). The amplitude of y_1 is obtained at the output of the block *MIN*, as the smallest of the three inputs (x_1, x_2, x_3). At the instants when the block switches between the inputs, the first derivative of the internal speed reference y_1 exhibits abrupt changes, reflected in the driving torque waveform given in Fig. 6.19. As shown in the phase plane in Fig. 6.14, one of the discontinuities occurs when the system abandons the square root boundary and enters linear mode with $\Delta\omega < \Delta\omega_{(\max)}$. Parameter change in the *KS* and *LIMITER* blocks of the model in Fig. 6.18 affects the torque waveform and alters the amplitude and form of the torque pulses. Note that the position controller designed in this section cannot provide a fully rectangular braking pulse. In its final stage, the system enters the linear operating mode, where the braking torque exponentially decays, according to the strictly aperiodic nature of the closed-loop poles.

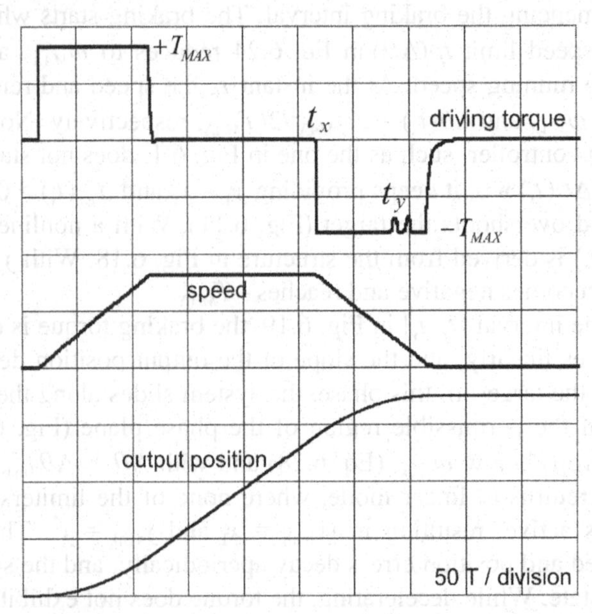

Fig. 6.19. Simulation traces of the output position and driving torque obtained for a large reference step, driving the system into nonlinear mode and making it reach both the speed and torque limits. The model with the PID position controller comprising the nonlinear speed limit is given in Fig. 6.17 and Fig. 6.18.

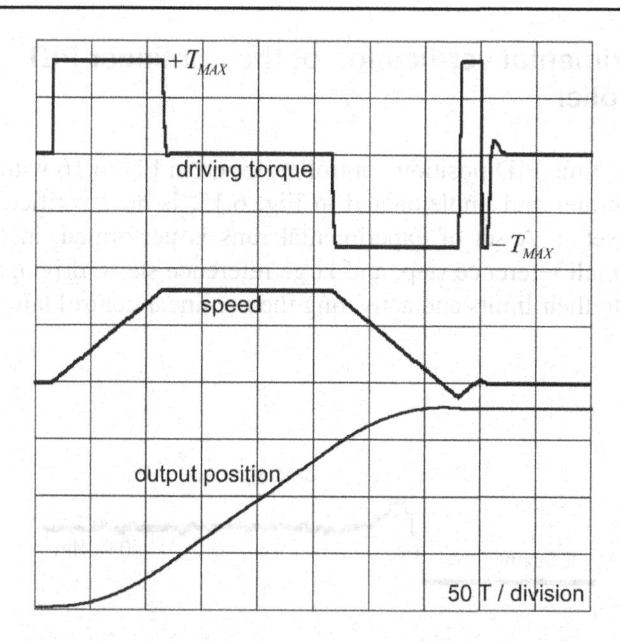

Fig. 6.20. The previous simulation run, given in Fig. 6.19, is repeated without compensating for the speed error of the internal speed controller. With reference to Fig. 6.18, blocks *S2* and *LIMITER* are removed, the signal x_2 is fed from the *KS* output to the *MIN* input, and the scaling coefficient is set to $K_S = 1$. Due to a finite error $\Delta y = y_1 - y_2$, the speed exceeds the braking limit $fp(\Delta\theta)$(Eq. 6.24). The overshoot in the output position can be inferred from the speed and torque waveforms.

It is of interest to investigate the impact of the internal speed controller imperfection on the large step response. In the previous simulation run, these were compensated for by subtracting the torque signal (block *Abs1*) from x_2 in Fig. 6.18, and by scaling down the square root speed limit *fp* by $K_S = 0.98$. The same simulation is then repeated, this time omitting the changes and implementing the saturable integrator as originally designed in Fig. 6.15. The Simulink subsystem in Fig. 6.18 is edited by removing blocks *S2* and *LIMITER*, and feeding x_2 straight from the *KS* output to the *MIN* input. At the same time, the scaling coefficient is set to $K_S = 1$. The simulation results are given in Fig. 6.20. While in the braking phase, the actual running speed exceeds the square root limit $fp(\Delta\theta)$, resulting in an overshoot. The excess speed is caused by the static error and a limited bandwidth of the internal speed controller. Due to a finite error $\Delta y = y_1 - y_2$, the speed exceeds the braking limit $fp(\Delta\theta)$. The overshoot in the output position is not obvious in Fig. 6.20, since it is much smaller than the position step. It can be inferred from the speed and torque waveforms.

6.7 Experimental verification of the nonlinear PID controller

The discrete-time PID position controller, given in Fig. 6.16 with a nonlinear speed limiter and implemented in Fig. 6.15, is now verified in an experimental setup. A set of experimental runs is performed, including the load step, small reference step, and large reference steps, driving the speed and torque to their limits and activating the nonlinear control law.

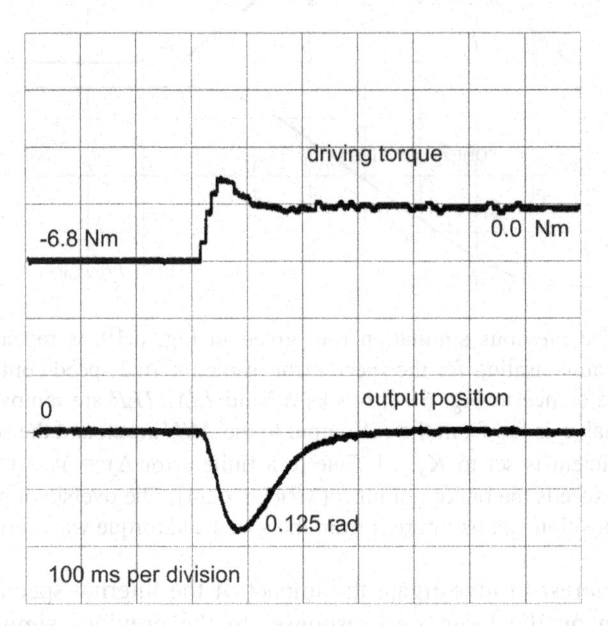

Fig. 6.21. Experimental traces of the driving torque and the position obtained for the step change in the load torque. The system is controlled by the PID controller with the nonlinear speed limiter (Fig. 6.16).

The experimental setup has been detailed in the previous chapter and is similar to the one described in Chapter 4. A four-pole, 1 HP induction motor is supplied from a CRPWM inverter and attached to an incremental encoder with 1250 pulses per revolution. The IFOC algorithm provides the driving torque response time of 250 μs. With the sampling time of $T =$ 10 ms, delays in the torque actuator are considered to be negligible. The motor is coupled to an inertial load. The equivalent inertia of the system comprising the motor, the load, and the inertial disk amounts $J = 0.032$ kgm^2. The peak torque capability of the torque actuator is determined by the inverter peak current and amounts to $T_{MAX} = 13.6$ Nm. The maximum permissible speed is $\omega_{MAX} = \omega_{nom} = 145$ rad/s. Control structures are implemented on a 16-bit fixed-point DSP platform and given in Appendix 1 and Appendix 2 of this book. The traces given in the subsequent figures are obtained by writing the digital words corresponding to the relevant torque and position samples on dedicated D/A converters and feeding their analog outputs to the oscilloscope.

The experimental traces plotted in Fig. 6.21 are obtained as the system response to the step change in the load torque. The PID controller with the nonlinear speed limiter is applied (Fig. 6.16). The load torque is changed from $T_L(0-) = -6.8$ Nm to $T_L(0+) = 0$ Nm. For practical reasons, load removal is applied instead of the load step, in order to achieve a change in the load torque resembling a Heaviside step. In approximately 15 T, the PID position controller suppresses the error and the output position returns to the reference. Hence, the presence of the load torque does not produce a position error in steady-state conditions.

In Fig. 6.22, the experiment is repeated with the PD position controller designed and described in Chapter 5 (Fig. 5.17). Following the load step, the PD controller provides the driving torque that counterbalances the load, preventing a further position drop and bringing the system to steady state. In the absence of integral action, the load step produces an error in the output position. The error is proportional to the load torque. The steady-state ratio $T_{LOAD}/\Delta\theta$, known as the *stiffness coefficient*, is given in Eq. 5.18. and Eq. 5.27.

The experimental traces of the driving torque and position, obtained with the PID controller for a small reference step, are shown in Fig. 6.23. With the step of 0.62 rad, the driving torque does not enter the limits and the system operates in linear mode. Within 15 sampling periods, the target position is reached and the system enters steady state. The torque and position transients are strictly aperiodic. For the reference steps inferior to $\Delta\theta_{(\max)}$ (Eq. 6.27), the system remains in linear mode, with $|T_{em}| \leq T_{MAX}$. While in linear mode, the speed does not exceed the threshold $\omega_{(\max)}$ (Eq. 6.28) designating the boundary between linear and nonlinear modes. In such conditions, the rise time and character of the step response does not change with the amplitude of the input disturbance.

In Fig. 6.24, the experimental traces are given for a large reference step. In less than 5 s, the output position changes by 500 rad. The system is controlled by the PID controller with the nonlinear speed limiter, as shown in Fig. 6.16. The driving torque reaches the limit T_{MAX} during acceleration and in the braking phase. The speed reaches the limit ω_{MAX} and remains in the limit for most of the transient, resulting in a linear change in the output position.

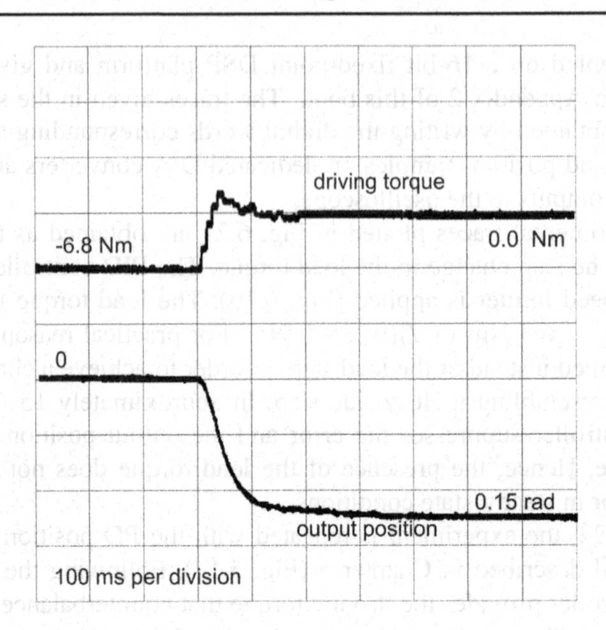

Fig. 6.22. The load step response obtained with the PD position controller designed in Chapter 5 and given in Fig. 5.17. Other settings are the same as in the previous experiment.

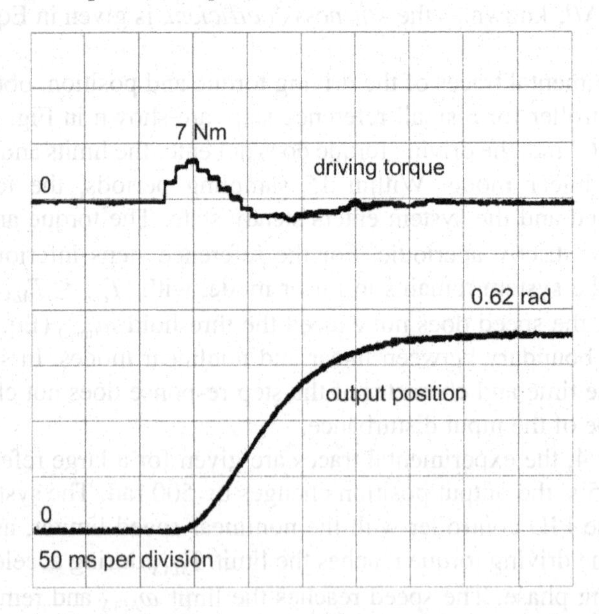

Fig. 6.23. Experimental traces of the driving torque and position obtained for a small step change in the position reference. The system is controlled by the PID controller with a nonlinear speed limiter (Fig. 6.16).

While running at a constant speed of $\omega = \omega_{MAX}$, the driving torque is rather small. With the load torque of $T_L = 0$, the driving torque T_{em} counter-balances a relatively small friction. In this phase, it is interesting to note the torque pulsations, caused by the quantization effects and the motor imperfections. At the end of the constant speed phase, the square root speed limit is activated, and the system enters the braking phase, where the maximum available deceleration of $d\omega/dt = -T_{MAX}/J$ is applied in order to reduce the speed and stop the system prior to exceeding the target.

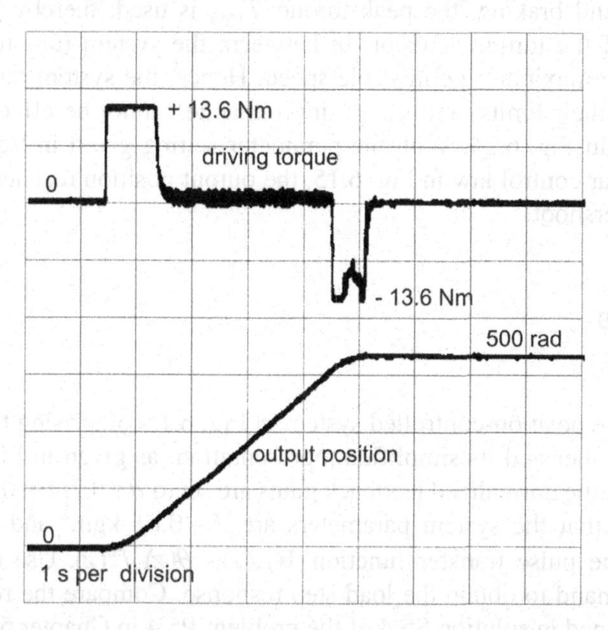

Fig. 6.24. Experimental traces of the driving torque and position obtained for a large step change in the position reference. The system is controlled by the PID controller with nonlinear speed limiter (Fig. 6.16). The speed and torque reach their limits, and the system operates in nonlinear mode.

If we compare the simulation traces in Fig. 6.19 with the experimental results in Fig. 6.24, slight differences are perceived and will be discussed. The simulation trace of the driving torque comprises the acceleration and braking pulses of the driving torque that are of the same width. The braking pulse in the experimental waveform is narrower. While the simulation model (Fig. 6.17) does not take into account any motion resistance and assumes that the control object is modeled as $W_P(s) = 1/Js^2$, the friction torque T_F in the experimental setup reduces the acceleration to $d\omega/dt = +T_{MAX}/J - T_F/J$ and helps the braking ($d\omega/dt = -T_{MAX}/J - T_F/J$). As shown by the braking pulse of the driving torque in Fig. 6.24, the notch where the braking torque deviates from $-T_{MAX}$ is more pronounced than in Fig. 6.19. This is because of a finite-wordlength, C-coded, real-time implementation of the saturable integrator with the square root speed limit given in Fig. 6.15 and Fig. 6.18.

It is worthwile to note that the torque and position responses given in Fig. 6.24 represent the fastest possible transition from the initial to the final position, given the speed and torque limits of the actual system. Both in acceleration and braking, the peak torque T_{MAX} is used, thereby getting the most out of the torque actuator. In between, the system runs towards the target at the maximum permissible speed. Hence, the system resources are used up to their limits, and the position change cannot be effectuated any faster than in Fig. 6.24. With the parameter setting given in Eq. 6.20 and the nonlinear control law in Fig. 6.15, the output position reaches the target with no overshoot.

Problems

P6.1

Consider the position-controlled system in Fig. 6.1, comprising the PID position controller and its simplified representation, as given in Fig. 6.2. Assuming that the normalized feedback gains are set to $d = 0.2$, $p = 0.05$, and $i = 0.005$, and that the system parameters are $J = 0.01$ kgm^2 and $T = 1$ ms, calculate the pulse transfer function $W_{LS}(z) = \theta(z)/T^L(z)$. Use the Matlab *dstep* command to obtain the load step response. Compare the result to the traces obtained in solution S5.4 of the problem P5.4 in Chapter 5.

P6.2

For the system described in the previous problem, derive the closed-loop transfer function $W_{SS}(z) = \theta(z)/\theta^*(z)$, and obtain the output position response to the step change of the reference. Estimate the speed and the driving torque traces by calculating the derivatives of the output position. Compare the rise time to the result obtained in P5.6/S5.6.

P6.3

For the system described in the previous problem, calculate the closed-loop zeros and poles in the *z*-domain, and find their equivalents in the *s*-domain.

P6.4

Consider the system described in P6.1. Assume that the normalized feedback gains are set according to Eq. 6.19, $d = 0.21609$, $p = 0.0516627$, $i = 0.0052195$. Obtain the output position response to the step change of the reference. Calculate the closed-loop poles.

P6.5

The characteristic polynomial $f_{PID}(z)$ of the system given in Fig. 6.2 is obtained in Eq. 6.11. Explain why the four closed-loop poles cannot be placed in an arbitrary way.

P6.6

Consider the closed loop system transfer function $W_{SS}(s)$:

$$W_{SS}(s) = \frac{1}{\left(1 + \frac{s}{\omega_p}\right)^n}.$$

Calculate the bandwidth frequency f_{BW} from the condition $|W_{SS}(j2\pi f_{BW})| = 1/\text{sqrt}(2)$. Given the value of ω_p, calculate the ratio between the bandwidth frequencies obtained with $n = 3$ and $n = 4$. Compare this ratio to $f_{BW}^{PD} / f_{BW}^{PID}$, obtained in Fig. 6.4.

P6.7

Consider the Simulink model of the linear discrete-time PID position controller, given in Fig. 6.5. The model takes into account a limited peak torque capability of the torque actuator. Assume that the load torque is equal to zero and that the position reference changes from 0 rad to 0.02 rad with an adjustable slope. The model is comprised within the file P6_7.mdl. Open the file by typing its name at the Matlab command prompt. Locate the block *repeating sequence* providing the position reference and note the parameter *tx* used to adjust the slope $d\theta^*/dt = 0.02/tx$. Run the model and observe the output position and the driving torque. Start with $tx = 0.05$ and decrease *tx* in small steps towards 0.01. Identify the maximum slope of the reference profile that provides the output position without an overshoot. Notice that the Matlab *m*-file *P6_7cmd.m* can be used to initialize the model parameters and plot the simulation traces. The search can be performed by the command below.

```
>> tx = 0.04;          % setting the desired value of tx
>> P6_7cmd             % invoking the command file that plots the sim. traces
```

P6.8

The system with the linear discrete-time PID position controller operates in linear mode, provided that the driving torque and speed do not reach the system limits T_{MAX} and ω_{MAX}. With larger input and load disturbances, the system limits are reached and the system enters nonlinear operating mode. Given the system parameters K_P, K_I, T, J, and T_{MAX}, determine the largest input step $\Delta\theta^*$ that does not drive the system into nonlinear mode.

7 Trajectory Generation and Tracking

In the majority of applications, position-controlled systems track predefined position reference profiles or trajectories. In this chapter, the tracking error is defined and expressed in terms of the reference profile derivatives and the position controller gains. Computer simulations are used in order to explore the error reduction achieved by the proper shaping of the profile. The reference profile generation is described and explained. The trajectory generation problem is defined as devising the function that changes between the given points. The time derivatives of the function are to be restricted, and with higher-order derivatives preferably equal to zero. The analytical considerations and simulation runs are included, relating the tracking error and the peak torque requirement to the profile time derivatives. In the closing section, interpolation of reference profiles is introduced, explained, and demonstrated.

7.1 Tracking of ramp profiles with the PID position controller

It is frequently required that position controlled systems track a position reference that has a ramp shape. In a system that tracks a reference profile with a constant slope $d\theta^*/dt$, the speed of motion is constant. In this section, the capability of the PID controller to track a constant-slope reference profile is analyzed. The tracking error is derived both analytically and from a Simulink model.

It is found that the implementation of the proportional gain influences the controller capability to track ramp-shaped reference profiles. With the proportional gain in the feedback path (Fig. 6.2), the system tracks the reference ramp with an error. The tracking error is proportional to the speed and depends on the feedback gains. In cases when the proportional action of the controller resides in the direct path, the gain K_P multiplies the error $\Delta\theta$ (Fig. 7.1). The analytical considerations and computer simulations in this section prove that such a controller is capable of tracking a ramp-shaped position profile without an error. Due to zeros in its closed-loop transfer function, the system with a PID controller, shown in Fig. 7.1, exhibits an overshoot in the step response, notwithstanding the strictly aperiodic character of the transient phenomena.

7.1.1 The steady-state error in tracking the ramp profile

Consider the PID controller represented in Fig. 6.2. With the proportional and derivative action relocated in the feedback path, the closed-loop system transfer function $W_{SS}(z) = \theta(z)/\theta^*(z)$ is derived in Eq. 6.10. The complex image of the tracking error is calculated as $\Delta\theta(z) = \theta^*(z) - \theta(z) = \theta^*(z)\,(1 - W_{SS}(z))$. It is interesting to examine the steady-state tracking of the position reference having a ramp shape with a constant slope $d\theta^*/dt$. If we assume that the samples $\theta^*_{(n)}$ of the position reference increment by R^* within each sampling period (i.e., $R^* = \theta^*_{(n+1)} - \theta^*_{(n)}$), the z-transform of the ramp-shaped reference profile $\theta^*_{(n)} = nR^*$ can be found as

$$\theta^*(z) = \sum_{j=0}^{+\infty} R^* j z^{-j} = R^* \sum_{j=0}^{+\infty} z^{-j}\left(\sum_{k=0}^{+\infty} z^{-k}\right) = \frac{R^*}{\left(1-z^{-1}\right)^2} = \frac{z^2 R^*}{(z-1)^2}\ . \tag{7.1}$$

From Eq. 6.10, the complex image $\Delta\theta(z)$ is calculated as:

$$\begin{aligned}\Delta\theta(z) &= \left(1 - W_{SS}(z)\right)\theta^*(z) = \frac{f_{PID}(z) - \left(z^3 + z^2\right)i}{f_{PID}(z)}\Delta\theta^*(z)\\
&= \frac{z^4 - (3-p-d)z^3 + (3-d)z^2 - (1+p+d)z + d}{z^4 - (3-p-i-d)z^3 + (3-d+i)z^2 - (1+p+d)z + d}\Delta\theta^*(z)\\
&= \frac{(z-1)\left(z^3 + (p+d-2)z^2 + (p+1)z - d\right)}{z^4 - (3-p-i-d)z^3 + (3-d+i)z^2 - (1+p+d)z + d}\,\frac{z^2 R^*}{(z-1)^2}\\
&= \frac{z^3 + (p+d-2)z^2 + (p+1)z - d}{z^4 - (3-p-i-d)z^3 + (3-d+i)z^2 - (1+p+d)z + d}\,\frac{z^2 R^*}{z-1}.\end{aligned}$$

The steady-state tracking error $\Delta\theta(\infty)$ is found from the final value theorem of the z-transform, expressed in Eq. 4.13. If we apply the final value theorem to the previous expression for $\Delta\theta(z)$, the tracking error is found as

$$\Delta\theta(\infty) = \lim_{z\to 1}\left(\frac{z-1}{z}\Delta\theta(z)\right) = R^*\frac{p}{i}\,. \tag{7.2}$$

Hence, the error in tracking the ramp profile with the controller structure given in Fig. 6.2 is proportional to the slope R^* (i.e., to the speed). It also depends on the ratio p/i between the normalized values of the proportional and integral gain. It is worth noting at this point that the tracking error in Eq. 7.2 corresponds to the criterion function Q, as defined in Eq. 6.17.

At this point, the block diagram in Fig. 6.1 and its signals Δy_1, y_1, y_2, and y_3 are going to be considered, in an attempt to establish the relation between the relocation of the proportional gain and the tracking error in Eq. 7.2. In the figure, the increment of the integral action $K_{FB}K_I\Delta\theta = K_{FB}K_I(\theta^*_{(n)} - \theta_{(n-1)})$ and the increment in the proportional action $-y_3 = -K_{FB}K_P(\theta_{(n)} - \theta_{(n-1)})$ are summed into the increment Δy_1. The derivative action $y_2 = K_{FB}K_D(\theta_{(n)} - \theta_{(n-1)})$ creates a feedback signal proportional to the speed $\omega = \mathrm{d}\theta/\mathrm{d}t$. The subsystem on the right in Fig. 6.1, comprising the derivative action and the control object, constitutes a local speed controller, with y_1 assuming the role of the speed reference and with the speed $y_2 = K_{FB}K_D(\theta_{(n)} - \theta_{(n-1)})$.

If the system tracks the reference profile $\theta^*_{(n)} = n\ R^*$ and the steady state is reached, the running speed ω has to be constant and equal to $\omega = R^*/T$. Therefore, the feedback signal $y_2 = K_{FB}K_D(\theta_{(n)} - \theta_{(n-1)})$ and the internal speed reference y_1 have to be constant. In such a case, the increment Δy_1 must be zero, leading to $K_{FB}K_I\Delta\theta - y_3 = K_{FB}K_I\Delta\theta - K_{FB}K_P(\theta_{(n)} - \theta_{(n-1)}) = 0$. With a constant speed ω, the position increments $\theta_{(n)} - \theta_{(n-1)}$ are equal to ωT. From these results, the steady-state tracking error can be determined to be $\Delta\theta = K_P\,\omega T/K_I = K_P R^*/K_I$. This result is in accordance with Eq. 7.2.

It will be shown that the tracking error disappears as the proportional action is restored into the direct path. In this case, the contribution of the proportional action to the driving torque is $K_{FB}K_P\Delta\theta$. The increment of such an action is calculated as $K_{FB}K_P(\Delta\theta_{(n)} - \Delta\theta_{(n-1)})$. In the structure in Fig. 6.1, and with the proportional action in the direct path, the increment Δy_1 is obtained as $K_{FB}K_I\Delta\theta_{(n)} + K_{FB}K_P(\Delta\theta_{(n)} - \Delta\theta_{(n-1)})$. In the steady state, $\Delta\theta_{(n)} = \Delta\theta_{(n-1)}$ and $\Delta y_1 = 0$. Thus, the steady-state tracking error is obtained as $\Delta\theta = 0$. It is of interest to confirm this result by evaluating the closed-loop system transfer function $W^+_{SS}(z) = \theta(z)/\theta^*(z)$, obtained with the PID controller that has the proportional action restored into the direct path. A simplified block diagram of such a system is given in Fig. 7.1. The integral and proportional control actions are located in the direct path, while the derivative action remains in the feedback path. The gain K_M of the torque actuator and the position sensor gain are assumed to be $K_{FB} = K_M = 1$. The closed-loop transfer function $W^+_{SS}(z)$ of the system is given in Eq. 7.3.

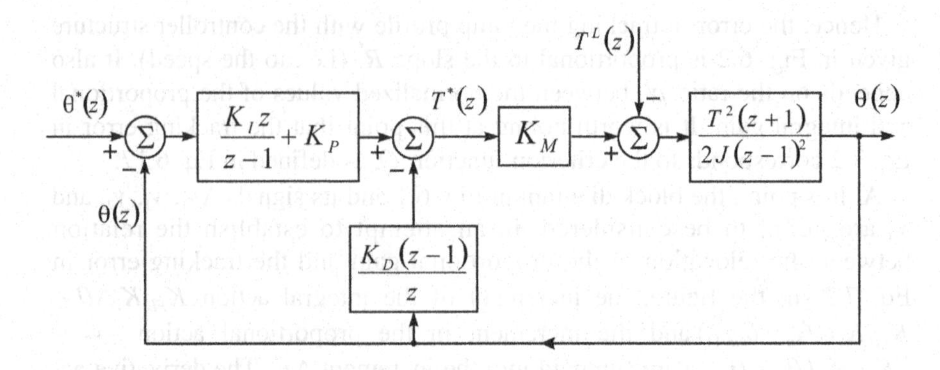

Fig. 7.1. Linear position controller with the proportional and integral gains in the direct path and with the derivative gain in the feedback path. It is assumed that the sensor gain K_{FB} and the torque actuator gain K_M are equal to one.

$$W_{SS}^{+}(z)=\frac{K_{FB}\left[\frac{K_M T^2(z+1)}{2J(z-1)^2}\right]\left[\frac{K_I z}{z-1}+K_P\right]}{1+\left[\frac{K_M T^2(z+1)}{2J(z-1)^2}\right]\left[K_I K_{FB}\frac{z}{z-1}+K_P K_{FB}+K_D K_{FB}\frac{z-1}{z}\right]}$$

$$=\frac{(p+i)z^3+iz^2-pz}{z^4-(3-p-i-d)z^3+(3-d+i)z^2-(1+p+d)z+d} \qquad (7.3)$$

Compared with $W_{SS}(z)$, given in Eq. 6.10, the polynomial in the numerator of the transfer function $W^{+}_{SS}(z)$ has one additional zero, $z_+ = p/(p+i)$. With an optimized setting of the feedback gains, $z_+ = 0.9096$. Given the reference profile $\theta^*_{(n)} = nR^*$, the complex image of the tracking error is derived in Eq. 7.4. In Eq. 7.5, the steady-state tracking error is obtained as $\Delta\theta(\infty) = 0$. Hence, the position controller in Fig. 7.1 is capable of tracking the ramp-shaped reference profiles without a steady-state error, provided that the slope R^* does not cause the system to exceed the system limit of ω_{MAX}.

$$\Delta\theta(z) = \left(1 - W_{SS}^{+}(z)\right)\theta^*(z) = \frac{f_{PID}(z) - (p+i)z^3 - iz^2 + pz}{f_{PID}(z)}\Delta\theta^*(z)$$

$$= \frac{(z-1)^2\left(z^2 - z + dz + d\right)}{z^4 - (3-p-i-d)z^3 + (3-d+i)z^2 - (1+p+d)z + d}\frac{z^2 R^*}{(z-1)^2} \tag{7.4}$$

$$= \frac{z^2 - z + dz + d}{f_{PID}(z)} z^2 R^*$$

$$\Delta\theta(\infty) = \lim_{z\to 1}\left(\frac{z-1}{z}\Delta\theta(z)\right) = \lim_{z\to 1}\left(\frac{z-1}{z}\frac{2d}{2i}R^*\right) = 0\,. \tag{7.5}$$

7.2 Computer simulation of the ramp-tracking PID controller

The position controlled system with the PID controller is simulated in the ramp tracking mode. The driving torque and the output position traces are obtained for the case with the proportional action in the feedback path, as well as with the proportional action in the direct path. A simplified Simulink model of the linear discrete-time PID controller is given in Fig. 7.2. The model is based on the previous one, given in Fig. 6.5. The system parameters K_{FB} and K_M are set to one. At the same time, it is assumed that the ramp tracking task does not involve the system limits ω_{MAX} and T_{MAX}. Within the model, a dedicated switch can be used to replace the proportional control action from the direct path into the feedback path. In the simulations to follow, the feedback gains are set in accordance with Eq. 6.19.

In Fig. 7.3, the simulation traces of the driving torque and the output position are given for the case when the reference trajectory has a relatively small slope $d\theta^*/dt$. The configuration switch (Fig. 7.2) is placed in a position where the proportional control action is placed in the feedback path. Within the interval denoted by C, the speed is constant, and the output position lags behind the reference with a constant tracking error $\Delta\theta$. In Fig. 7.4, similar traces are given with the slope of the ramp profile $\theta^*(t)$ doubled with respect to the previous case. In accordance with Eq. 7.2, the tracking error observed in this figure is doubled as well.

The simulation run presented in Fig. 7.5 is obtained with the proportional gain replaced into the direct loop. The switch *Relocate proportional action* (Fig. 7.2) is placed in the upper position. Initially, the output position stays behind the reference. In approximately 15–20 T, the tracking error decays and reaches zero. Hence, during the constant speed interval, the system tracks the reference ramp without an error.

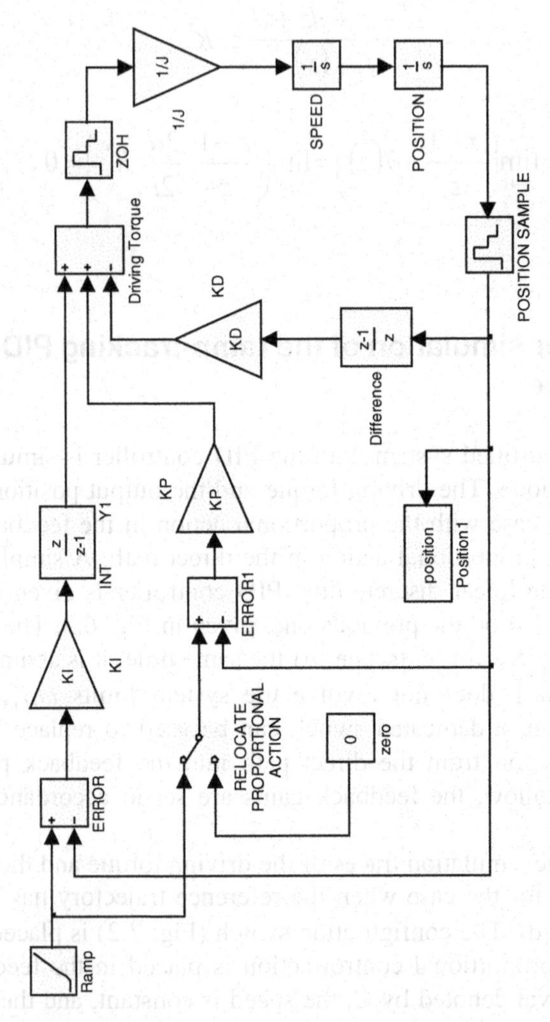

Fig. 7.2. Simplified Simulink model of linear discrete-time PID position controller. Parameters K_{FB} and K_M are assumed to be equal to one. The switch entitled *Relocate proportional action* can be operated to replace the proportional control action from the direct path into the feedback path.

When the position reference stops increasing and settles at the target value (interval D), the output position exhibits an overshoot. The overshoot is present regardless of the strictly aperiodic character of the transient response. As already discussed in the previous chapters, the overshoot originates from the closed-loop zeros.

The closed-loop system transfer function $W^{+}_{SS}(z)$, obtained with the proportional gain in the direct path, has one additional zero, $z_{+} = p/(p+i)$, giving rise to overshoots. In Sections 2.2.4, 4.6, and 5.3.2, it has been proved that the relocation of the control actions into the feedback path removes the closed-loop zeros, suppressing, in this way, the overshoot in the step response. When the proportional gain is kept in the direct path (Fig. 7.1), the output position in Fig. 7.5 goes beyond the setpoint. Within the interval D in the same figure, the driving torque becomes negative and reduces the speed. When the position error becomes zero, the speed retains a small positive value and the output position overshoots the target. The application of a negative torque lasts until instant D in Fig. 7.5, resulting in a brief interval with a negative speed, wherein the output position is pulled back toward the target. Eventually, the torque becomes positive and suppresses the negative speed. At the same time, the output error $\Delta\theta$ decays exponentially.

Although the transient response retains a strictly aperiodic nature, it is not acceptable in a number of cases, since overshooting the target position may cause collision and damage the tool or the work piece. In cases when the input changes in a stepwise manner, the overshoot in the output position is even more emphasized. In Fig. 7.6, the transient response of the same system (Fig. 7.1) is given for a small step change in the reference position. Although the gains are set to provide real closed loop poles and a strictly aperiodic response, the output overshoots the target by 30%.

The previous analysis indicates that the implementation of the proportional control action has a decisive impact on the tracking error. With the action in the feedback path (Fig. 6.2), the output position reaches the setpoint without an overshoot, but it tracks ramp profiles with a finite error (Eq. 7.2). Relocation of the proportional action into the direct path (Fig. 7.1), suppresses the tracking error (Eq. 7.5), but the output position overshoots the target (Fig. 7.4, Fig. 7.5). In most position-controlled systems, overshooting the target position is not acceptable. Therefore, it is important to discuss the options that may lead to error-free tracking of the ramp profiles, yet providing an output position response having no overshoots.

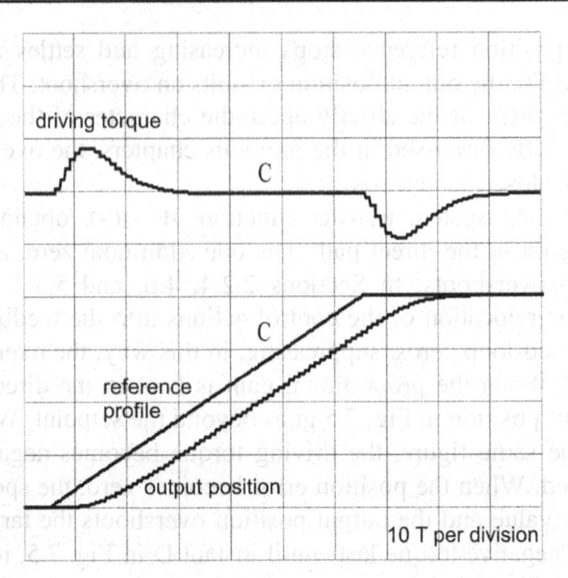

Fig. 7.3. Simulation traces of the driving torque, the output position, and the reference profile obtained from the model in Fig. 7.2. The proportional action is located in the feedback path. The input disturbance θ^* is a ramp function. During the constant speed interval (labeled C), the tracking error $\Delta\theta$ is constant.

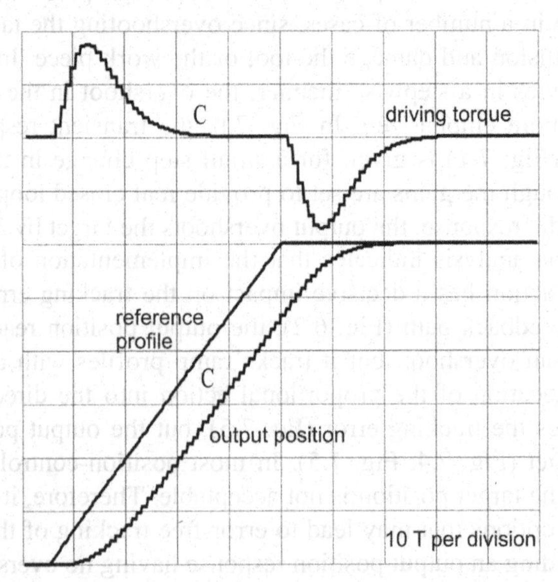

Fig. 7.4. Simulation traces of the driving torque, the output position, and the reference profile obtained from the model in Fig. 7.2. The slope of the ramp profile $\theta^*(t)$ is doubled with respect to Fig. 7.3. Other settings retain the same value as in the previous case.

If we renounce the optimized setting of the feedback gains (Eq. 6.19), the step response of the system with K_P gain in the direct path can be made slower, suppressing and eventually eliminating the overshoot. On the other hand, this decision will result in a significant decrease in the closed loop bandwidth, reducing as well the system capability to suppress the impact of load torque fluctuations on the output.

The tracking error can be suppressed by means of the feedforward control action. The analysis in Section 2.2.8 proves that the errors in the system output, obtained with a feedback controller, can be further reduced by devising feedforward compensation signals and adding these to the reference signals. In cases where the control object parameters are known and where the input reference profile is available prior to execution time, the changes in reference signals can be anticipated and applied to reduce or eliminate residual errors in the system output. The additional signals are referred to as the *feedforward compensation signals.*

Consider the PID controller structure with K_P gain in the feedback path (Fig. 6.2). The tracking error (Eq. 7.2) can be suppressed by introducing an additional signal, proportional to the slope R^* of the ramp profile. In the block diagram in Fig. 6.1, in the steady-state tracking of the reference profile $\theta^*_{(n)} = nR^*$, the summation point with Δy_1 at the output receives $K_{FB}K_I\Delta\theta$ and $y_3 = K_{FB}K_P(\theta_{(n)} - \theta_{(n-1)})$ as inputs. In the steady state, Δy_1=0, and the signal $y_3 = K_{FB}K_PR^*$ contributes to the tracking error $\Delta\theta = K_PR^*/K_I$. In order to suppress the error, the feedforward compensation signal $y_{FB} = K_{FB}K_PR^*$ has to be added to the summation point, resulting in $K_{FB}K_I\Delta\theta = 0$. In cases where the reference profile has a form other than the ramp, the feedforward compensation can be calculated as $y_{FB} = K_{FB}K_P(\theta^*_{(n)} - \theta^*_{(n-1)})$. In other words, the feedforward compensation must be proportional to the first derivative of the reference position.

Note at this point that this sample application of feedforward control suppresses the errors caused by the first derivative of the reference position and does not require knowledge of control object parameters, specifically, the equivalent inertia J. This outcome is due to the motion speed remaining constant during the ramp tracking, resulting in the absence of the acceleration component ($J\,\mathrm{d}\omega/\mathrm{d}t$) in the driving torque. In cases where the reference profile comprises acceleration intervals with $\mathrm{d}^2\theta^*/\mathrm{d}t^2 \neq 0$, the feedforward compensation requires that changes in the equivalent inertia J be known in advance.

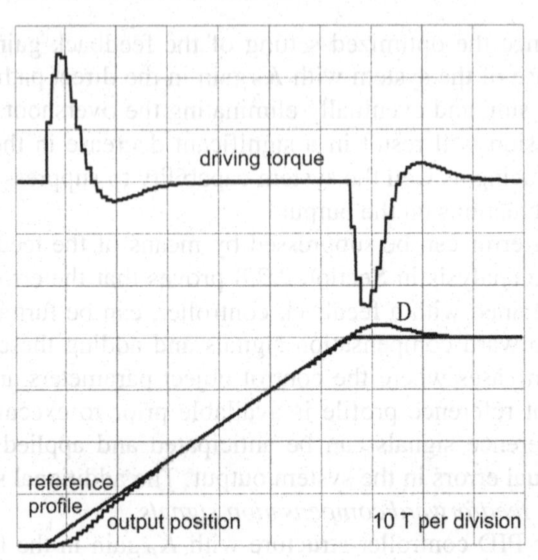

Fig. 7.5. Simulation traces of the driving torque, the output position, and the reference profile obtained from the model in Fig. 7.2. The proportional action is restored into the direct feedback path. The input disturbance θ^* is a ramp function. During the constant speed interval, the tracking error $\Delta\theta$ is equal to zero. When reaching the target (interval D), the output position exhibits an overshoot.

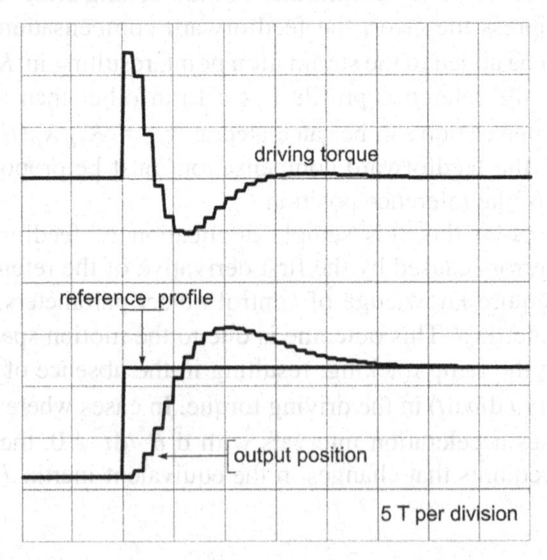

Fig. 7.6. Response to a small step in position reference obtained with the linear discrete-time position controller given in Fig. 7.1. Proportional and integral actions are placed in the direct path. The feedback gains are tuned to provide a strictly aperiodic response.

In cases where the feedforward compensation is not feasible, and where the overshoot in the output position (Fig. 7.5) is not acceptable, there is a possibility of adopting the PID structure with K_P gain in the direct path (Fig. 7.1) and modifying the reference profile in a way that suppresses the overshoot. In Fig. 7.7, the Simulink model given in Fig. 7.2 is used to produce the simulation traces of the driving torque and position, obtained with a modified ramp profile and with the proportional control action placed in the direct path (Fig. 7.1). In the final stage of the ramp profile, prior to reaching the target position, the slope $d\theta^*/dt$ is gradually reduced. The overshoot in the output position is not eliminated completely; however, it is an order of magnitude smaller than the overshoot experienced with the same system at instant D in Fig. 7.5. It is concluded that the overshoot in the system output can be significantly reduced by the proper shaping of the reference profile.

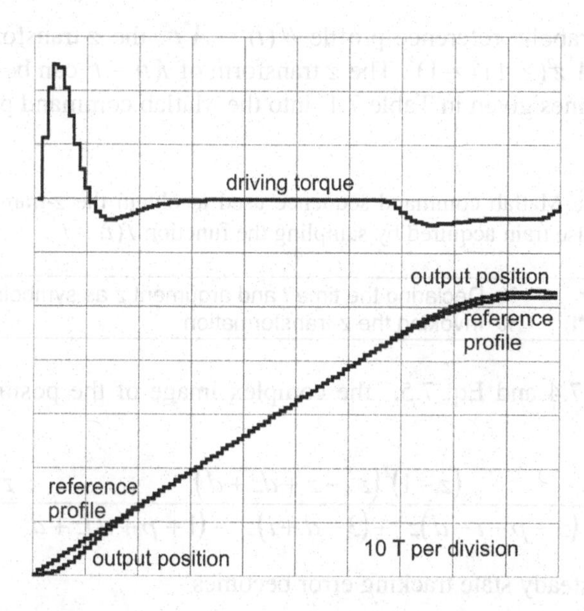

Fig. 7.7. Reduction of the overshoot in the output position obtained with the PID controller with K_P gain in the direct path. In its final stage, the ramp profile is modified. Compared with Fig. 7.5, the slope $d\theta^*/dt$ reduces as the reference approaches its steady state value. On the right in the figure, the output position overshoots the target by a small amount. The simulation traces are obtained from the Simulink model in Fig. 7.2.

7.3 Generation of reference profiles

In the previous section, the discussion focused on the capability of the PID controller to track changes in the reference position. The PID position controller in Fig. 7.1 has the proportional and integral actions in the direct path, and it is capable of tracking the ramp-shaped position profile without an error in the steady state. In practical applications, the position reference rarely assumes the form of a Heaviside step or a ramp. In most cases, the function $\theta^*(t)$ is composed from segments that are sinusoidal, parabolic, or expressed as a polynomial function of time. Hence, in most cases, the second derivative of the position reference is not equal to zero. This causes the PID controller to operate with a tracking error. If we consider the PID controller in Fig. 7.1, it is noteworthy to show that the steady-state error in the output position is proportional to the second derivative $d^2\theta^*/dt^2$ of the reference.

With a parabolic reference profile $\theta^*(t) = A^*t^2$, the z-transform $\theta^*(z)$ is found to be $A^*z^2(z+1)/(z-1)^3$. The z-transform of $f(t) = t^2$ can be derived by entering the lines given in Table 7.1 into the Matlab command prompt.

Table 7.1. The Matlab command sequence used to obtain the z-transform of the pulse train acquired by sampling the function $f(t) = t^2$.

```
>> syms t z        % Declaring the time t and argument z as symbolic constants
>> ztrans(t*t)     % Invoking the z-transformation
```

From Eq. 7.4 and Eq. 7.5, the complex image of the position error is found as

$$\Delta\theta(z) = \frac{(z-1)^2(z^2-z+dz+d)}{z^4-(3-p-i-d)z^3+(3-d+i)z^2-(1+p+d)z+d}\,\frac{z^2(z+1)A^*}{(z-1)^3}$$

while the steady state tracking error becomes

$$\Delta\theta(\infty) = \lim_{z\to 1}\left(\frac{z-1}{z}\Delta\theta(z)\right) = \frac{d}{i}A^*. \tag{7.6}$$

Hence, even with the proportional action restored in the direct path, as in Fig. 7.1, the PID controller cannot track a parabolic reference without an error. If we consider profiles with $d^3\theta^*/dt^3 > 0$ and apply the results given in Eq. 7.4 and Eq. 7.5, it is possible to prove that the reference tracking is unstable, with a continuous increase in the tracking error $\Delta\theta$. With $\theta^*(t) = B^* t^3$, the z-transform $\theta^*(z)$ has the factor $(z-1)^4$ in the denominator, leading to $\Delta\theta(\infty) = \infty$. It must be noted that the references with the third derivative strictly positive ($d^3\theta^*/dt^3 > 0$) are not considered for use. Although the error $\Delta\theta(\infty) = \infty$ is not to be expected, the PID controller will exhibit an intermittent error $\Delta\theta$ in tracking the reference profiles with sinusoidal or polynomial segments. The error amplitude $|\Delta\theta|$ depends on the closed-loop gains and the magnitude of higher-order derivatives of the profile $\theta^*(t)$. Tracking errors may present a problem in both single-axis and multiaxis systems.

7.3.1 Coordinated motion in multiaxis systems

A motion-control system where the tool or the work piece has one degree of freedom is referred to as a *single axis system*. The motion consists of translation along one axis or revolution around a predefined rotation axis. A typical motion cycle consists of moving the object from the initial position $\theta(A)$ to the target $\theta(B)$. In most cases, the reference trajectory $\theta^*(t)$ is provided, with $\theta^*(t_{START}) = \theta(A)$ and $\theta^*(t_{END}) = \theta(B)$. The output position is supposed to track the reference, presumably without an error. The responses given in Fig. 7.4 illustrate the case of a single-axis positioner tracking the ramp profile with a finite tracking error. When the tracking of reference profiles involves position errors, (see the traces in Fig. 7.4) the error may be acceptable, provided that the output position does not overshoot the target. In a single-axis positioner, the error $\Delta\theta = \theta^* - \theta$ does not drive the moving part away from the predefined path. Instead, it results in the attainment of specific points on the trajectory, with a small delay.

In cases where the motion of multiple axes is coordinated, the tracking error in one of the axes may cause collision or other undesired outcome. An example of single-axis positioning is the drill-carrying spindle with automatic tool exchange. In order to approach the toolbox, deposit the tool, and seize a new one, the rotor has to assume the proper angular position. The motion of the rotor towards the desired position can tolerate the tracking error as long as the target position is reached accurately and in a timely fashion. Another example of where the tracking error is sustainable is the drives that feed the work pieces into the work area of the production machine (i.e., under the tools). Similarly, when the drives are removing the work pieces after completion of the operation and are moving them towards storage, a certain error in tracking the reference profiles is acceptable in a number of cases. The motion of work pieces can be rotation (revolving tables) or translation (conveyers). Performance is measured in terms of the speed and precision in assuming the target position, while accuracy in tracking the reference profiles is of lower importance.

In multiaxis drives, the moving part follows the reference trajectory in space, thus requiring synchronized motion of several servo motors. In the case where the moving object has to be positioned in the *x*-*y*-*z* coordinate system, the reference trajectory is given by $[x^*(t), y^*(t), z^*(t)]$. In cases when the three servo motors track their references with finite errors Δx, Δy, and Δz, the moving object location is defined by the coordinates $[x^* - \Delta x, y^* - \Delta y, z^* - \Delta z]$. Depending on the tracking error, the object drifts away from the predefined path $[x^*, y^*, z^*]$ to a smaller or larger extent. In some cases, the moving part may collide with an obstacle. Delays in tracking the profile may result in the part reaching undesired positions and colliding with other elements and supports. When the moving object is a tool, the tracking error impairs the accuracy and quality of generating the work-piece surface. Therefore, deviation from the reference trajectory has to be suppressed in order to avoid damage to the work-piece.

An example of a multiaxis system is the plotter. Attached to the computer, the plotter moves the pen in the *x*-*y* plane, providing $y = f(x)$ curves plotted on the paper. The position of the pen is controlled by two electric motors, controlling the motion in the *x* and *y* direction. The motion must be coordinated in order to provide the proper shapes $y = f(x)$. Consider a plotter drawing a circle: the two motors follow trajectories $x^*(t) = X_0 + R \sin(\omega t)$ and $y^*(t) = Y_0 + R \cos(\omega t)$, where R is the circle radius while X_0 and Y_0 are the coordinates of its center. The two degrees of motion have to be coordinated. In cases when the axis x tracks the reference $x^*(t)$ with a delay, the resulting figure will be distorted: instead of a circle, an ellipse. An illustration of such a case can be obtained by entering the sequence of commands in Table 7.2 at the Matlab command prompt.

Table 7.2. The Matlab command sequence used to evaluate the effects of delays in tracking the reference profile in two-axis *x*–*y* positioning systems.

```
>> t=[1:1000]; delay = 0.5;      % Consider a time span of 1:1000; initialize delay
>> y = cos(2*pi*t/1000);         % y-axis follows trajectory with no delay
>> x = sin(2*pi*t/1000-delay);   % x-axis tracks the reference with a certain delay
>> plot(x,y)                     % Plot the figure
```

Industrial manipulators and robots used in automotive, metal forming, wood, glass, plastics, and other production lines involve a number of motors, which are required to perform coordinated motion and track predefined trajectories with the smallest inevitable tracking error. In Fig. 7.8, the effects of the tracking error on work piece quality are illustrated on a two-axis system.

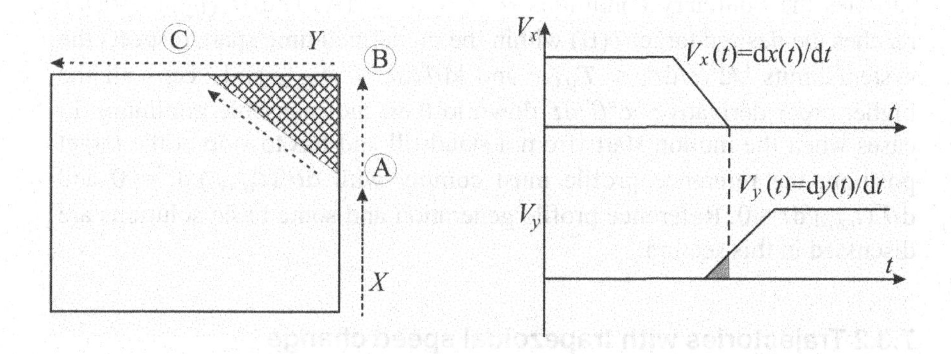

Fig. 7.8. The two-axis system is supposed to move the tool along the path A-B-C, resulting in a rectangular work pattern on the work piece. Due to delays and errors in tracking the reference profile, the tool sweeps over the shaded area and damages the piece. The speed profiles are given on the right.

It is assumed that the tool is moved in two-dimensional x-y space, powered by a pair of two servo motors controlling the position (x, y) and speed (dx/dt, dy/dt) in each axis. In Fig. 7.8, it is assumed that the objective is to move the tool along the path denoted by X in the figure, then stop at vertex B, and then proceed by following path Y. Provided that the position control is error free, the tool passes through points A, B, and C in the figure. Consider the case when the coordinate $x(t)$ tracks the reference $x^*(t)$ with an error, and observe the instant when the reference x^* arrives at point B. Then the actual positioning of the tool may be delayed and found in the position A. At this time, the motion commences in axis y. Instead of moving over the path A-B-C, the tool departs from A and moves towards point C, cutting through the shaded area in the Figure. Hence, the work piece is damaged, as it does not assume the desired rectangular form.

In order to prevent collision and damages, reference profiles are planned and coordinated in a way that reduces the tracking error. Smooth trajectories are obtained by controlling the amplitude of higher-order derivatives $d^n\theta^*/dt^n$. Regarding the first- and the second-order derivative, the constraints $|Jd^2\theta^*/dt^2| < T_{MAX}$ and $|d\theta^*/dt| < \omega_{MAX}$ have to be met, where T_{MAX} and ω_{MAX} represent the torque and speed limits of the system. In Eq. 7.6, it is shown that the tracking error obtained with a PID controller is proportional to the second derivative of the reference profile. Moreover, the presence of a strictly positive third derivative $d^3\theta^*/dt^3$ results in a permanent increase of the tracking error. For that reason, the task of generating the reference profile can be described as devising the function $\theta^*(t)$ that satisfies the boundary conditions $\theta^*(t_{START}) = \theta(A)$ and $\theta^*(t_{END}) = \theta(B)$, reaches the desired target $\theta(B)$ within the predefined time span, respects the system limits $|Jd^2\theta^*/dt^2| < T_{MAX}$ and $|d\theta^*/dt| < \omega_{MAX}$, and keeps all the higher order derivatives $d^n\theta^*/dt^n$ down to their indispensable minimum. In cases when the motion starts from a standstill and has to stop at the target position, the reference profile must comply with $d\theta^*(t_{START})/dt = 0$ and $d\theta^*(t_{END})/dt = 0$. Reference profile generation and some basic solutions are discussed in this section.

7.3.2 Trajectories with trapezoidal speed change

In the previous section, the ramp-shaped reference profiles were analyzed. Now consider the ramp $\theta^*(t)$ that starts from $\theta(A)$, increases with a constant slope, and stops in $\theta^*(t_{END}) = \theta(B)$. It is of interest to note that the constraint $|Jd^2\theta^*/dt^2| < T_{MAX}$ is not met: at instants t_{START} and t_{END}, the first derivative (the speed) changes instantly, while the second derivative (the torque) is infinite. Hence, whatever the controller structure and parameter setting, the ramp-shaped profiles cannot be tracked without a transient error in starting and stopping.

In Fig. 7.9, the reference profile $\theta^*(t)$ is obtained with a limited second-order derivative. Along with the reference trajectory, given at the top of the figure, the first derivative $d\theta^*/dt = \omega^*$ is shown in the middle. The second derivative $d^2\theta^*/dt^2 = d\omega^*/dt = T^*_{em}/J$ is given as the bottom trace, proportional to the driving torque required for the system to track the reference profile. In a position-controlled system with negligible friction and with $T_L = 0$, the application of the driving torque $T_{em} = Jd^2\theta^*/dt^2$ would result in $\theta(t) = \theta^*(t)$, provided that the initial conditions $\theta(0) = \theta^*(0)$ and $\omega(0) = \omega^*(0)$ are met. With the profile given in Fig. 7.9, the system accelerates with a constant acceleration (T_{em}/J), reaches the top speed, runs at a constant speed towards the target, decelerates at a rate $-T_{em}/J$, reaches the target, and stops. The required driving torque has a pulse form: namely, at $t = 0$, it changes instantly from $T_{em}(0^-) = 0$ to $T_{em}(0^+) = Jd^2\theta^*/dt^2$. A similar change in the driving torque takes place at the end of the acceleration interval, as well as on the edges of the deceleration pulse. At instants where the torque exhibits sudden jumps, the third-order derivative of the reference profile assumes an infinite value. According to the findings in Eq. 7.6, the third-order derivative of the reference profile increases the tracking error. Additionally, sudden changes in the driving torque may give rise to mechanical resonance phenomena, explained briefly in the following section.

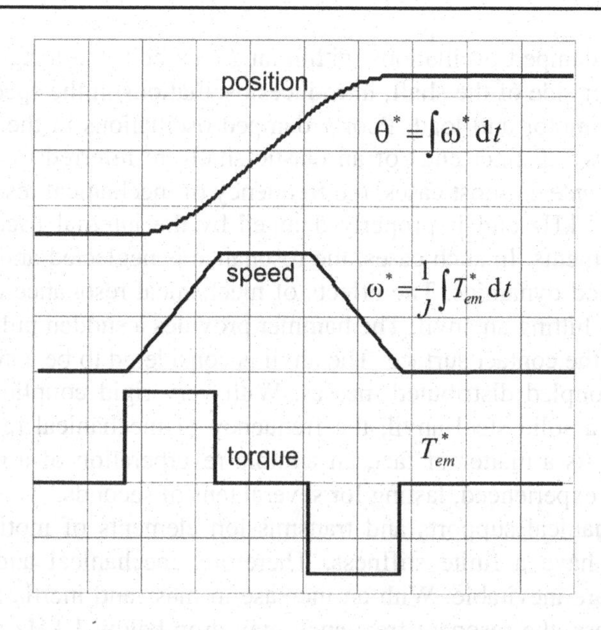

Fig. 7.9. The reference profile with the second-order derivative $Jd^2\theta^*/dt^2$, limited to T_{MAX}. At instants where the torque exhibits sudden jumps, the third-order derivative of the reference profile assumes an infinite value.

7.3.3 Abrupt torque changes and mechanical resonance problems

The mechanical subsystem consists of a number of elements such as the motor, supports, and tools, coupled by means of mechanical transmission devices. The motor-tool coupling is considered to be rigid: any change of the rotor shaft position is assumed to have an immediate effect on the tool (load) position, in proportion to the transmission ratio of the mechanical coupling. This assumption appears to be unquestionable in cases when the rotor and load are connected by means of a short, rigid shaft. Even in such a case, it must be noted that the rotor and the load masses are connected by a flexible coupling. Although the shaft stiffness can be very high, it is finite, and it reduces as the shaft becomes longer and thinner. The two masses, with their elastical coupling, make a poorly damped resonant subsystem, with the resonant frequency proportional to the shaft stiffness and inversely proportional to mass (inertia J). An electrical dual of the two-mass subsystem is the circuit with two capacitors connected by a series inductance, wherein C corresponds to J while the inductance L is inversely proportional to the stiffness. Since a Heaviside step of the supply voltage results in undamped oscillations within an LC circuit, the torque step, applied to either side of the shaft, introduces oscillations in the speed and position of the motor and load. Poorly damped oscillations in the speed and position at two distinct ends of an elastic shaft are referred to as the *torsional resonance.* In most cases, the frequency of mechanical resonance exceeds several kHz and is properly damped by the internal friction of the coupling elements. In such cases, the resonance is neglected and treated as the unmodeled dynamics. The effects of mechanical resonance can be observed when hitting an anvil. The hammer provides a sudden pulse of force exercised at the contact surface. The anvil is considered to be a system with elastically coupled distributed masses. With very rigid coupling between the ends of a solid steel anvil, the frequency of mechanical resonance is rather high. As a matter of fact, an audible reverberation of a rather high pitch can be experienced, lasting for several tens of seconds.

The mechanical supports and transmission elements of motion-control systems all have a finite stiffness. Therefore, mechanical and torsional resonances are inevitable. With an increase in mass and inertia and with a lower stiffness, the resonant frequency may drop below 1 kHz and impair the response of the system. The energy required to set the resonant modes into oscillation is obtained from the torque. The oscillation amplitude depends on the spectral contents of the torque signal at the resonant frequency. A pulse-shaped torque with abrupt changes is rich in high-frequency components. Since the reference profile in Fig. 7.9 requires instant changes in the driving torque, the servo system that is tracking such profiles provides a significant excitation to mechanical resonance modes. The spectral energy in the high-frequency range is reduced by smoothing the edges of the torque pulses. Hence, the resonant modes of an imperfect mechanical structure can be alleviated by devising position reference profiles that require smooth changes in the driving torque.

7.3.4 '*S*' curves

The mechanical structure, supports, and transmission elements of contemporary motion-control systems have a finite stiffness. The requirement to reduce the weight and cost of mechanics results in lighter and more elastic structures, thus emphasizing the mechanical resonance phenomena and torsional oscillations. The resonant frequency of existing machines lies between 100 Hz and 5 kHz. The resonant modes can be suppressed by avoiding sudden changes in the driving torque and reducing the slope dT_{em}/dt of the torque waveform. Given $|T_{em}| >> |T_L|$, the first derivative of the torque waveform is proportional to the third derivative of the output position. In cases when the position-controlled system tracks the reference trajectory, the torque slope $\mathrm{d}T_{em}/\mathrm{d}t$ is proportional to $\mathrm{d}^3\theta^*/\mathrm{d}t^3$. Hence, the design of the reference profile must ensure that the third derivative of the reference profile is limited.

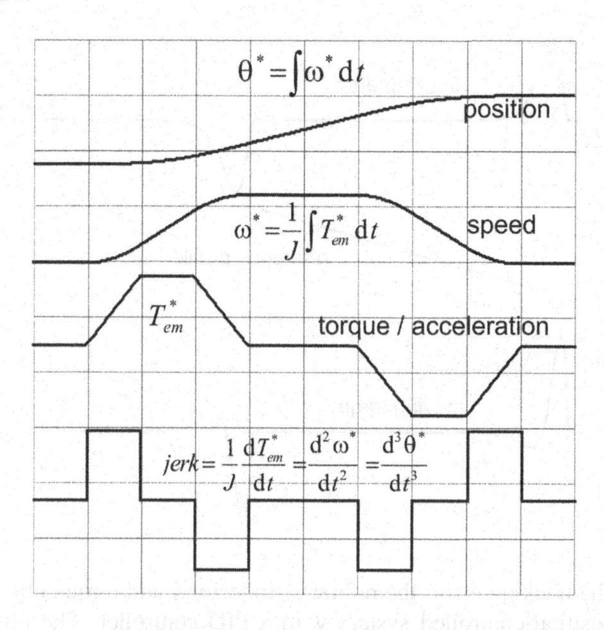

Fig. 7.10. The reference profile with limited third-order derivative. The first-order derivative of the torque (acceleration) is frequently referred to as the *jerk*. Due to the shape of the velocity profile, the waveform is known as the *S-curve*.

The profile with limited third derivative is given in Fig. 7.10. It is assumed that the system starts from a standstill, runs towards the target position, and arrives at the target position with zero speed. The torque waveform is proportional to the second derivative of the reference profile $\mathrm{d}^2\theta^*/\mathrm{d}t^2$ and represents the acceleration $a = \mathrm{d}\omega/\mathrm{d}t$. The speed waveform is obtained as the integral of the torque. Rather than having a trapezoidal waveform, as in Fig. 7.9, the speed exhibits a smooth profile resembling the letter *S*. For that reason, the waveform in Fig. 7.10 is known as the *S-curve*. Notice that the torque and acceleration have a limited first derivative. The waveform $\mathrm{d}a/\mathrm{d}t = \mathrm{d}^3\theta^*/\mathrm{d}t^3$ is known as the *jerk*, and it is given at the bottom of Fig. 7.10. In order to obtain a smooth torque waveform while tracking the reference trajectory $\theta^*(t)$, the jerk $\mathrm{d}^3\theta^*/\mathrm{d}t^3$ has to include one positive and one negative pulse in acceleration, followed by an inverted sequence of jerk pulses in deceleration. The use of *S* profiles reduces the excitation of mechanical resonance modes in motion control systems. *S* profiles are frequently used in elevators, reducing the subjective sensation of acceleration and stress exercised on passengers.

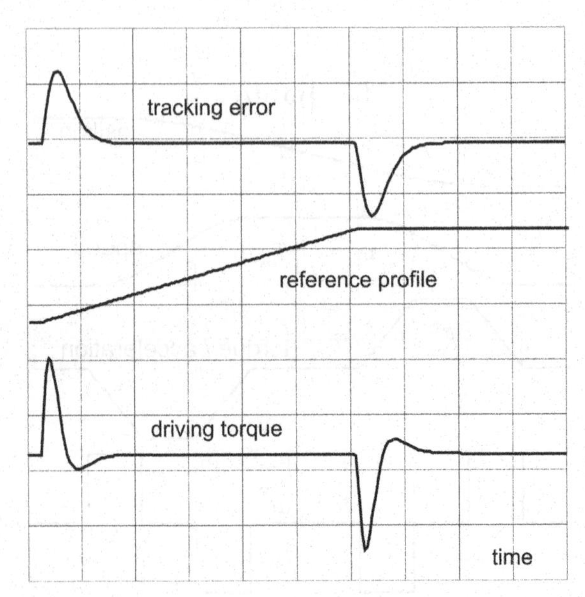

Fig. 7.11. The tracking error, the reference trajectory, and torque obtained from a position-controlled system with a PID controller. The position reference starts from the initial position and increases with a constant slope towards the target.

The smooth reference profile obtained with the *S*-profile (Fig. 7.10) has the potential to reduce the tracking error. In order to verify the impact of the reference profile on the tracking error, the Simulink model, given in Fig. 6.17, is run with the ramp-shaped and *S*-shaped reference profiles. In Fig. 7.11, the driving torque and tracking error are given for the position-controlled system tracking the ramp profile. Due to an abrupt change in $d\theta^*/dt$ at the startup and in the final stage, the output position falls away from the reference. At the same time, substantial torque pulses are obtained, with the potential of exciting the resonance modes within the mechanical structure of the motion-controlled system.

In Fig. 7.12, the system travels the same path, this time tracking the *S*-profile defined in Fig. 7.10. The required torque is proportional to the second derivative of the profile and has a limited slope. The third derivative $d^3\theta^*/dt^3$ (jerk) of the reference profile is limited. The peak tracking error is halved, with respect to the previous case. The peak torque is four times lower, having smooth edges and a significant reduction in dT_{em}/dt. The torque waveform obtained in Fig. 7.12 has reduced high-frequency content and reduces the excitation of the mechanical resonance modes. A smooth change in the driving torque has beneficial effects on the transmission elements. Besides *S*-profiles, trajectory generators use a range of functions with limited higher-order derivatives. In practice, trajectories are frequently composed from fragments of a sinusoidal function sin(kt). These shapes are known as *cycloids*. Other solutions include polynomial and other functional approximations of position references, in an attempt to control the amplitude of the higher-order derivatives $d^n\theta^*/dt^n$ up to the order of $n = 7$. In the next section, the reference trajectory composed from fragments obtained by polynomial approximation is discussed, analyzed, and simulated.

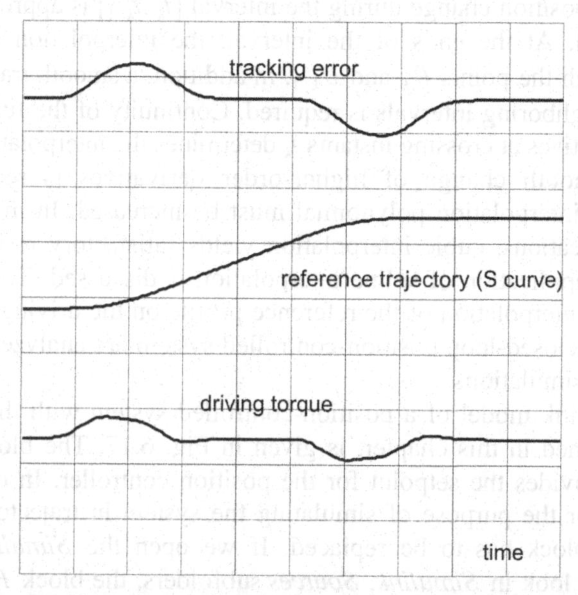

Fig. 7.12. The tracking error, reference trajectory, and torque obtained with the reference trajectory given in Fig. 7.10 (the *S-curve*).

7.4 Spline interpolation of coarse reference profiles

In a number of motion-control applications, the desired trajectory of the moving part or tool is defined by a limited number of points. In such cases, the position reference profiles have to be generated on the basis of the coordinates of such points. In a single-axis system, the reference profile $\theta^*(t)$ has to be generated on the basis of several position targets θ_1 .. θ_N to be reached at given time instants t_1 .. t_N. A simple and straightforward approach to reference profile generation consists of stepping the reference $\theta^*(t)$ from θ_k to θ_{k+1} at any given instant t_{k+1}. Where the number of data points is limited and the time span $t_{k+1} - t_k$ exceeds the sampling interval T by an order of magnitude, the resulting profile may be too coarse, involving sudden changes in the position reference and the consequential torque spikes. In such cases, it is interesting to interpolate the position reference. In practice, interpolation on the interval $[t_k..t_{k+1}]$ consists of devising additional data points describing a smooth transition of the position reference from the initial value of θ_k to the value θ_{k+1}, assumed at the end of the interval. A frequently-encountered approach is spline interpolation [18], wherein the position change during the interval $[t_k..t_{k+1}]$ is approximated by a polynomial. At the ends of the interval, the interpolation polynomial passes through the points θ_k and θ_{k+1}. In addition, a smooth transition between the neighboring intervals is required. Continuity of the function $\theta^*(t)$ and its derivatives at crossing instants t_k determines the interpolation quality. Where a smooth change of higher-order derivatives is required, the order of the interpolation polynomial must be increased. In most motion-control applications, cubic interpolation yields satisfactory results. In this section, the application of spline interpolation is discussed. The effects of cubic spline interpolation of the reference profile on the driving torque and position of a closed-loop position-controlled system are analyzed by means of computer simulations.

The Simulink model of a position-controlled system with the PID controller, designed in this chapter, is given in Fig. 6.17. The block *position reference* provides the setpoint for the position controller. In order to use the model for the purpose of simulating the system in trajectory tracking mode, this block has to be replaced. If we open the *Simulink Library Browser* and look in *Simulink*, *Sources* subfolders, the block *From Workspace* is found. It must be inserted into the model and connected as the position reference. At simulation time, the block supplies individual position values from a predefined reference trajectory stored in Matlab workspace. The sampling time of the block has to be set to $T_S = 1$ ms, which is the sampling period of the PID position controller comprised within the model.

A coarse reference profile comprising seven data points can be defined by entering the commands shown in Table 7.3 at the Matlab command prompt.

Table 7.3. The Matlab command sequence used to define a coarse position reference profile and prepare the array with *time–position* pairs.

```
>> stim = [0 0.05 0.1 0.15 0.2 0.25 0.3];           % Definition of time instants t1..t7
>> sdat = [ 0 .002 .006 .011 .014 .016 .017]; % Setpoints teta1 .. teta7
>> simin = [stim(:) sdat(:)];                        % Array with [time, data] points
```

The coarse reference profile, stored in the array *simin* has time stamps in its first column and position data points in the second column. The successive position references are spaced by five sampling periods (50 T_S = 50 ms). The simulation traces obtained with such a reference are given in Fig. 7.13A.

Due to a coarse reference trajectory, the output position changes in a stepwise manner, while the driving torque exhibits a series of spikes. Sudden changes in the driving torque give rise to mechanical resonance and stress the transmission elements. In order to obtain a smooth motion and suppress the peaks in the driving torque, the coarse reference position must be interpolated. Specifically, between each pair of successive position setpoints, spaced by $t_{k+1} - t_k = 50\ T_S = 50$ ms, another 50 setpoints are devised, determined in a way that describes a smooth transition of the reference position from the initial value of θ_k to the value θ_{k+1}, to be reached at the end of the interval.

A simple solution consists of incrementing the reference 50 times by $(\theta_{k+1} - \theta_k)/50$. This approach, referred to as *linear interpolation*, results in a linear change of the reference position for the interval $[t_k..t_{k+1}]$. At the edges of the interval, the slope of the reference trajectory exhibits an abrupt change, as the increments $(\theta_{k+2} - \theta_{k+1})/50$, applied in the neighboring interval, may be different. With linear interpolation, the first derivative of the reference profile exhibits sudden changes at the ends of the interpolation interval, while the second derivative assumes an infinite value. The driving torque T_{em}, required to track the reference profile under the no-load condition, is proportional to the second derivative $d^2\theta^*/dt^2$. Therefore, linear interpolation results in large torque spikes and cannot be tracked without a relatively large error in the output position.

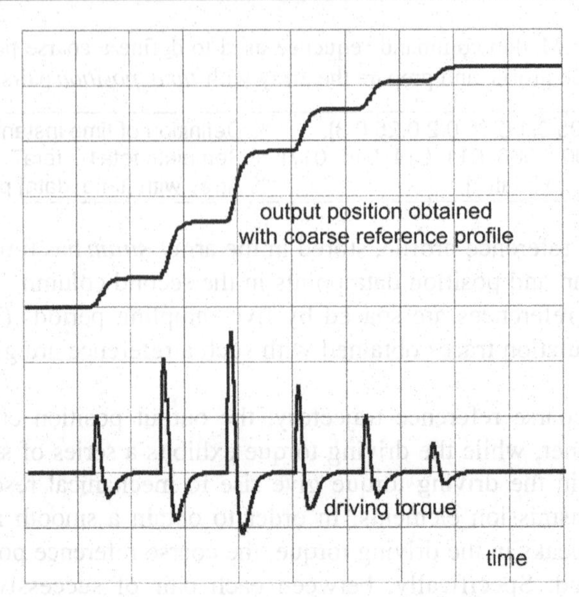

Fig. 7.13A. Simulation traces of the output position and driving torque obtained by running the model in Fig. 6.17 supplied with the coarse reference profile, with successive position setpoints spaced by 50 sampling periods ($50T$).

The change of the reference position from θ_k at the beginning of the interval to the value θ_{k+1}, achieved at the end, can be approximated by a polynomial. The polynomial coefficients can be adjusted in a way that results in a smooth transition between the intervals. Increasing the order of the interpolation polynomial results in the reference profile, with more higher-order derivatives $d^n\theta^*/dt^n$ having finite values at the interval boundaries. The Matlab function *spline* implements the cubic interpolation. The smooth reference profile obtained with the cubic polynomial interpolation is obtained by entering the commands shown in Table 7.4 at the Matlab command prompt.

Table 7.4. The Matlab command sequence used to perform *spline* interpolation of a coarse position reference profile and to plot the result.

```
>> stim1 = [1:300]*Ts                    % Defining Ts = 1 ms spaced time stamps
>> sdat1 = spline(stim,sdat,stim1)       % Performing cubic spline interpolation
>> stairs(stim,sdat,'b');                % Plotting the coarse trajectory
>> hold on;                              % Holding the figure
>> stairs(stim1,sdat1,'r')               % Plotting the interpolated trajectory
>> simin = [stim1(:) sdat1(:)]           % Preparing array for the Simulink model
```

In Fig. 7.13B, both the coarse and interpolated trajectories are given. The waveform contained in the vector *sdat*1 can be processed further to obtain the first, second, third, and fourth derivative. The derivatives can be obtained by using the Matlab function *diff.* It can be shown that the first derivative is a smooth function, the second derivative has a continuous change, the third derivative changes in a stepwise manner with a restrained amplitude, and the fourth derivative exhibits very large peaks. An interpolation polynomial of a higher order would result in a smooth change in the second-, third-, or fourth-order derivative. According to Eq. 7.6, the third-order derivative of the reference profiles increases the tracking error. Therefore, trajectories with a restrained third order derivative result in an acceptable tracking error in most of the cases. Therefore, the cubic spline interpolation of coarse reference profiles gives satisfactory results, rendering higher-order interpolation polynomials unnecessary. In Fig. 7.14, the simulation traces of the output position and the driving torque are given, obtained from the model of Fig. 6.17. The model is fed by the interpolated reference profile, obtained by the cubic spline interpolation. Compared with Fig. 7.13A, the output position is smooth, while the driving torque peaks are at least 20 times smaller.

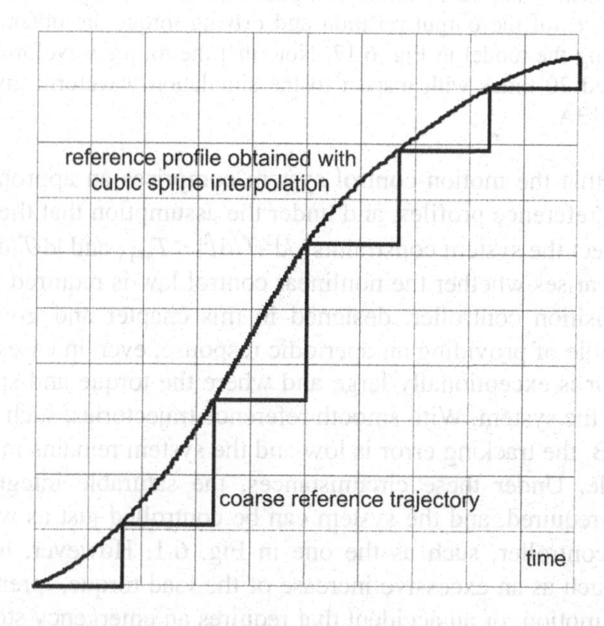

Fig. 7.13B. The coarse reference trajectory and the interpolated curve obtained by the cubic spline interpolation.

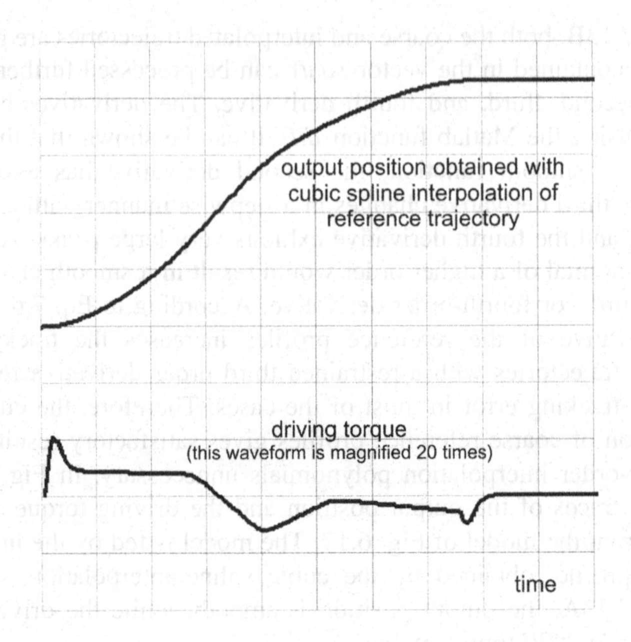

Fig. 7.14. Tracking the cubic spline interpolated reference trajectory. Simulation traces of the output position and driving torque are obtained by running the model in Fig. 6.17. Note that the torque waveform is magnified 20 times with respect to the simulation waveform given in Fig. 7.13A.

Provided that the motion-control system comprises an appropriate generator of the reference profiles, and under the assumption that the reference profiles respect the system constraints $|Jd^2\theta^*/dt^2| < T_{MAX}$ and $|d\theta^*/dt| < \omega_{MAX}$, the question arises whether the nonlinear control law is required at all. The nonlinear position controller, designed in this chapter and given in Fig. 6.16, is capable of providing an aperiodic response, even in cases when the tracking error is exceptionally large and where the torque and speed reach the limits of the system. With smooth reference trajectories, such as the one in Fig. 7.13B, the tracking error is low and the system remains in linear operating mode. Under these circumstances, the saturable integrator (Fig. 6.15) is not required, and the system can be controlled just as well from a linear PID controller, such as the one in Fig. 6.1. However, unexpected conditions such as an excessive increase of the load torque, premature termination of motion, or an accident that requires an emergency stop may result in tracking errors exceeding the nonlinear operation threshold $\Delta\theta_{(max)}$ in Eq. 6.27. In such cases, the use of a linear controller results in overshoot and instability, similar to those given in Fig. 6.10. Hence, the use of nonlinear elements (Fig. 6.15) within the PID position controller is a prerequisite for securing a stable, aperiodic response of the output position, even in exceptional operating conditions.

Problems

P7.1

Consider the linear PID position controller given in Fig. 6.2, with proportional and derivative actions replaced into the feedback path. Assume that the reference profile has a ramp shape, with the position reference samples $\theta^*_k = kR^*$. Calculate the steady-state position error $\Delta\theta(\infty)$. Calculate this error for the case when the integral and proportional gains are in the direct path, while the derivative gain resides in the feedback path.

P7.2

Consider the simplified Simulink model of a linear discrete-time PID position controller, given in Fig. 7.2, where the parameters K_{FB} and K_M are assumed to be equal to one. The reference profile has a ramp shape, incrementing at a constant slope. The proportional action can be switched from the direct path into the feedback path. Run the model for both locations of the proportional gain. Plot the output error and output position, verify the results from P7.1/S7.2, and note the overshoot in the output position. Note that Matlab files *P7_2cmd.m*, *P7_2F.mdl*, and *P7_2D.mdl* can be used to obtain the desired traces. In order to initialize the model parameters and plot the simulation traces, type *P7_2cmd* at the Matlab command prompt.

P7.3

In the previous problem, the gains of the PID position controller are set to provide a strictly aperiodical response. The model does not contain nonlinear elements and operates in linear mode. Notwithstanding linear operation and aperiodic settings, the output position overshoots the target when K_P is in the direct path. Provide the sample transfer function $W_{SS}(s)$ without conjugate complex poles or zeros, yet resulting in a step response comprising an overshoot.

P7.4

Use the previous Simulink model with the proportional gain in the direct path. Replace the position reference generator with a *repeating sequence* block and make an attempt to reduce the overshoot in the output position by smoothing the ramp profile at its starting and ending regions. Use Matlab files *P7_4cmd.m* and *P7_4.mdl.*

P7.5

Use the previous Simulink model and replace the proportional gain into the feedback path. Introduce the incremental implementation of the proportional control action, in accordance with Fig. 6.1. Add the feedforward compensation $y_{FB} = K_{FB}K_P(\theta^*_n - \theta^*_{n\text{-}1})$ to the summation junction providing the signal Δy_1 (refer to the discussion in Chapter 6). Run the model and investigate the impact of the feedforward gain on the tracking error. Use Matlab files *P7_5cmd.m* and *P7_5.mdl.* Hint: enter the string

```
>> ff = KP/2; p7_5cmd
```

in order to obtain the output position and the tracking error obtained with the feedforward gain of $K_P/2$.

P7.6

The position reference profile is given by seven time-position data pairs: $\theta(0)$ = 0 rad, $\theta(0.05)$ = 0.002 rad, $\theta(0.1)$ = 0.006 rad, $\theta(0.15)$ = 0.011 rad, $\theta(0.2)$ = 0.014 rad, $\theta(0.25)$ = 0.016 rad, and $\theta(0.3)$ = 0.017 rad. By using linear interpolation, calculate the reference trajectory with the time resolution of 1 ms. Use the Matlab command *interp1*(). Analyze the first and second derivative of the generated trajectory.

P7.7

Consider the position reference profile given in the previous problem. Use the Simulink model of the PID position controller, contained within the Matlab file *P7_7.mdl,* to obtain the output position and torque responses. Compare the traces obtained with the profile defined in seven points (*stim, sdat*) and the traces obtained with linear interpolation (*stim1, sdat1*). Hint: use the Matlab command file *P7_7cmd.m.*

P7.8

Repeat the simulation described in the previous problem with the reference profile obtained with cubic spline interpolation. Compare the output position and torque waveforms obtained with linear and spline interpolation. Hint: use the Matlab command file *P7_8cmd.mdl.*

P7.9

Consider the reference profile *sdat1* obtained by cubic spline interpolation in P7.8. Use the Matlab *diff*() command to probe the first, second, and third derivatives.

8 Torsional Oscillations and the Antiresonant Controller

This chapter explains the mechanical resonance and torsional oscillations within mechanical structures of the motion-control systems. Their impact on closed-loop performance is predicted and evaluated. The cases are distinguished where the lowest resonance frequency remains well beyond the desired bandwidth and where the resonant modes can be neglected as secondary phenomena. For applications where the resonance phenomena overlap with the frequency range of interest, passive and active antiresonant control actions are devised and evaluated. An insight is given into designing and using antiresonant controllers by means of simulation and experiments.

In Chapter 1, the mechanical part of a motion-controlled system has been modeled as a concentrated inertia J with friction coefficient B and with an external load torque disturbance T_L. When the servo motor is coupled to the load by means of a rigid shaft, the motor and load positions are the same, and the inertia coefficient corresponds to the sum of the load inertia J_L and the inertia of the rotor J_M ($J = J_L + J_M$). In cases when a stiff shaft connects several revolving objects, the equivalent inertia J_{EQ} is obtained as a sum, while the transfer function of the control object remains $W_P(s) = 1/(J_{EQ}s + B)$. The analysis and discussion in the preceding chapters assume a rigid connection between mechanical elements. Therefore, the control object is considered and modeled as a concatenated inertia, comprising the rotor, load, and the equivalent inertia of all the moving elements.

Mechanical structures, joints, and couplings within a motion-control system have a finite stiffness: that is, transmission elements such as shafts do not ensure equal positions at the shaft ends. Even a small flexibility results in certain torsion $\Delta\theta$ of the shaft, proportional to the applied torque. In cases when the shaft couples two revolving parts, each one with a distinct inertia J, the two inertias and the shaft constitute a resonant subsystem. For example, when a step in the driving torque is applied, the speed and position at both ends of the shaft exhibit poorly damped oscillations. Oscillating phenomena involving the speed, torque, and position of revolving objects are referred to as *torsional oscillations*. In cases where the translation of the tool is obtained from a linear motor, providing the driving force F [N] and performing a linear motion along a predefined direction, an elastic coupling between the motor and tool results in a resonant system where the linear speed and displacement of both objects oscillate. Such a system resembles two masses connected by a spring. The oscillating phenomena are referred to as the *mechanical resonance.* The resonance caused by a finite stiffness is also encountered in cases where a transmission element converts the rotation of a conventional servo motor into linear motion.

The oscillation frequency depends on the stiffness in transmission and decreases with an increase in inertia on both ends of the transmission. In many cases, the frequency of the mechanical resonance exceeds several kHz, and it decays quickly. In such cases, the effects of elastic coupling can be neglected. The mechanical subsystem can be treated as a single, concatenated equivalent inertia with $W_P = 1/Js$. Where the resonant frequency is lower and the oscillation damping is insufficient, the presence of a resonant mode impairs the step response of a servo system and may result in sustained oscillations or instability.

In this chapter, the effects of resonance modes on the closed loop performance of motion-controlled systems are considered. The transfer function of a control object comprising resonance modes is derived and its impact on the closed-loop poles is discussed. Furthermore, the principle of operation and the design of series antiresonant compensators is given. Insight into improvements of the closed-loop dynamic performance, obtained with IIR and FIR [22] *notch* filters, is obtained from computer simulations. The torsional resonance problem and remedies are demonstrated by the experimental traces obtained on a contemporary speed-controlled servo drive.

8.1 Control object with mechanical resonance

In this section, the transfer function $W_P(s)$ is derived for a mechanical subsystem with torsional resonance. The system under consideration is given in Fig. 8.1. The rotor of a servo motor is coupled to the inertial load by means of a flexible shaft. The shaft has a finite stiffness K_K. When the shaft transmits the torque T_S from the servo motor to the load, the shaft position at the motor end will differ from the position of the load. The shaft torsion $\Delta\theta_S$ is proportional to the torque transmitted and inversely proportional to the stiffness coefficient ($\Delta\theta_S = T_S/K_K$).

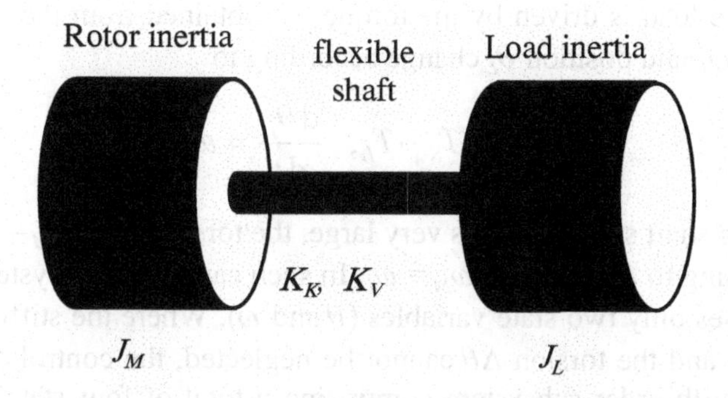

Fig. 8.1. The rotor inertia J_M and load inertia J_L are coupled by a flexible shaft, described by a finite stiffness K_K and an internal viscous friction, characterized by the coefficient K_V.

In cases where the subsystem in Fig. 8.1 is part of a speed-controlled system, the shaft sensor is attached either to the motor or to the load, acquiring the speed signal and closing the feedback loop. With the feedback device on the motor end of the shaft, the motor speed is the output of the system. Cases where the speed-sensing device is attached to the load are also encountered. Motion-control applications where the torsional resonance is pronounced and the closed-loop performance is critical may have shaft sensors on both ends of the flexible coupling.

The mechanical subsystem in Fig. 8.1 is driven by the torque T_{em}, supplied by the servo motor. The load torque T_L acts at the other end of the shaft, while the motor and load speed and position are the outputs. When the shaft ends are displaced by the torsion angle $\Delta\theta = \theta_M - \theta_L$, the torque T_S, transmitted by the shaft, is obtained as

$$T_S = K_K \Delta\theta + K_V \frac{d(\Delta\theta)}{dt} = K_K \Delta\theta + K_V \Delta\omega . \quad (8.1)$$

With reference to the rotor of the servo motor, the driving torque T_{em}, produced through the electromagnetic interaction between the stator and rotor, accelerates the inertia J_M, while the shaft torque T_S acts in the opposite direction. If we neglect the friction and mechanical power losses within the motor, the change in the rotor speed ω_M and position θ_M are described by

$$J_M \frac{d\omega_M}{dt} = T_{em} - T_S, \quad \frac{d\theta_M}{dt} = \omega_M . \quad (8.2)$$

The load is modeled as a concentrated inertia J_L with a negligible friction coefficient B and with all the motion resistances represented as an external load T_L. The load is driven by the torque T_S, obtained from the shaft. The load speed ω_L and position θ_L change according to

$$J_L \frac{d\omega_L}{dt} = T_S - T_L, \quad \frac{d\theta_L}{dt} = \omega_L. \tag{8.3}$$

When the shaft stiffness K_K is very large, the torsion $\Delta\theta = \theta_M - \theta_L$ is negligible, leading to $\theta_M = \theta_L$ and $\omega_M = \omega_L$. In such cases, the subsystem in Fig. 8.1 comprises only two state variables (θ and ω). Where the stiffness has a finite value and the torsion $\Delta\theta$ cannot be neglected, the control object becomes a fourth-order subsystem comprising a total of four state variables (θ_M, θ_L, ω_M, ω_L). Equations 8.2 and 8.3 are presented in Fig. 8.2 in the form of a block diagram. The two integrators in the upper part of the diagram output the motor side variables, while the two bottom integrators provide the load-side speed and position. The shaft torque T_S is obtained from Eq. 8.1. The block diagram considers an elastic shaft coupling, but it can be advantageously applied in all the cases where a flexible mechanical coupling transmits the torque (or force) from a servo motor to the load.

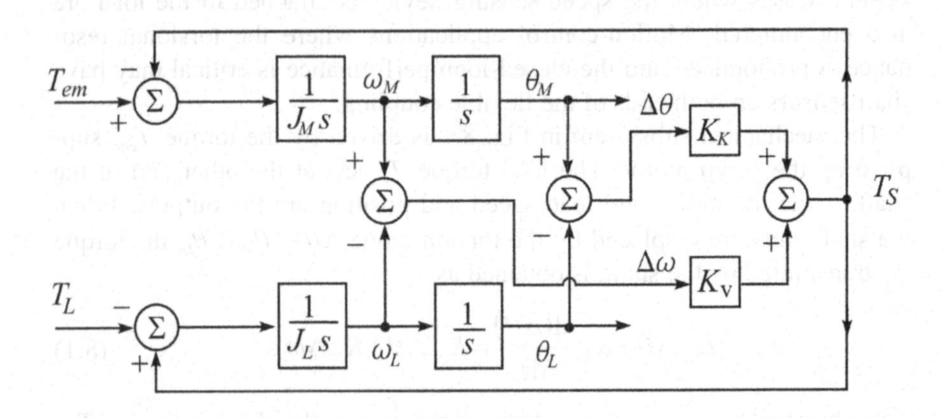

Fig. 8.2. A mechanical subsystem with torsional resonance mode comprises four state variables (θ_M, θ_L, ω_M, ω_L). The input to the system is the driving torque T_{em}, while the outputs are the rotor and load speed and position.

Given that $K_K \to \infty$, the transfer function $W_P(s) = \omega(s)/T_{em}(s)$ is obtained as $1/Js^2$. With a finite stiffness, the transfer functions $W_{P1}(s) = \omega_R(s)/T_{em}(s)$ and $W_{P2}(s) = \omega_L(s)/T_{em}(s)$ are more involved. The transfer function W_{P1} is of interest in cases when the shaft sensor is attached to the motor side. From Fig. 8.2, it is obtained as

$$W_{P1}(s)=\frac{1}{(J_M+J_L)s}\frac{1+\frac{K_V}{K_K}s+\frac{J_L}{K_K}s^2}{1+\frac{K_V}{K_K}s+\frac{J_LJ_M}{K_K(J_M+J_L)}s^2}=\frac{1}{J_{EQ}s}\frac{1+\frac{2\zeta_z}{\omega_z}s+\frac{1}{\omega_z^2}s^2}{1+\frac{2\zeta_p}{\omega_p}s+\frac{1}{\omega_p^2}s^2}. \tag{8.4}$$

In Eq. 8.4, ω_p represents the natural frequency of the conjugate complex pair of poles, while the coefficient ζ_p corresponds to the damping factor. The damping is proportional to the viscous friction K_V. The transfer function has a pair of conjugate complex zeros, with their natural frequency ω_z and the damping factor of ζ_z. From Eq. 8.4, the relevant frequencies and coefficients are expressed in terms of K_K and K_V:

$$\omega_p=\sqrt{\frac{K_K(J_M+J_L)}{J_MJ_L}},\qquad \omega_z=\sqrt{\frac{K_K}{J_L}},$$
$$\zeta_p=\sqrt{\frac{K_V^2(J_M+J_L)}{4K_KJ_MJ_L}},\qquad \zeta_z=\sqrt{\frac{K_V^2}{4K_KJ_L}}. \tag{8.5}$$

Note in Eq. 8.4 that with $K_K\to\infty$, $W_{P1}(s)$ becomes $1/J_{EQ}s$. With the shaft sensor attached to the load side, the feedback signal and the output of the system are the load speed ω_L. The corresponding transfer function $W_{P2}(s)=\omega_L(s)/T_{em}(s)$ is given in Eq. 8.6. It has one real zero, a pair of conjugate complex poles, attributed to the torsional resonance, and one pole at the origin, deriving from the factor $1/J_{EQ}s$ of $W_{P2}(s)$.

The resonant frequency (Eq. 8.5) is directly proportional to the shaft stiffness and inversely proportional to the equivalent inertia. Larger objects with flexible coupling exhibit torsional resonance phenomena at lower frequencies. The only damping to oscillations is obtained from a relatively low viscous friction K_V. Where the friction B (neglected in Fig. 8.2 and in Eq. 8.5) has meaningful values, it contributes to damping and suppresses torsional resonance. In a number of cases, the motion resistances $B\omega$ and the shaft viscous friction K_V are rather small, resulting in a very low damping. In such cases, the torsional resonance results in sustained oscillations. Nested within the closed-loop speed or position control, poorly damped resonant dynamics in Eq. 8.4 may result in a lack of stability and limit the range of applicable gains.

$$W_{P2}(s)=\frac{1}{(J_M+J_L)s}\frac{1+\dfrac{K_V}{K_K}s}{1+\dfrac{K_V}{K_K}s+\dfrac{J_L J_M}{K_K(J_M+J_L)}s^2}$$
$$=\frac{1}{J_{EQ}s}\frac{1+\dfrac{2\zeta_z}{\omega_z}s}{1+\dfrac{2\zeta_p}{\omega_p}s+\dfrac{1}{\omega_p^2}s^2} \tag{8.6}$$

8.2 Closed-loop response of the system with torsional resonance

Torsional resonance introduces a pair of weakly-damped conjugate complex poles into the transfer function of the control object. With the feedback device connected to the motor, the pair of poles is accompanied by a pair of conjugate complex zeros (Eq. 8.4). When the feedback sensor is attached to the load, the torsional resonance mode adds one real zero (Eq. 8.6) to the pair of conjugate complex poles. In both cases, the damping of the poles may be insufficient, due to a limited viscous friction K_V.

In a majority of motion-control applications, the stiffness K_K of mechanical transmission is high, while the equivalent inertia of the system is low. Therefore, the resonance frequency $\omega_p \sim K_K/J$ exceeds the desired closed-loop bandwidth and resides well-above the frequency range of interest. With $\omega_p > 1$ kHz, the torsional resonance phenomena are classified as the unmodeled dynamics (Chapter 4, [1]). In such cases, analysis of the system dynamics and design of the closed-loop speed and position control may proceed as described in the preceding chapters, under the assumption that the inertial elements within the control objects are concatenated with all mechanical couplings and transmission elements being rigid ($K_K \to \infty$). In applications such as rolling mills, comprising long, elastic shafts and a mechanical subsystem of a considerable mass, the ratio K_K/J is low, turning the control object into a low-frequency mechanical resonator with insufficient damping. In all the cases with a large inertia and flexible coupling, the torsional resonance phenomena affect the closed-loop performance and have to be taken into account in the controller design. In this section, the impact of the torsional resonance on the closed-loop performance of a speed controller is investigated by means of computer simulation. Considerations are based on the digital PI speed controller, designed in Chapter 4, and the Simulink model given in Fig. 4.18.

The Simulink model of a speed controlled system with torsional resonance is given in Fig. 8.3. The model comprises the PI speed controller with the proportional action placed in the direct path. For simplicity, the nonlinear elements such as the torque limiter are omitted. The mechanical subsystem with torsional resonance is represented by the subsystem on the extreme right of the figure. This subsystem is given in Fig. 8.4. It comprises the rotor inertia J_M with the state variables θ_M and ω_M, and the load inertia J_L with the state variables θ_L and ω_L. The shaft is assumed to have a finite stiffness K_K and the internal friction K_V.

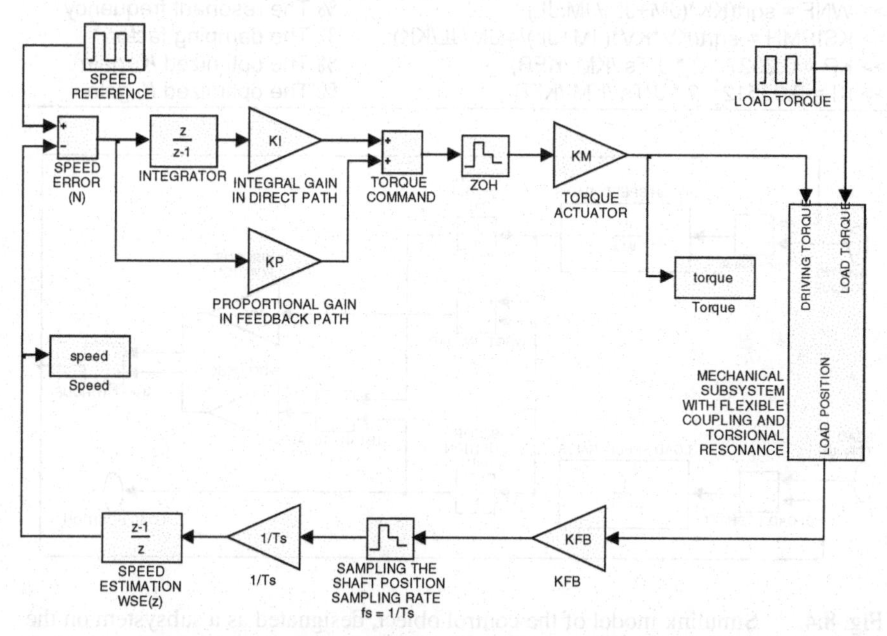

Fig. 8.3. Simulink model of a speed-controlled system with the PI controller and elastic coupling between the motor and load. The transfer function of the mechanical subsystem $1/Js^2$ in Fig. 4.18 is replaced by the resonant plant, given in Fig. 8.4. The PI controller has both control actions in the direct path.

The model is used for the purpose of investigating the impact of flexible coupling on the closed-loop performance. In order to set the model parameters, it is necessary to enter the sequence of commands given in Table 8.1 at the Matlab command prompt.

Table 8.1. The Matlab command sequence used to initialize the model parameters of the mechanical resonator given in Fig. 8.4.

```
>> Tsim = 0.000001;      % The simulation step is set to 1 μs (required for
>>                       % modeling the high frequency torsional phenomena)
>> Ts = 0.0003;          % The sampling time of the speed controller is 300 μs
>> JM= 0.0007;           % The motor inertia, [kgm^2]
>> JL= 0.0007;           % The load inertia
>> KK = 9300;            % The shaft stiffness, [Nm/rad]
>> KV = 0.15;            % Viscous (internal) friction of the shaft, [Nm/(rad/s)]
>> J = JM + JL;          % Equivalent inertia
>> KFB = 1;              % The gain of the position sensor
>> KM = 1;               % The torque amplifier gain
>>
>> WNF = sqrt(KK*(JM+JL)/JM/JL);               % The resonant frequency
>> KSIPMH = sqrt(KV*KV*(JM+JL)/4/JM/JL/KK);    % The damping factor
>> KP = 0.2027 * 2 * J/Ts /KM /KFB;            % The optimized K_P gain
>> KI = 0.03512 * 2 * J/Ts /KM /KFB;           % The optimized K_I gain
```

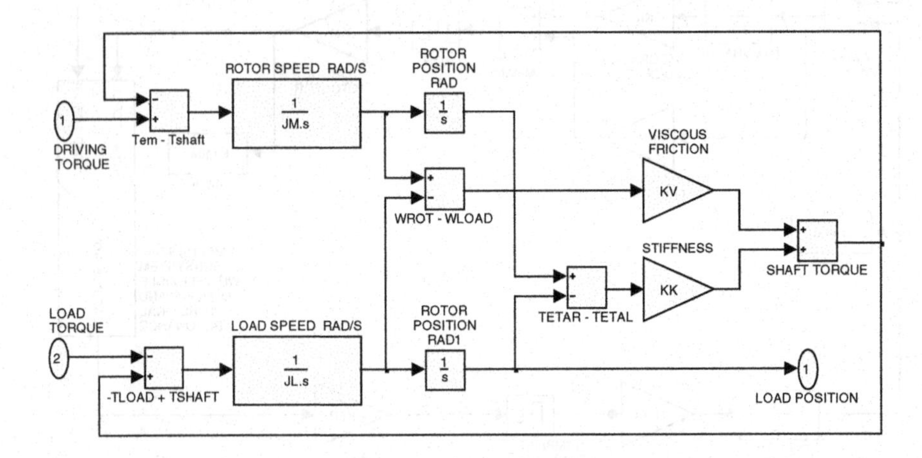

Fig. 8.4. Simulink model of the control object, designated as a subsystem on the right in Fig. 8.3. The motor and load are coupled by an imperfect shaft, characterized by the stiffness K_K and viscous friction K_V. The mechanical subsystem under consideration has a total of four state variables (θ_M, θ_L, ω_M, ω_L).

For the given speed-controlled system, it is interesting to determine the minimum frequency f_{TRmin} of the torsional resonance that does not interfere with the closed-loop dynamics of the PI speed controller. When the frequency of torsional oscillations exceeds f_{TRmin}, the torsional resonance phenomena can be neglected and treated as the unmodeled dynamics. In such cases, the controller design can proceed as prescribed in the preceding chapters. In this section, the analytical approach to finding f_{TRmin} is omitted. Instead, a series of computer simulations is performed with different torsional resonance frequencies. In Fig. 8.5, the driving torque response to the reference step and the load step is given for the resonant frequency f_{TR} ranging from 820 Hz to 1.6 kHz. The traces are obtained with K_K= 35000 Nm/rad, K_K= 15000 Nm/rad, K_K = 9500 Nm/rad, and K_K = 9300 Nm/rad, resulting in f_{TR} of 1600 Hz, 1000 Hz, 830 Hz, and 820 Hz, respectively. With f_{TR} = 1.6 kHz (the bottom trace in the figure), the torque response does not have any conspicuous effects of the torsional resonance and is similar to the traces obtained in Chapter 4. With f_{TR} = 1 kHz, the torque waveform has relatively small oscillations at the resonant frequency. The two upper traces represent the stability limit: with f_{TR} = 820 Hz, the step disturbance results in sustained oscillations. For the given parameter setting, any further decrease in the resonant frequency drives the system into instability. Note at this point that a stable operation with f_{TR} < 820 Hz can be achieved due to lowering the feedback gain K_P and K_I. In Fig. 8.6, the speed and torque responses to the reference and the load step are given, obtained with $K_P = K_P^{OPT}/2$ and $K_I = K_I^{OPT}/2$. With f_{TR} = 650 Hz, the response is stable. Damped torsional oscillations are seen in both the speed and torque traces. With f_{TR} = 650 Hz, the amplitude of torsional oscillations is increased and their damping is lower.

It is helpful to express the limit f_{TRmin} in terms of the closed-loop bandwidth f_{BW}. According to the analysis given in Section 4.8, and given that the speed-controlled system in Fig. 8.3 has the optimized setting of the feedback gains K_P and K_I , the bandwidth frequency is found as $f_{BW} = 1/21/T_S$ = 1/21/300 μs = 158 Hz. From this result, a rule of thumb is devised, stating that the elasticity in mechanical coupling and the associated torsional resonance phenomena can be neglected in cases where the resonant frequency exceeds the target bandwidth by a factor of f_{TR}/f_{BW} > 1000/158 ≈ 7.

In cases where $f_{TR} < 7f_{BW}$, the closed-loop gains have to be reduced in order to suppress the speed and torque oscillation and provide an acceptable damping. The gain reduction leads to a lower closed-loop bandwidth, reduced stiffness, and a sluggish response. Therefore, in motion-control systems where the mechanical resonance is pronounced, the control structure and the setting of the adjustable feedback parameters have to be changed in order to accommodate the antiresonant features. Practicable control solutions to the torsional resonance are reviewed in the following sections.

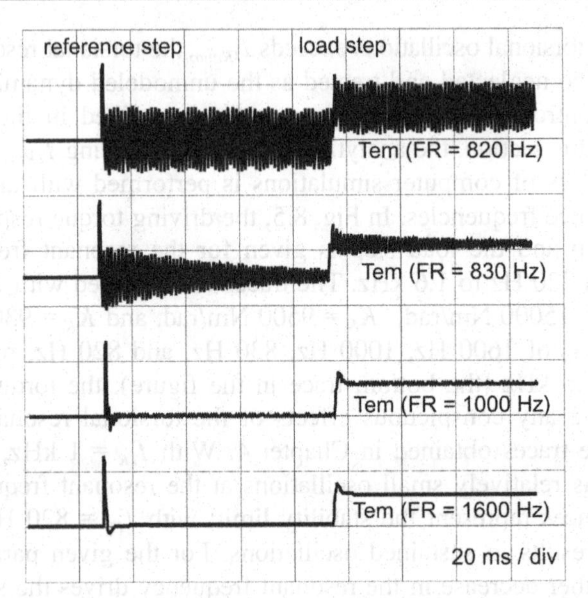

Fig. 8.5. Simulation traces of the driving torque obtained during the reference step (left) and the load step (right) transients. The resonant frequency is altered by varying the shaft stiffness K_K.

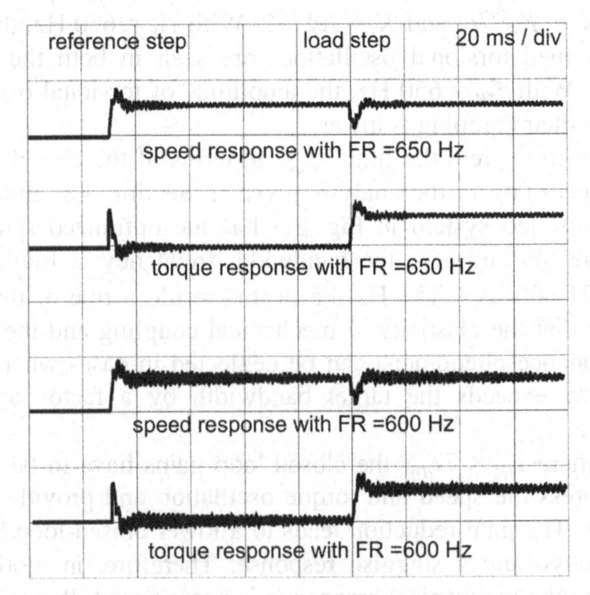

Fig. 8.6. The speed and torque response to the reference and the load step. The feedback gains K_P and K_I are halved with respect to their optimized setting. With reduced gains, the system provides an acceptable response for f_{TR} = 650 Hz. Further decrease in the resonant frequency emphasizes the torsional oscillations.

8.3 The ratio between the motor and load inertia

The amplitude and damping factor of torsional oscillations depend on the ratio between the motor inertia J_M and load inertia J_L. In the resonant subsystem in Fig. 8.1, with its transfer function given in Eq. 8.4, the natural frequency of weakly damped poles ω_p and zeros ω_z is expressed in terms of the inertia coefficients J_M and J_L:

$$\omega_p = \sqrt{\frac{K_K(J_M + J_L)}{J_M J_L}}, \quad \omega_z = \sqrt{\frac{K_K}{J_L}} .$$

The ratio $R_R = \omega_p/\omega_z$ is known as the resonance ratio:

$$R_R = \frac{\omega_p}{\omega_z} = \sqrt{1 + \frac{J_L}{J_M}} . \tag{8.7}$$

In cases where $J_L \ll J_M$, the resonance ratio R_R is close to one. It is worthwile to consider the impact of this resonant ratio on the transfer function $W_{P1}(s)$ (Eq. 8.4), obtained with the shaft sensor attached to the motor side. The natural frequency of resonant poles is very close to the frequency of zeros. The damping coefficient ζ_p of the conjugate complex poles is given in Eq. 8.5, along with the damping of zeros ζ_z. With $R_R \approx 1$, $(J_L + J_M)/J_L/J_M \approx 1/J_L$, leading to $\zeta_p \approx \zeta_z$. Hence, in cases where the resonance ratio is close to unity, the resonant poles and zeros tend to cancel out. The transfer function $W_{P1}(s) = \omega_M(s)/T_{em}(s)$ reduces to $1/J_{EQ}s \approx 1/J_M s$. The design and parameter setting of the closed-loop speed and position controllers can be done according to the procedures developed in the preceding chapters. With the control object transfer function $W_{P1}(s) \approx 1/J_M s$, the presence of a flexible mechanical coupling does not affect the transient response of the rotor speed ω_M, nor does it affect the driving torque T_{em}. With large feedback gains, the stiffness of the servo system is high, and the motor speed tracks the reference with a negligible error ($\omega_M \approx \omega^*$). Note at this point that the control objective is to regulate the load speed ω_L and position θ_L. Although the rotor speed tracks the reference with no error, the speed on the load side may differ due to a finite stiffness of the mechanical coupling.

In the system where the shaft sensor is placed on the rotor side and stiff control of the rotor speed is assumed with $\omega_M \approx \omega^*$ and $R_R \approx 1$, the control object in Fig. 8.1 can be envisaged as the subsystem having rotor speed $\omega_M \equiv \omega^*$ at the input and load speed at the output. The shaft torque is obtained as

$$T_S = K_K \Delta\theta + K_V \frac{\mathrm{d}(\Delta\theta)}{\mathrm{d}t} = K_V\left(\omega^* - \omega_L\right) + K_V \int\left(\omega^* - \omega_L\right)\mathrm{d}t . \quad (8.8)$$

From Eq. 8.3 and Eq. 8.8, the complex images $\omega_L(s)$ and $\omega^*(s)$ are related by:

$$J_L s^2 \omega_L(s) = K_V s\left(\omega^*(s) - \omega_L(s)\right) + K_K\left(\omega^*(s) - \omega_L(s)\right)$$

leading to the transfer function $W_{RR1}(s) = \omega_L(s)/\omega^*(s)$, obtained as

$$W_{RR1}(s) = \frac{1+\frac{K_V}{K_K}s}{1+\frac{K_V}{K_K}s+\frac{J_L}{K_K}s^2} = \frac{1+\frac{2\zeta_z}{\omega_z}s}{1+\frac{2\zeta_z}{\omega_z}s+\frac{1}{\omega_z^2}s^2} . \quad (8.9)$$

Hence, in cases with $J_L \ll J_M$, where the speed controller secures stiff control of the rotor speed ($\omega_M \approx \omega^*$), the load speed and position may exhibit torsional oscillations. According to Eq. 8.9, these oscillations are determined by the natural frequency and damping factor of the resonance zero, given in Eq. 8.5. In cases where the load inertia is small and the shaft has reasonable stiffness, the frequency ω_z of the load side oscillations may exceed several kilohertz. Properly damped, the oscillations have a negligible effect on the overall performance. On the other hand, systems with $J_L \ll J_M$ and with significant load-side resonance are controlled with great difficulty. The oscillations in the load-side speed and position can hardly be detected with the shaft sensor attached on the rotor side, measuring the rotor speed and position. Due to a large ratio J_M/J_L, any signal contained in the rotor speed, that is related to the load-side resonance, is filtered out by a large rotor inertia. As the attenuation suppresses the load-side signals down to the noise level, they cannot be efficiently used for feedback purposes. In such cases, it is necessary to install a shaft sensor on the load side as well.

In cases where $J_L \gg J_M$, the poles frequency is well beyond the frequency of zeros, and the resonance ratio is $R_R \gg 1$. With $\omega_p \gg \omega_z$, and given that high-frequency resonant phenomena in the mechanical subsystem are negligible, the transfer functions $W_{P1}(s)$ and $W_{P2}(s)$ in Eq. 8.4 and Eq. 8.6 retain their zeros, while their poles are of secondary importance. The transfer function $W_{P1}(s)$ is obtained with the shaft sensor installed on the rotor side, and it has a pair of conjugate complex zeros. The function $W_{P2}(s)$ is obtained with the shaft sensor attached to the load, and it has one real zero. In order to avoid having poorly damped conjugate complex zeros, it is advisable to fasten the feedback device to the load side in all cases where $J_L \gg J_M$.

With a large load inertia, the contribution of J_M to the equivalent inertia J_{EQ} of the subsystem is negligible. The servo motor provides the driving torque T_{em}, required to control the load-side speed ω_L and to make it track the reference profile $\omega^*(t)$ with an error $\Delta\omega$ as small as possible. The torque is transmitted over a flexible shaft having a finite stiffness K_K. It is of interest to consider the system dynamics under the assumption that the shaft sensor is attached to the load side, while the speed controller manages to drive the load speed along the reference profile with a negligible error ($\omega_L \approx \omega^*$). If we assume that $T_L \approx 0$, $\Delta\omega \approx 0$, and $\omega_L(0) = 0$, the shaft torque T_S has to be equal to

$$T_S = J_L \frac{\mathrm{d}\omega^*}{\mathrm{d}t} \Rightarrow T_L(s) = J_L \omega^*(s).$$

The same torque can be expressed in terms of the motor speed ω_M and the load speed $\omega_L \approx \omega^*$:

$$T_S = K_K \Delta\theta + K_V \frac{\mathrm{d}(\Delta\theta)}{\mathrm{d}t} = K_V(\omega_M - \omega^*) + K_V \int(\omega_M - \omega^*)\mathrm{d}t.$$

From the previous, the complex images $\omega_M(s)$ and $\omega^*(s)$ are related by

$$J_L s^2 \omega^*(s) = K_V s(\omega_M(s) - \omega^*(s)) + K_K(\omega_M(s) - \omega^*(s)),$$

while the transfer function $W_{RR2}(s) = \omega_M(s)/\omega^*(s)$ is obtained as

$$W_{RR2}(s) = \frac{1 + \frac{K_V}{K_K}s + \frac{J_L}{K_K}s^2}{1 + \frac{K_V}{K_K}s} = \frac{1 + \frac{2\zeta_z}{\omega_z}s + \frac{1}{\omega_z^2}s^2}{1 + \frac{2\zeta_z}{\omega_z}s}. \tag{8.10}$$

The function $W_{RR2}(s)$ has one pair of conjugate complex zeros and one real pole. Note in Eq. 8.9 that $W_{RR2}(s) = 1/W_{RR1}(s)$. With more zeros than poles, $W_{RR2}(s)$ has a derivative nature and emphasizes the high-frequency content of the reference signal ω^*. From Eq. 8.2, the driving torque T_{em} is obtained as $T_{em} = T_S + J_M\, \mathrm{d}\omega_M/\mathrm{d}t$. In cases where $T_L \neq 0$, the rotor speed and driving torque comprise the high-frequency content of the load torque as well.

If we assume that the load speed corresponds to the reference, and with $\omega_M(s) = W_{RR2}(s)\omega^*(s)$, the driving torque and rotor speed dynamics do not comprise resonant oscillations at the frequency ω_z of the conjugate complex zeros in Eq. 8.10. On the other hand, the presence of poorly damped conjugate complex poles in $W_{P2}(s)$ (Eq. 8.6) may reduce the range of applicable gains and make the condition $\omega_L \approx \omega^*$ unachievable.

In order to support the above considerations, the performance of the speed-controlled systems with torsional resonance is examined by means of computer simulations given in Section 8.8. The simulation waveforms are compared for the cases where $J_M \gg J_L$, $J_M \ll J_L$, and $J_M \approx J_L$.

8.4 Active resonance compensation methods

The mechanical subsystem in Fig. 8.2 comprises four state variables: θ_M, θ_L, ω_M, and ω_L. Suppression of torsional resonance phenomena can be achieved by closing the state feedback [1, 3]. Implementation of the state feedback consists of devising a control law where the driving torque T_{em} is calculated as the weighted sum of the state variables and the output error. This approach provides the designer with the possibility of selecting the closed-loop poles freely and without restrictions. The placement of the closed-loop poles is determined by the feedback gains attributed to each of the system states. Hence, poorly damped resonance phenomena, described by conjugate complex poles and zeros in Eq. 8.4, can be compensated for and cancelled out by appropriate selection of the feedback gains in a state feedback controller.

Implementation of such a controller that requires all the system states is readily available within the digital controller memory. Hence, the states θ_M, θ_L, ω_M, and ω_L have to be either measured or estimated. Direct measurement of the state variables requires that the shaft sensors be installed at both the motor and load sides of the shaft. At the same time, the parameters J_M, J_L, K_K, and K_V of the control object have to be known to ensure proper placement of the closed-loop poles.

When the state variables θ_M, θ_L, ω_M, and ω_L cannot be obtained by an explicit measurement, they can be reconstructed from the available signals. In most cases, the servo motor encloses a shaft sensor, providing the rotor speed ω_M and position θ_M. According to Eq. 8.1, the torsion displacement $\Delta\theta = \theta_M - \theta_L$ can be calculated from the shaft torque T_S. From Eq. 8.2, the shaft torque can be estimated as

$$T_S = T_{em} - J_M \frac{\mathrm{d}}{\mathrm{d}t}\omega_M = T_{em} - J_M \frac{\mathrm{d}^2}{\mathrm{d}t^2}\theta_M \,. \tag{8.11}$$

Calculation of the second derivative of the shaft position, implied in Eq. 8.11, amplifies the high-frequency noise. The shaft feedback θ_M is obtained either from an optical encoder or electromagnetic resolver [2]. High-frequency noise components include the quantization noise, errors caused by the sensor imperfection, and electromagnetic noise originating from the power converter. Brought into formula 8.11, where the shaft position is differentiated in order to obtain the torque estimate, the noise components increase in amplitude, impairing the reconstruction of T_S, ω_L, and θ_L. In a number of cases, the frequency of parasitic signals coincides with the resonance, further involving the noise filtering and the signal reconstruction. Further information on antiresonance measures based on the torsional torque estimation can be found in [19].

Inaccessible state variables can be reconstructed by means of the state observer [20], achieving, in this way, better suppression of parasitic noise components. The observer output can be used to close the state feedback [21] and similarly suppress the resonant phenomena. In cases with the resonant poles at a frequency f_{TR} of several kilohertz, and when their damping coefficient is exceptionally low, the desired accuracy of the state reconstruction is achieved, provided that the sampling frequency f_S of the observer is sufficiently high ($f_S > 50\ f_{TR}$).

8.5 Passive resonance compensation methods

The suppression of torsional resonance based on state feedback requires either measurement or reconstruction of the speed and position signals on both sides of the elastic coupling. Acquiring the feedback signals from a state estimator or an observer implies relatively high sampling frequencies and advanced filtering measures. The noise filtering must be capable of separating the resonance dynamics from the background noise, even in cases when the noise frequencies coincide with those of the resonance modes. In a number of cases, the required states of the resonant subsystem can be neither measured nor reconstructed. In cases when the subsystem model, given in Fig. 8.2, does not represent the resonant processes with sufficient accuracy, the state feedback approach cannot properly suppress the resonance modes. The differences between the model and the control object may include the backlash, the nonlinear viscous friction $K_V(\Delta\omega)$, the variable stiffness coefficient $K_V(\Delta\theta)$, and other nonlinear effects. Moreover, even the two-mass representation of the resonant subsystem given in Fig. 8.1 is an apprehensible representation of a system comprising a number of elastically coupled masses.

When an active suppression of torsional oscillations is not feasible, the effects of the resonant modes can be reduced or eliminated by passive measures. In Fig. 8.7, the torque reference T^*, obtained from the speed or position controller, is brought to the input of the antiresonant series compensator. The compensator acts as a filter, attenuating certain frequency components from the torque reference and providing the signal T_{em} at the output. The compensator design has to ensure that the torque T_{em} supplied to the control object does not comprise any frequency components in the vicinity of the resonant frequency f_{TR}. The mechanical resonator does not oscillate unless supplied with a certain amount of energy. The energy sources to the mechanical subsystem as a whole are the servo motor, feeding the driving torque T_{em}, and the load torque T_L. In cases when the antiresonant filter removes all of the driving torque frequency components in the neighborhood of f_{TR}, and where the load torque does not comprise an excitation to the torsional resonance, the system in Fig. 8.7 ensures operation without torsional oscillations.

The series antiresonant filter in Fig. 8.7 has to suppress or remove a narrow frequency range around the resonant frequency f_{TR}, while leaving other spectral components of the driving torque intact. Such filters are known as the *notch* filters. In the following sections, the IIR and FIR implementations [22] of the notch antiresonant filter are considered.

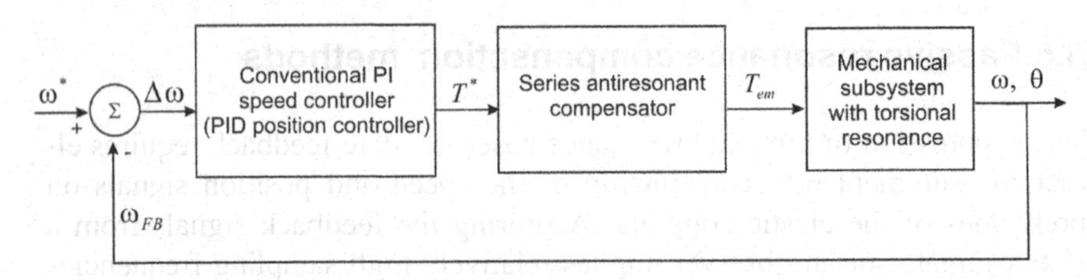

Fig. 8.7. Passive suppression of the torsional resonance by means of an antiresonant series compensator.

8.6 Series antiresonant compensator with a notch filter

In this section, the antiresonant series compensator based on the IIR notch filter with two conjugate complex poles and zeros is considered. The filter is to be connected in series with the speed (position) controller, as indicated in Fig. 8.7. The purpose of the filter is suppression of the driving torque components in the proximity of the resonant frequency f_{TR}. The application of the notch filter is expected to alleviate the torsional resonance problems and to increase the range of applicable gains.

In Section 8.2, the closed-loop response of the speed-controlled system with torsional resonance is considered. In order to obtain an acceptable response, the feedback gains are halved with respect to their optimized values (Fig. 8.6). The conclusions drawn in this section indicate that in all cases where the resonance frequency f_{TR} is low relative to the closed-loop bandwidth f_{BW}, the loop gains have to be reduced in order to suppress the oscillations. Without the antiresonant control measures, the largest bandwidth cannot exceed $f_{BW} < f_{TR}/7$. The notch filter, outlined in this section, is expected to extend the range of applicable gains and to increase the bandwidth.

The transfer function of the notch filter with one pair of conjugate complex poles and one pair of zeros is given in Eq. 8.12. The poles and zeros have the same natural frequency ω_{NF}, referred to as the *notch* frequency. For the frequencies $\omega \ll \omega_{NF}$ and $\omega \gg \omega_{NF}$, the transfer function $W_{NOTCH}(j\omega)$ is close to unity. Hence, all the frequency components of the input signal away from the notch frequency are passed without changes in their amplitude and phase. If we introduce $s = j\omega_{NF}$ in Eq. 8.12, the transfer function is calculated as $W_{NOTCH}(j\omega_{NF}) = \zeta_Z/\zeta_P$. Hence, in order to attenuate the frequencies in the region of ω_{NF}, the damping ζ_P of the poles must be large, while the damping ζ_Z of the zeros should be as small as possible.

$$W_{NOTCH}(s) = \frac{s^2+2\zeta_Z\omega_{NF}s+\omega_{NF}^2}{s^2+2\zeta_P\omega_{NF}s+\omega_{NF}^2} \tag{8.12}$$

8.6.1 The notch filter attenuation and width

The fundamental characteristics of the notch filters can be investigated by using Matlab. In order to obtain the amplitude and phase characteristics of the transfer function $W_{NOTCH}(j\omega)$ for $\omega_{NF} = 1$ rad/s, $\zeta_Z = 0.1$, and $\zeta_P = 1$, it is sufficient to type in the sequence given in Table 8.2 at the Matlab command prompt.

The impact of the ζ_Z/ζ_P ratio on the amplitude characteristics of the notch filter is given in Fig. 8.8. The plots are obtained by using the sequence of the Matlab commands in Table 8.2. The pole damping is kept constant at $\zeta_P = 1$, while the zero damping assumes the values of $\zeta_Z = 0.5$, $\zeta_Z = 0.25$, and $\zeta_Z = 0.1$. The largest attenuation is achieved when the excitation frequency coincides with ω_{NF}. The peak attenuation of 20 dB (i.e., ten times) is obtained with $\zeta_Z/\zeta_P = 0.1/1$. While ratio ζ_Z/ζ_P determines the peak attenuation of the notch filter, the damping ζ_P of the conjugate complex poles determines the width of the notch in the amplitude characteristics. This is illustrated in Fig. 8.9, where the amplitude characteristic $|W_{NOTCH}(j\omega)|$ is given for the case where the ratio ζ_Z/ζ_P is kept constant, with ζ_P ranging from 0.2 to 1.

Table 8.2. The Matlab command sequence used to obtain the amplitude characteristics for the recursive notch filter.

```
>> wnf = 1;                           % Setting the notch frequency to 1 rad/s
>> xp  = 1;                           % The damping of  poles
>> xz  = xp/10;                       % The damping of  zeros
>> den = [1 2*xp*wnf wnf*wnf];        % Denominator of WNOTCH transfer function
>> num = [1 2*xz*wnf wnf*wnf];        % Numerator of WNOTCH
>>
>> [mag,phase,ww] = bode(tf(num,den));
>>
>>                                    % Using the bode command, the amplitude
>>                                    % characteristics are obtained in mag, the
>>                                    % phase characteristics in phase
>>
>> plot(log10(ww(:)),20*log10(mag(:)));
>>
>>                                    % Plotting of the amplitude characteristics
>>                                    % in logarithmic scale
```

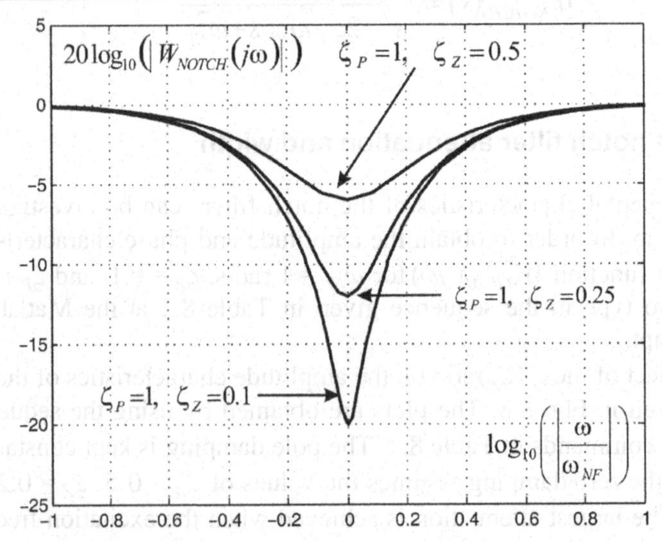

Fig. 8.8. The amplitude characteristics of the notch filter obtained with pole damping of $\zeta_P = 1$ and zero damping ranging from $\zeta_Z = 0.5$ to $\zeta_Z = 0.1$.

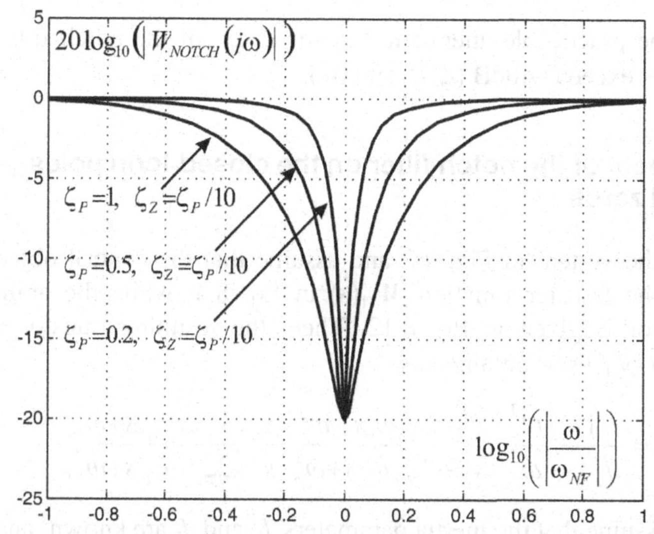

Fig. 8.9. The amplitude characteristics of the notch filter obtained with pole damping ranging from $\zeta_P = 1$ to $\zeta_P = 0.2$ and zero damping of $\zeta_Z = \zeta_P/10$.

A constant ζ_Z/ζ_P ratio results in the maximum attenuation of 20 dB. The width of the notch (i.e., the range of frequencies where the amplitude $|W_{NOTCH}(j\omega)|$ has a meaningful reduction) is proportional to ζ_P. With $\zeta_P = 0.1$, the amplitude characteristic goes below –3dB for $\omega = \omega_{NF} \pm 0.2\omega_{NF}$.

Note at this point that the notch filter given in Eq. 8.12 attenuates the signals at the notch frequency ω_{NF} but cannot remove them completely. The attenuation ζ_Z/ζ_P is limited due to practical limits in setting the pole and zero damping. The width of the notch in the amplitude characteristic $|W_{NOTCH}(j\omega)|$ is determined by ζ_P. Therefore, the pole damping has to be sufficiently low to keep the frequency components away from ω_{NF} intact. On the other hand, the damping of zeros is lower bounded. A practical limit in discrete-time implementation of $\zeta_Z << 1$ is the requirement to use very low sampling times. In cases with $\zeta_Z \approx 0.001$, the sampling frequency f_S has to be several hundred times larger than the notch frequency ω_{NF} in order to obtain a stable implementation of the filter. In order to obtain the implementation with an acceptable error between the actual and desired damping, the sampling time of the digital notch filter has to be $f_S > 1000/(2\pi\omega_{NF})$. The purpose of the notch filter in Fig. 8.7 is the suppression of torsional resonance. Therefore, the notch frequency is set to $\omega_{NF} = 2\pi f_{TR}$. In cases where $f_T \approx 1$ kHz, the required sampling frequency may reach 1 MHz. The sampling rates and interrupt periods of that kind are incompatible with the existing motion-control processors [10, 11]. For the above reasons, the practicable attenuation factor ζ_Z/ζ_P of the notch filter in Eq. 8.12 cannot exceed –40dB ($\zeta_Z/\zeta_P \geq 0.01$).

8.6.2 Effects of the notch filter on the closed-loop poles and zeros

Consider the system in Fig. 8.7 and assume that the control object is defined by the transfer function $W_{P1}(s)$ in Eq. 8.4, while the antiresonant compensator is given in Eq. 8.12. Then, the resulting transfer function $W_P(s) = \omega(s)/T^*(s)$ is obtained as

$$\frac{\omega(s)}{T^*(s)} = \frac{1}{J_{EQ}s}\frac{\omega_p^2}{\omega_z^2}\frac{s^2+2\zeta_z\omega_z s+\omega_z^2}{s^2+2\zeta_p\omega_p s+\omega_p^2}\frac{s^2+2\zeta_Z\omega_{NF}s+\omega_{NF}^2}{s^2+2\zeta_P\omega_{NF}s+\omega_{NF}^2}. \tag{8.13}$$

If we assume that the inertia parameters J_M and J_L are known, and we are provided with information on the shaft stiffness K_K and viscous friction K_V, the object parameters ω_p, ω_z, ζ_p, and ζ_z can be calculated from Eq. 8.5. With $\omega_{NF} = \omega_p$ and $\zeta_Z = \zeta_p$, the notch filter zeros cancel the conjugate complex poles of the torsional resonance mode. At the same time, the torsional resonance zeros cannot be canceled with the notch poles due to $\omega_z \neq \omega_p = \omega_{NF}$. Therefore, the resulting transfer function $\omega(s)/T^*(s)$ is

$$\frac{\omega(s)}{T^*(s)} = \frac{1}{J_{EQ}s}\frac{\omega_p^2}{\omega_z^2}\frac{s^2+2\zeta_z\omega_z s+\omega_z^2}{s^2+2\zeta_P\omega_{NF}s+\omega_{NF}^2}. \tag{8.14}$$

In Eq. 8.14, the weakly-damped conjugate complex poles ω_p of the mechanical resonator are replaced by the notch poles of the same frequency, having a sufficient damping $\zeta_P >> \zeta_p$.

It is noteworthy to analyze the effects of the series antiresonant compensator (Fig. 8.7), implemented according to Eq. 8.12, with $\omega_{NF} = \omega_p$, $\zeta_Z = \zeta_p$, and $\zeta_P >> \zeta_p$. In order to compare the range of stable gains prior to and after the insertion of the notch filter, the Evans root locus is constructed by using the Matlab function *rlocus*. The function plots the closed-loop poles of the system comprising the transfer function $W_S(s) = num(s)/den(s)$ and the loop gain K_P. The poles are plotted in the s-plane for the gain values ranging from $K_P = 0$ to $K_P = \infty$. A stability limit is reached for K_{PMAX} that results in the closed-loop poles reaching the imaginary axis. With $K_P > K_{PMAX}$, the characteristic polynomial $f(s) = den(s) + K_P\,num(s)$ has zeros residing in the right half of the s-plane.

The closed-loop system under consideration is given in Fig. 8.10. In order to keep the root locus readable, the system is simplified by the assumption that the speed controller comprises only the proportional action ($T^*= K_P\,\Delta\omega$). Delays in the torque actuator and feedback acquisition are modeled with the first-order transfer function $1/(1+s\tau_{HF})$. The resulting open-loop transfer function $W_S(s)$ of the system is given in Eq. 8.15, with the assumption that $W_{P1}(s)$ and $W_{NF}(s)$ are given in Eq. 8.4 and Eq. 8.12, respectively.

$$W_S(s)=\frac{K_P\ \omega_p^2\ s^2+2\zeta_z\omega_z s+\omega_z^2\ \ s^2+2\zeta_Z\,\omega_{NF}\,s+\omega_{NF}^2}{J_{EQ}s\,\omega_z^2\ s^2+2\zeta_p\omega_p s+\omega_p^2\ \ s^2+2\zeta_P\,\omega_{NF}\,s+\omega_{NF}^2}\frac{1}{\left(1+s\tau_{NF}\right)^2} \tag{8.15}$$

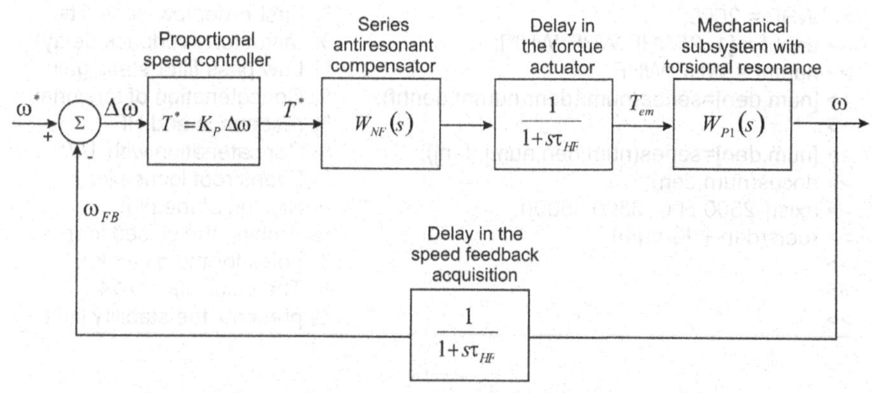

Fig. 8.10. Proportional speed controller with resonance in the mechanical subsystem. The notch filter cancels the resonance poles. The torque actuator delays and feedback acquisition are modeled as the first order lag with $1/\tau_{HF}= 2000$ rad/s. The root locus of the system is given in Fig. 8.11 and Fig. 8.12.

To begin, the root locus is constructed for the system without any antiresonant compensator ($W_{NF}= 1$). In order to construct the root locus, it is necessary to enter the Matlab commands given in Table 8.3.

The root locus obtained from the sequence in Table 8.3 is shown in Fig. 8.11. The system is stable for $K_P< K_{PMAX}= 3.14$. As the loop gain K_P exceeds K_{PMAX}, the root locus branches originating from the resonance poles pass into the right half of the s–plane. Hence, the range of applicable gains for a system without any antiresonant compensator is $0 < K_P< 3.14$.

Table 8.3. The Matlab command sequence used to obtain the root locus of the speed-controlled system with a mechanical resonator and without any antiresonant measures.

```
>> JM= 0.0007; JL= 0.0007;                    % Inertia coefficients
>> KK = 317; KV = 0.12;                       % Shaft parameters
>> WP = sqrt(KK*(JM+JL)/JM/JL);               % Res.pole natural frequency
>> WZ = sqrt(KK/JL);                          % Res.zero natural frequency
>> XP = sqrt(KV*KV*(JM+JL)/4/JM/JL/KK);        % Resonant pole damping
>> XZ = sqrt(KV*KV/4/JL/KK);                   % Resonant zero damping
>> numr = [1  2*XZ*WZ  WZ*WZ];                % Numerator (Eq. 8.4)
>> denr = [1  2*XP*WP  WP*WP];                % Denominator (Eq. 8.4)
>> denj = [ (JM+JL)  0];                      % Pole at the origin (Eq. 8.4)
>> numj = WP*WP/WZ/WZ;                        % Static gain of the function
>> WHF = 2000;                                % First-order low-pass filter
>> denhf = [1  2*WHF WHF*WHF];                % (actuator&feedback delay)
>> numhf = WHF*WHF;                           % Low pass filter static gain
>> [num,den]=series(numr,denr,numhf,denhf);   % Concatenation of torsional
>>                                            % resonance and HF
>> [num,den]=series(num,den,numj,denj);       % Concatenation with 1/Js
>> rlocus(num,den);                           % Evans root locus plot
>> axis([-2500 500 -3500 3500]);              % Scaling of the plot
>> roots(den + Kp num);                       % Probing the closed loop
>>                                            % poles for the given Kp.
>>                                            % The value Kp = 3.14
>>                                            % presents the stability limit
```

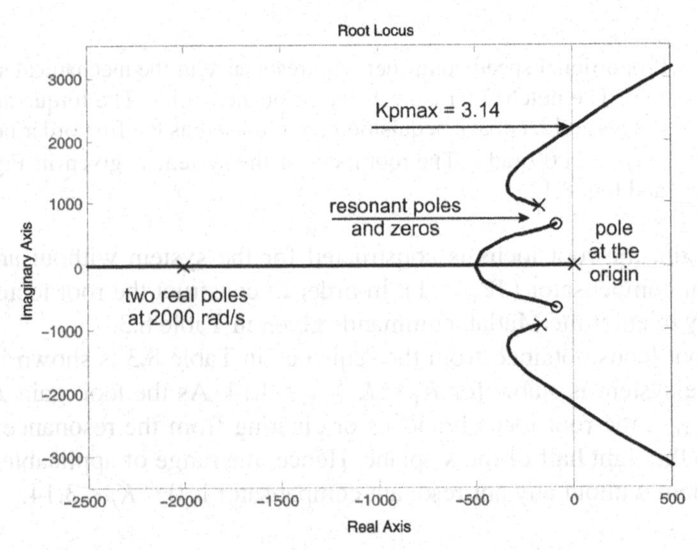

Fig. 8.11. Evans root locus obtained for the closed loop system with a proportional speed controller, a resonant mechanical subsystem, and the torque actuator and feedback acquisition delays modeled as $1/(1+s/\omega_{NF})^2$.

The purpose of the series antiresonant compensator is to cancel the resonance poles and extend the range of applicable gains. The notch filter zeros have to be tuned to $\omega_{NF} = \omega_p$, while their damping ζ_Z should correspond to the damping ζ_p of the resonance poles. With ζ_p depending on the viscous friction (Eq. 8.5), the condition $\zeta_Z = \zeta_p$ can hardly be expected. The friction may change with the running speed and position of the machine. At the same time, it tends to change with the ambient temperature and to increase with wear in the transmission elements. Therefore, it is assumed that an error $\Delta\zeta = \zeta_Z - \zeta_p$ exists, leading to a mismatch of 0.1 rad in the argument of the notch filter zeros. In order to obtain the root locus for the system in Fig. 8.7, comprising the notch filter with detuned zeros, a subsequent set of Matlab instructions has to be entered at the Matlab command prompt. For the correct result, the command sequence in Table 8.4 has to follow that in Table 8.3 without quitting Matlab or altering its variables.

Table 8.4. The Matlab command sequence used to obtain the root locus of the speed-controlled system with a mechanical resonator, compensated with a series antiresonant notch filter.

```
>> WNF = sqrt(KK*(JM+JL)/JM/JL);                % Notch filter design, ωNF=ωp
>> XZNF = cos(acos(XP)-0.1);                    % Detuning of zeros
>>                                              % by 0.1 rad
>> XPNF = 1;                                    % Notch poles damping = 1
>> numnf = [1 2*XZNF*WNF WNF*WNF];              % Notch filter numerator
>> dennf = [1 2*XPNF*WNF WNF*WNF];              % Notch filter denominator
>> [numc,denc]=series(num,den,numnf,dennf);     % Insertion of the notch
>>                                              % numnf(s)/dennf(s) in
>>                                              % series with the transfer
>>                                              % function num(s)/den(s)
>>                                              % of the uncompensated
>>                                              % system
>> close all;                                   % Close all previous figures
>> rlocus(numc,denc);                           % Printing the root locus of
>> axis([-2500 500 -3500 3500]);                % the resulting system and
>>                                              % adjusting the scaling
```

The root locus obtained in the prescribed way is shown in Fig. 8.12. Due to the notch filter zeros being displaced by 0.1 rad with respect to the resonance poles, the pole-zero cancellation is not perfect. Two pole-zero couples (dipoles) are formed in the figure, defining the two branches of the root locus, each one originating from one resonance pole, moving away as the gain K_P increases, and ending in the neighboring notch filter zero.

Compared with the locus in Fig. 8.11, the branch that starts from the resonance pole is absorbed by the zero and is thus prevented from sliding towards the right half-plane. There is a total of five real poles in Fig. 8.12. The pole at the origin corresponds to the factor $1/J_{EQ}s$ in Eq. 8.13. The two notch poles are real due to $\zeta_P = 1$, and they reside on the real axis next to $\omega \approx -1000$ rad/s. The two branches of the root locus, departing from the real axis between the notch poles and the origin, are absorbed by the resonance zeros.

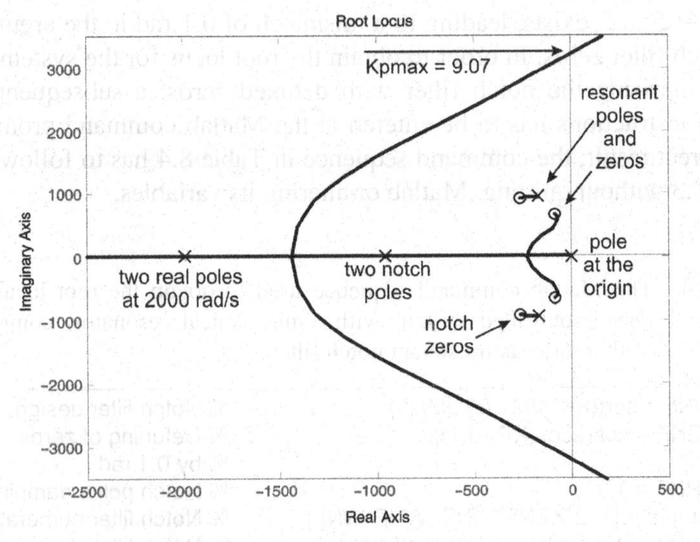

Fig. 8.12. Evans root locus obtained for the system in Fig. 8.10 equipped with a series antiresonant compensator $W_{NF}(s)$. It is assumed that the resonant poles cancellation is impaired by a mismatch of 0.1 rad in the argument of the notch filter zeros.

At $\omega = -2000$ rad/s, the open-loop transfer function has one pair of real poles. Between these poles and the notch poles, another two branches depart from the real axis and sway towards the imaginary axis. For $K_P = 9.07$, the closed-loop poles, sliding along this pair of branches, reach the stability limit. Hence, even a detuned notch filter increases the range of applicable gains from $K_P = 3.14$ (Fig. 8.11) to $K_P = 9.07$.

The implementation aspects of the notch filter and further investigation of its impact on the closed-loop performance are discussed in subsequent sections of this Chapter.

8.6.3 Implementation aspects of the notch antiresonant filters

The notch antiresonant filter has to be implemented in discrete-time form, within the same digital controller that executes the speed controller and other motion control tasks. In most cases, the sampling period of the notch filter T_{SNF} is set to be equal to the sampling period T_{SSC} of the speed controller. In cases with the resonant frequency ω_{TR} exceeding 1 kHz, where the required sampling rate $1/T_{SNF}$ exceeds $1/T_{SSC}$, the sampling period of the filter is set to $T_{SNF} = T_{SSC}/p$, where $p > 1$ is an integer.

The notch transfer function $W_{NF}(s)$, given in Eq. 8.12, has to be converted in the *z*-domain. The corresponding pulse transfer function $W_{NF}(z)$ has second-order polynomials in both numerator and denominator:

$$W_{NF}(z) = \frac{K_3 z^2 - K_4 z + K_5}{z^2 - K_1 z + K_2}.$$

The coefficients K_1, K_2, K_3, K_4, and K_5 of the pulse transfer function can be obtained from its poles $z_{p1/2}$ and zeros $z_{z1/2}$. The notch poles and zeros in the *s*-domain are obtained from Eq. 8.12 as $s_{p1/2} = -\xi_P\omega_{NF} \pm j\,(1-\xi_P^2)^{1/2}\,\omega_{NF}$ and $s_{p1/2} = -\xi_p\omega_{NF} \pm j\,(1-\xi_p^2)^{1/2}\,\omega_{NF}$. In order to cancel the resonance poles, the notch frequency ω_{NF} and damping of zeros must be set to $\xi_Z = \xi_p$ and $\omega_{NF} = \omega_p$ (Eq. 8.5). According to the analysis given in previous sections, the damping coefficient ξ_P of the notch poles determines the width of the frequency band involved. If we convert $W_{NF}(z)$ polynomials in terms of $(z - z_{z1})(z - z_{z2})$ and $(z - z_{p1})(z - z_{p2})$ and calculate $z_{p1/2} = \exp(s_{p1/2}T_{SNF})$ and $z_{z1/2} = \exp(z_{z1/2}T_{SNF})$, the pulse transfer function coefficients are found as

$$\begin{aligned}
K_1 &= 2\exp(-\zeta_P\omega_{NF}T_{SNF})\cos\left(\omega_{NF}T_{SNF}\sqrt{1-\zeta_P^2}\right)\\
K_2 &= \exp(-2\zeta_P\omega_{NF}T_{SNF})\\
K_3 &= \exp[-(\zeta_P-\zeta_Z)\,\omega_{NF}T_{SNF}]\\
K_4 &= 2\exp(-\zeta_P\omega_{NF}T_{SNF})\cos\left(\omega_{NF}T_{SNF}\sqrt{1-\zeta_Z^2}\right)\\
K_5 &= \exp[-(\zeta_P+\zeta_Z)\,\omega_{NF}T_{SNF}]\,.
\end{aligned}$$

In addition, the discrete-time notch filter can be designed by using Matlab. The sample command sequence resulting in the $W_{NF}(z)$ structure and coefficients is given in Table 8.5.

Table 8.5. The Matlab command sequence used to obtain the pulse transfer function of the discrete-time recursive notch filter.

```
>> wnf = 1000;                        % Notch frequency set to 1000 rad/s
>> xp  = 0.5;                         % Notch pole damping set to 0.5
>> xz  = 0.01;                        % Notch zero damping set to 0.01
>> Ts  = 0.0001;                      % Sampling rate set to 10 kHz
>> num = [ 1 2*wnf*xz wnf*wnf];       % Numerator of the s-domain transfer function
>> den = [ 1 2*wnf*xp wnf*wnf];       % Denominator, s-domain
>> sysd = c2d(tf(num,den),Ts)         % Conversion into z-domain and printing WNF
>> [numd,dend] = tfdata(sysd,'v')     % Obtaining the WNF(z) polynomials
```

Discrete-time implementation requires conversion of the transfer function $W_{NF}(z)$ into a difference equation, expressing the next sample of the filter output in terms of previous inputs, outputs, and coefficients. The difference equation can be obtained by using the time-shift property of the z transform (Eq. 4.7):

$$Y(z)\left(z^2 - K_1 z + K_2\right) = X(z)\left(K_3 z^2 - K_4 z + K_5\right)$$

$$zY(z) = K_1 Y(z) - K_2 z^{-1} Y(z) + K_3 zX(z) - K_4 X(z) + K_5 z^{-1} X(z)$$

$$y_{n+1} = K_1\, y_n - K_2\, y_{n-1} + K_3\, x_{n+1} - K_4\, x_n + K_5\, x_{n-1}\,.$$

In order to simplify calculations, the coefficients K_2 and K_4 can be stored in the digital controller memory as $K^*_2 = -K_2$ and $K^*_4 = -K_4$:

$$y_{n+1} = K_1\, y_n + K^*_2\, y_{n-1} + K_3\, x_{n+1} + K^*_4\, x_n + K_5\, x_{n-1}\,.$$

It is worthwile to explore the execution time required to implement an antiresonant notch filter. Therefore, the implementation of the above difference equation is considered for existing motion-control processors [10, 11], and in particular for the TI TMS320LF240x family of fixed-point DSP. A total of five multiplications has to be performed. The products have to be summed in order to obtain the torque reference.

It is assumed that coefficients, inputs and outputs are represented as 16-bit signed integers residing in the internal DSP memory. The numbers are written in Q12 format, where the bits b15–b12 keep the sign and integer part of the number, while the remaining bits b11–b0 keep the fractional part. In Q12 format, the value of 1.00 is written as 1000h.

For proper access to RAM using *direct addressing* mode, it is necessary to set the *data page pointer.* In the above code, this is taken care of by the instruction LDP. Notice that the variables *Xnew*, *Xn* and *Xold*, corresponding to x_{n+1}, x_n, and x_{n-1}, have to be written in the successive memory cells. In this way, the operation $x_n = z^{-1}\, x_{n+1}$ requires the value of *Xnew* to be copied into the next memory cell *Xn*. The same requirement holds for *Yn* and *Yold* (y_n and y_{n-1}).

Table 8.6. The implementation of the antiresonant notch filter on a 16-bit fixed-point DSP platform, coded in TMS320LF240x assembly language.

```
LACC  PI_output          ; Load the output of the speed
                         ; controller into the accumulator
                         ; this is the new sample at the
                         ; input of the notch filter
LDPK  #relevant_page     ; Load data page pointer;
                         ; relevant variables must reside
                         ; within the same 128-word-long
                         ; page of the internal RAM,
                         ; determined by DP field of the DSP
                         ; status, this instruction sets DP
SACL  Xnew               ; Store PI_output as X(n+1)
ZAC                      ; Clear 32-bit accumulator
                         ; ACCH:ACCL = 0x0000:0000
LT    Xold               ; Load T register with X(n-1)
MPY   K5                 ; P register (32 bit) becomes
                         ; X(n-1) * K5
LTD   Xn                 ; Load T register with X(n),
                         ; copy X(n) into X(n-1),
                         ; and add P = X(n-1) * K5 to the
                         ; accumulator (32 bit)
MPY   K4                 ; P register (32 bit) becomes
                         ; X(n) * K4
LTD   Xnew               ; Load T register with X(n+1),
                         ; copy X(n+1) into X(n),
                         ; and add P = X(n) * K4
                         ; to 32-bit ACC
MPY   K3                 ; P reg. = X(n+1) * K3
LTA   Yold               ; Load T register with Y(n-1)
                         ; and add P register to ACC
MPY   K2                 ; P reg. = Y(n-1) * K2
LTD   Yn                 ; T reg. = Y(n), copy Y(n)
                         ; into Y(n-1), and add P reg.
                         ; to ACC
MPY   K1                 ; P reg. = Y(n) * K1
APAC                     ; Add P reg. = Y(n) * K1 to
                         ; the accumulator. At this point,
                         ; the result Y(n+1) is stored in
                         ; the 32-bit accumulator. It has to
                         ; be scaled, converted into a
                         ; 16-bit format and stored.
SACH  Yn,4               ; The shift of 4 goes with Q12
                         ; format. The new output
                         ; Y(n+1) will be used as Y(n)
                         ; at the next sampling instant
SACH  Tem_Ref,4          ; The output of the filter is the
                         ; torque reference
```

The input to the filter is received from the PI speed controller (PI_output). The output is used as the driving torque reference (Tem_Ref). The program flow implements the notch filter difference equation, wherein the specific samples of the input and output are multiplied by the filter parameters K_1–K_5, and accumulated in the 32-bit accumulator. The hardware multiplier provides the 32-bit result. It multiplies the 16-bit contents of the dedicated T register and the addressed RAM variable. Therefore, prior to each MPY instruction, the T register must be loaded with one of the multiplication factors. It is interesting to note that the instruction LDP performs three operations at a time. It loads the T register, adds the P register contents to the accumulator, and moves the addressed memory cell to the subsequent location. Upon completion, the output Tem_Ref is obtained by shifting the 32-bit accumulator four times towards the left and storing the accumulator upper 16 bits (ACCH). This operation is required in all cases when the numbers are represented in Q12 format. In the multiplication 1.00 × 1.00, where both values are represented in Q12 format, the multiplication produces 1000h × 1000h = 0100 0000h. The proper result is obtained by left shifting the result by four and storing the upper 16 bits.

The execution of the code given above requires 18 instruction cycles. With an instruction cycle of t_C = 1/40 MHz = 25 ns [10, 11], the algorithm is completed in 450 ns. In cases where the damping coefficient ξ_Z has to be very low, a 16-bit implementation is not sufficient, and the filter variables have to be represented as 32-bit signed integers. This prolongs the execution at least four times. Even with 4·450 ns = 1.8 μs, the execution time is only a fraction of the shortest conceivable sampling period.

8.7 Series antiresonant compensator with the FIR filter

The application of notch antiresonant filters such as the one in Eq. 8.12 is limited by their sensitivity to parameter changes, their inability to compensate both the resonant poles and zeros, and an insufficient attenuation ζ_P/ζ_Z of the resonance frequencies.

The notch filter sensitivity to changes in the system parameters comes from the need to cancel the resonance poles with the notch zeros, wherein both the natural frequency ω_{NF} and damping ζ_Z of such zeros have to correspond with the control object parameters ω_p and ζ_p. Any change in the viscous friction, speed-dependent motion resistances, and temperature, as well as the wear of mechanical parts, results in the notch filter detuning.

Even in cases where the notch zeros cancel the resonance poles in full, the notch poles cannot remove the resonance zeros from the open-loop transfer function (Eq. 8.15) due to $\omega_p \neq \omega_z$.

The notch filter attenuation ζ_P/ζ_Z cannot exceed –40 dB due to the reasons detailed in the previous section. Therefore, the filter W_{NF} in Fig. 8.10 cannot eliminate the resonant frequency components in the signal T^* from passing into the driving torque T_{em}. Therefore, a certain amount of energy is supplied to resonance modes of the control object $W_{P1}(s)$.

In this section, the design of a Finite Impulse Response (*FIR*) antiresonant filter is considered, with the goal of securing a complete removal of the resonant frequencies from the torque reference signal in Fig. 8.10 and reducing the filter sensitivity to changes in the system parameters. The filter has to replace the block W_{NF} in Fig. 8.10.

8.7.1 IIR and FIR filters

The salient features of discrete-time Finite Impulse Response (*FIR*) filters and Infinite Impulse Response (*IIR*) filters, explained in the literature [22], are briefly summarized below. The input pulse, supplied at $t=kT$, affects the output of an IIR filter in the interval $[kT .. +\infty]$, where T stands for the sampling period. Hence, the effects of an input pulse are not limited in time (*infinite*). With the same excitation fed to a FIR filter, the effects of the input pulse are visible in an interval $[mT .. nT]$, while the output of the filter remains unaffected by the input for $t > nT$. In the latter case, the input to the filter has effects that are limited in time (*finite*). The transport delay $(m-k)T$ and duration of the response $(n-m)T$ depend on the FIR filter order and design.

Linear discrete time filters can be described by their transfer function $W(z) = num(z)/den(z)$, where the roots of equations $num(z) = 0$ and $den(z) = 0$ determine zeros and poles of the filter, respectively. The difference equations and pulse transfer functions $W(z)$ describing linear discrete-time IIR and FIR filters are given in Eq. 8.16 and Eq. 8.17. In difference equations, x_p stands for the input sample at $t = pT$, y_p represents the corresponding sample of the filter output, and the integers n and m determine the number of filter poles and zeros.

In Eq. 8.16, the IIR filter output sample y_p is calculated from the actual input x_p, past input samples $x_{p-1}...x_{p-m}$, and past output samples $y_{p-1}...y_{p-n}$. The previous input and output samples are stored in the memory of the digital controller. The FIR output sample y_p is obtained in Eq. 8.17 from samples x_p, $x_{p-1},...x_{p-m}$ of the input, and it does not depend on the past outputs. Hence, the output of the FIR filter is calculated as a weighted sum of the present and past input samples, and it represents a moving average of the input pulse train. The filter does not involve any recursion, since the output y_p does not depend on the output history. Therefore, FIR filters do not have polynomial $den(z)$ in their denominator. With $den(z) = 1$, their transfer function is described as $W(z) = num(z)$.

$IIR:$

$$
\begin{aligned}
& y_p = (a_0 x_p + a_1 x_{p-1} + \ldots + a_m x_{p-m}) - (b_1 y_{p-1} + b_2 y_{p-2} + \ldots + b_n y_{p-n}) \\
& Y(z)(1 + b_1 z^{-1} + b_2 z^{-2} .. + b_n z^{-n}) = X(z)(a_0 + a_1 z^{-1} + \ldots + a_m z^{-m}) \\
& W(z) = \frac{num(z)}{den(z)} = \frac{a_0 + a_1 z^{-1} + \ldots + a_m z^{-m}}{1 + b_1 z^{-1} + b_2 z^{-2} .. + b_n z^{-n}}
\end{aligned}
\tag{8.16}
$$

$FIR:$

$$
\begin{aligned}
& y_p = (a_0 x_p + a_1 x_{p-1} + \ldots + a_m x_{p-m}) \\
& Y(z) = X(z)(a_0 + a_1 z^{-1} + \ldots + a_m z^{-m}) \\
& W(z) = \frac{num(z)}{1} = \frac{a_0 + a_1 z^{-1} + \ldots + a_m z^{-m}}{1}
\end{aligned}
\tag{8.17}
$$

A sample first-order IIR filter $W(z) = (1-\alpha)/(z-\alpha)$ has a single pole $z = \alpha$ and no zeros. With $\alpha = \exp(-T/\tau)$, the filter has its s-domain equivalent $W_E(s) = 1/(1+s\tau)$. Supplied with a unity impulse $\delta(t)$, the filter $W_E(s)$ produces an output $y(t) = \exp(-t/\tau)$. Although the effects of the input supplied at $t = 0$ exponentially decay, they do affect the filter output in the interval $[0 .. +\infty]$.

An example of a simple FIR filter is $W(z) = 1 + z^{-1} + z^{-2}$. With the input excitation defined as $x_i = 1$ for $i = p$ and $x_i = 0$ for $i \neq p$, the output samples are calculated as $y_p = y_{p+1} = y_{p+2} = 1$, with $y_i = 0$ in all the remaining sampling instants. Hence, the input sample x_p affects the output during a limited interval of time.

The FIR filters are capable of providing infinite attenuation at desired frequencies. This feature is crucial for designing an efficient series antiresonant compensator.

8.7.2 FIR antiresonant compensator

The design of a FIR antiresonant compensator with unlimited attenuation of resonant oscillations is based on considerations illustrated in Fig. 8.13. It is assumed that the input signal, given at the bottom of the figure, is fed to the resonator having the transfer function $W_R(s) = \omega_{TR}^2/(s^2 + \omega_{TR}^2)$. It is of interest to obtain the output of the resonator $y_{OUT}(t)$, shown at the top of Fig. 8.13. The electrical dual of the system under consideration is an LC circuit where the input voltage is fed through series inductance L while the output voltage is obtained across parallel capacitor C. The transfer function of this circuit is $u_{OUT}(s)/u_{IN}(s) = 1/(1+s^2LC) = \omega_{TR}^2/(s^2+\omega_{TR}^2)$, with $\omega_{TR} = 1/\mathrm{sqrt}(LC)$. Supplied with $u_{IN}(t) = Ah(t)$, the dual circuit output becomes $u_{OUT}(t) = A\,(1-\cos(\omega_{TR}t))$. The waveform corresponds to the trace $r_1(t)$ in Fig. 8.13.

In order to suppress undamped oscillations, the input step in Fig. 8.13 is split into two equal half-steps. One of these is passed without delay, while the other is delayed by one half of the resonance period $T_{TR} = 1/f_{TR}$. Each of these steps excites undamped oscillations of the resonator $W_R(s)$, given as $r_1(t)$ and $r_2(t)$ in Fig. 8.13. The resulting output $y_{OUT}(t) = r_1(t) + r_2(t)$ does not contain oscillations at the resonance frequency. Given the assumptions that the half-step amplitude is $A/2$ and that the input step commences at t_0, responses $r_1(t)$ and $r_2(t)$ to individual half steps are obtained as $r_1(t) = A/2\,(1-\cos(\omega_{TR}t - \omega_{TR}t_0))$ and $r_2(t) = A/2\,(1-\cos(\omega_{TR}t - \omega_{TR}t_0 - \pi))$. Due to $\cos(\alpha) + \cos(\alpha+\pi) = 0$, the sum $y_{OUT}(t) = r_1(t) + r_2(t)$ does not have any residual oscillations.

The transfer function of an antiresonant FIR filter, designed on the basis given in Fig. 8.13, is shown in Eq. 8.18. The number q is the closest integer representation of the ratio $T_{TR}/2/T$ between the resonance half period and the sampling time T. The transfer function $W_{FIR}(z)$ has to replace the block $W_{NF}(z)$ in Fig. 8.10. Hence, it must suppress the signals at the resonant frequency and prevent them from entering the control object. The torque reference T^*, generated by the speed controller, is fed to the filter input. The filter output T_{em} is fed to the torque actuator and supplied to the control object as the driving torque.

$$W_{FIR}(z) = \frac{1}{2} + \frac{z^{-q}}{2}, \quad q \approx \frac{T_{TR}}{2T} \tag{8.18}$$

Each step in the reference torque T^* (Fig. 8.10) is split into two equal half steps. One of these is passed to the torque actuator (T_{em}) without delay, while the other is delayed by one half of the resonance period $T_{TR} = 1/f_{TR}$. Each of these steps alone would excite torsional oscillations. The interaction of responses to the two successive half steps cancels the resonance phenomena and produces an output waveform similar to $y_{OUT}(t)$ in Fig. 8.13.

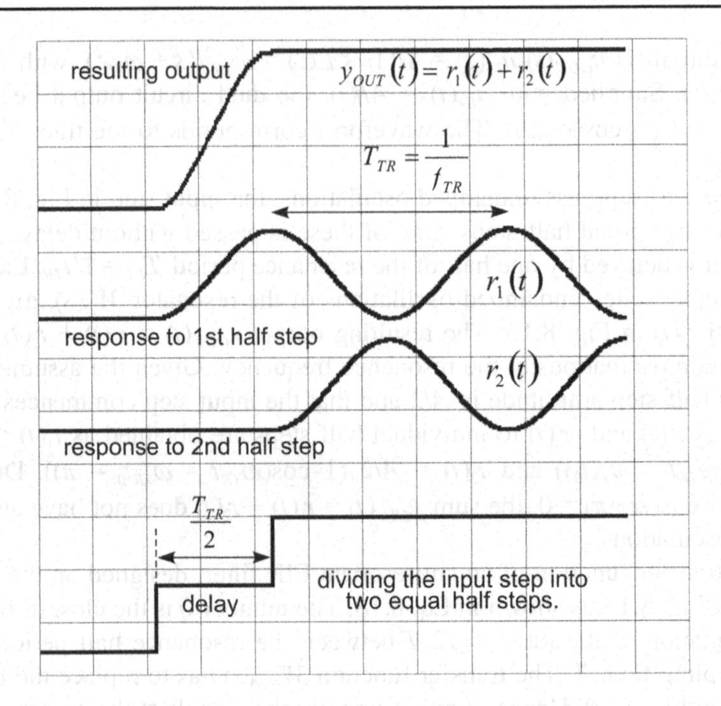

Fig. 8.13. Principles of a FIR antiresonant filter. The input step (lower trace) is split into two half steps, one of them applied instantly and the other delayed by one half of the resonance period ($T_{TR}/2 = 1/2/f_{TR}$). Both half steps incite resonant oscillations (traces in the middle) in the control object (the block $W_{PI}(s)$ in Fig. 8.10). The oscillations are canceled out in the resulting output, being the sum of the former two (upper trace).

Note at this point that tuning the FIR antiresonant filter requires only one system parameter. The integer q in Eq. 8.18 is set according to the resonance period T_{TR}. With q determined as the integer value closest to the ratio $T_{TR}/T/2$, the rounding error depends on the ratio between the resonance period T_{TR} and the sampling period T. With $T_{TR} > 1$ ms in most cases, and with the motion-control processors [10, 11] capable of achieving sampling intervals of several tens of microseconds, the parameter setting of the FIR compensator can be accomplished with sufficient accuracy. While the notch filter attenuation (ζ_p/ζ_z) cannot exceed 40 dB, the FIR antiresonant filter with $qT = T_{TR}/2$ offers an infinite attenuation at the frequency $\omega_{FIR} = 2\pi/T_{TR}$, and removes completely all such frequency components from the input signal T^*.

8.7.3 Implementation aspects of FIR antiresonant compensators

It is important to discuss the implementation aspects of FIR series antiresonant compensators. The PI speed controller presented in Fig. 4.10 and Fig. 4.25 calculates the increments of the proportional and integral control actions. The increments are summed in order to obtain the driving torque reference. Insertion of the FIR filter in Eq. 8.18 into an incremental PI speed controller is explained in Fig. 8.14. The increment of the proportional action $K_P(\Delta\omega_p - \Delta\omega_{p\text{-}1})$ and the increment of the integral action $K_I\,\Delta\omega_p$ are fed to the input of the filter. The design parameter q has to be set to ensure $qT \approx T_{TR}/2$. The filter splits each torque increment ΔT_{em} into two half-steps and delays the second half-step by qT. The discrete-time integrator on the right of the figure accumulates the increments and generates the output. With $qT = T_{TR}/2$, the output T_{em} of the structure in Fig. 8.14 does not have any spectral energy at the resonant frequency. Therefore, the torsional resonance modes do not receive any energy from the servo motor. In such conditions, any torsional oscillations would eventually decay due to a lack of energy supply and a small but finite damping of the resonance poles. Under these assumptions, an increase in feedback gains does not augment the energy supplied to the mechanical resonator, due to the FIR filter having an infinite attenuation at $\omega = \omega_{TR}$. For this reason, the range of applicable gains is extended, increasing, in this way, the closed-loop bandwidth and the response speed of the system. Further performance evaluation is carried out by means of computer simulations and experimentally.

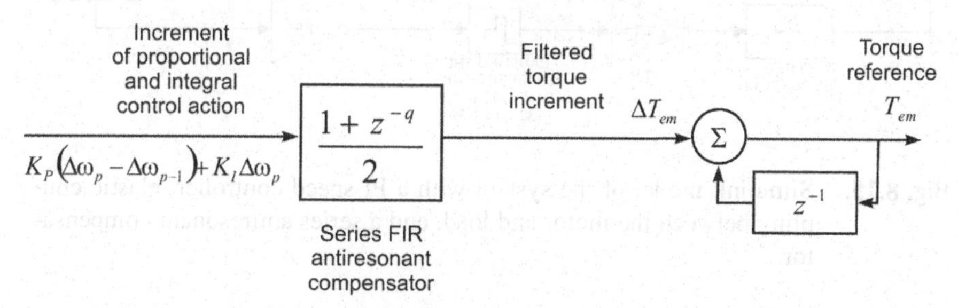

Fig. 8.14. Implementation of the antiresonant FIR compensator in conjunction with the PI speed controller in its incremental form.

8.8 Computer simulation of antiresonant compensators

The impact of antiresonant series compensators on the input and load step response of the system with a PI speed controller and torsional resonance in the control object is investigated here by means of computer simulations. In Section 8.2, the Simulink model of the system has been developed and given in Fig. 8.3. The model of the mechanical resonator is contained within the Simulink subsystem given in Fig. 8.4. The model used in this section is shown in Fig. 8.15 and is obtained from the previous one by adding the series antiresonant compensator between the speed controller and the torque actuator.

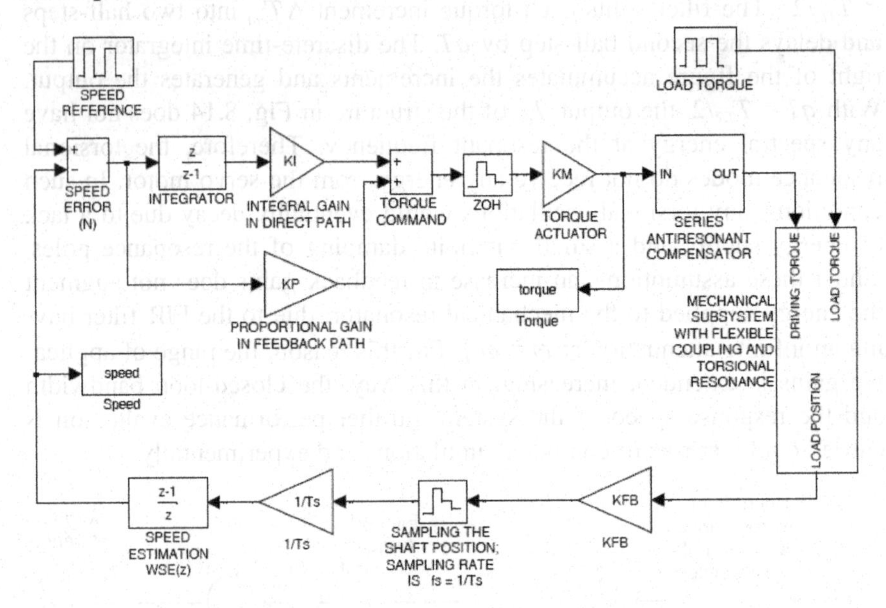

Fig. 8.15. Simulink model of the system with a PI speed controller, elastic coupling between the motor and load, and a series antiresonant compensator.

In this section, the notch filter and the FIR filter are used as series antiresonant compensators. The Simulink subsystem with the notch filter is given in Fig. 8.16. The antiresonant series compensator with the FIR filter is given in Fig. 8.17. The Matlab command sequence used to set the simulation parameters is given in Table 8.1. In this section, the stiffness coefficient K_K is set to 310 Nm/(rad/s), resulting in a resonant frequency of 150 Hz.

Fig. 8.16. Simulink subsystem with a notch filter.

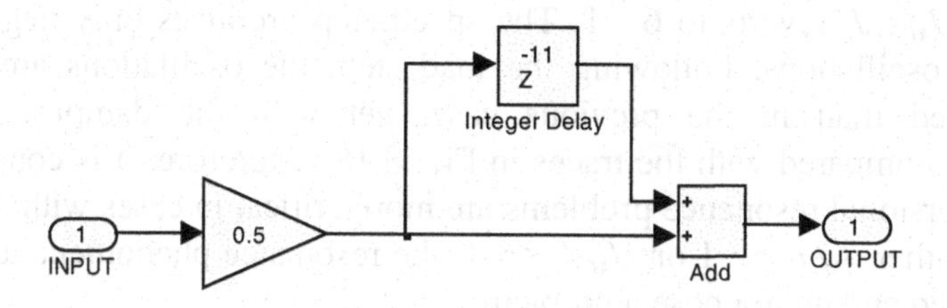

Fig. 8.17. Simulink subsystem with a FIR antiresonant filter.

For the purpose of comparison, the first simulation run is performed without any antiresonant compensator. In order to achieve a stable response and reduce the torsional oscillations, the closed-loop gains are reduced with respect to their optimized values, as suggested in Eq. 4.59, and set to $K_P = K_{POPT}/4$ and $K_I = K_{IOPT}/20$. These gains are maintained through all the simulation runs in this section. The simulation traces of the driving torque and speed are given in Fig. 8.18. The speed reference step (left) produces torsional oscillations that decay with a time constant of several hundreds of milliseconds. The load step results in oscillations of even larger magnitude. The response obtained in the figure is stable, but it is not acceptable due to the excessive amplitude of torsional oscillations. In the absence of an antiresonance compensator, the gains would have to be further reduced in order to obtain an acceptable response quality.

The simulation traces obtained with the notch filter are given in Fig. 8.19. The filter is properly tuned, and it removes the resonance phenomena. An overshoot in the step response is present due to the fact that the feedback gains K_P and K_I differ from their optimized values (Eq. 4.59). The simulation traces obtained with the FIR filter are given in Fig. 8.20. The torque and speed responses are similar to those given for the previous case. Hence, it is concluded that, with the proper tuning, both the notch and FIR filter remove the resonant frequencies from the driving torque. Therefore, the torsional resonance modes do not receive any energy from the servo motor. Further verification of antiresonant filters is performed experimentally in the next section.

It is of interest to investigate the effects of the resonant ratio R_R (Eq. 8.7) on the amplitude and damping of torsional oscillations. In Fig. 8.18, the simulation traces are given for $J_M = J_L$, resulting in the resonance ratio of $R_R = 1.41$. In Fig. 8.21, the torque and speed responses are obtained with $J_M : J_L = 1 : 3$. The model does not include any antiresonant filter; nevertheless, the speed oscillations are relatively small and well damped. In Fig. 8.22, the ratio $J_M : J_L$ reverts to 6 : 1. The speed step produces only negligible torsional oscillations. Following the load step, the oscillations are more pronounced than in the previous case, yet with the damping much improved compared with the traces in Fig. 8.18. Therefore, it is concluded that the torsional resonance problems are more critical in cases with $J_M/J_L \approx 1$. With either $J_M/J_L >> 1$ or $J_M/J_L << 1$, the resonance phenomena are less emphasized and do not pose a problem.

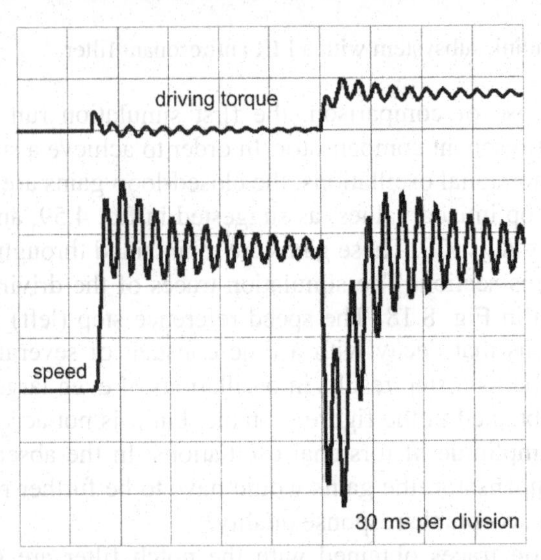

Fig. 8.18. Simulation traces of the driving torque and speed obtained from the model in Fig. 8.15 during the reference step (left) and the load step (right) transients. This model does not include any antiresonant filter.

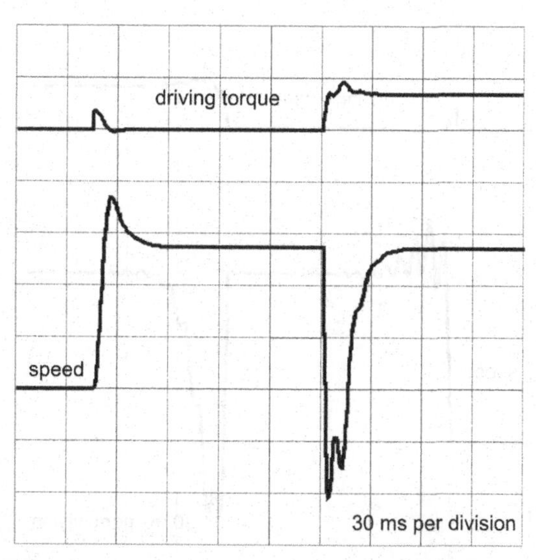

Fig. 8.19. The simulation given in Fig. 8.18 is repeated with a notch antiresonant filter. Other settings and scalings remain unaltered.

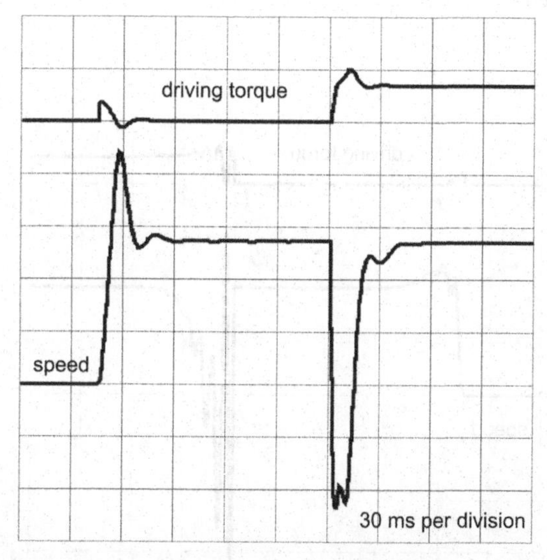

Fig. 8.20. The simulation given in Fig. 8.18 is repeated with a FIR antiresonant compensator. Other settings and scalings remain unaltered.

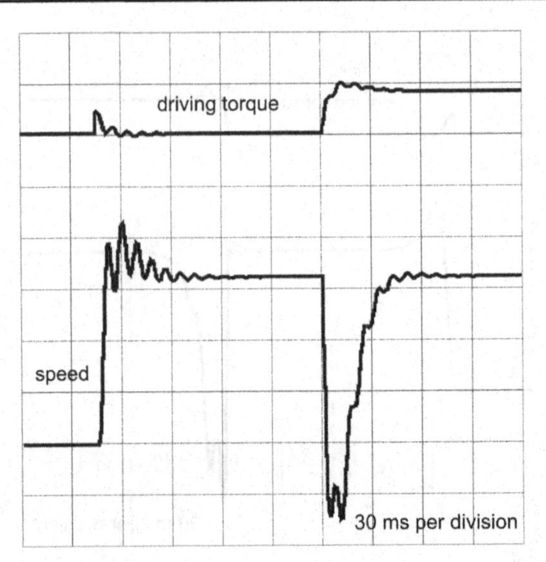

Fig. 8.21. The impact of the resonant ratio R_R (Eq. 8.7) on torsional oscillations. Simulation traces are obtained with $J_M : J_L = 1 : 3$. The model does not include any antiresonant filter. Other settings and scalings remain unaltered.

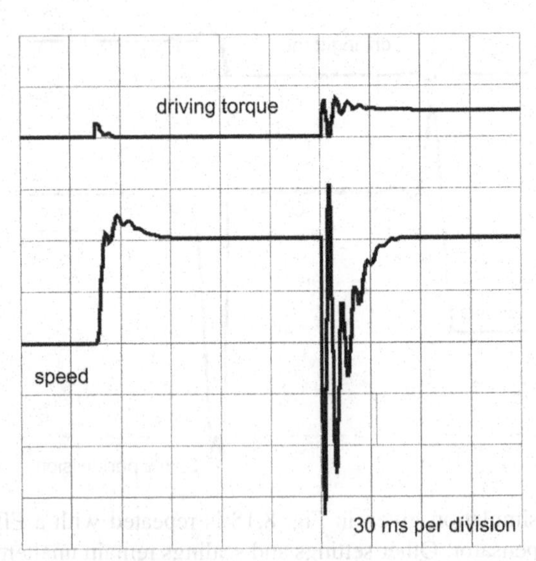

Fig. 8.22. The simulation given in Fig. 8.21 is repeated with $J_M : J_L = 6 : 1$. Other settings and scalings remain unaltered.

8.9 Experimental evaluation

The analytical considerations and simulation traces obtained in the preceding sections may differ from the experimental results obtained from a practical resonant servo system. The presence of quantization noise, cogging torque, speed-dependent motion resistances, and nonlinear friction is not taken into account in the preceding analysis. The two-mass system in Fig. 8.1 represents a simplified representation of a system comprising a number of elastically coupled masses, and this assumption may result in differences between simulated and experimental traces. In the vast majority of practical cases, where the range of applicable gains is limited due to mechanical resonance, the problem reveals itself in the form of sustained oscillations. For that reason, the conclusions derived from analysis and simulation throughout this chapter are verified experimentally.

The resonant control object in the experimental setup consists of two identical synchronous permanent magnet motors. One of the motors is used as the torque actuator, while the other serves as the load. The motors are coupled by a flexible hollow shaft. Both motors are equipped with electromagnetic resolvers. Hence, it is possible to close the loop by using either the motor- or the load-side variables. At the same time, both the load and the motor speed and position can be stored and printed. The R/D converter bandwidth is 1 kHz, and its resolution is set to 12 bits [2].

In order to decouple the mechanical resonance phenomena from the speed and torque oscillations caused by the cogging torque of the permanent magnet motors, the motor shafts are coupled so as to minimize the sum of the cogging torques coming from the motor side and the load side of the experimental setup. Furthermore, all the experiments are performed at low speeds, keeping the measurement results free from cogging torque disturbances. The low-speed behavior is of greater interest, since the problems of sustained torsional oscillations and instability are more pronounced at lower speeds, and, in particular, at standstill.

Detailed information on the servo motor FAST1M6030 is given in manufacturer's catalog [23]. The basic parameters of the motor and the shaft are given in Table 8.7.

Table 8.7. Relevant parameters of the experimental setup, used to compare the notch and FIR antiresonant compensators.

Rated torque:	$T_{nom} = 5.7$ Nm
Peak torque:	$T_{max} = 24$ Nm
Rated speed:	$\omega_{nom} = 3000$ rpm
Rated power:	$P_{nom} = 1.49$ kW

(continued)

Table 8.7. *(continued)*

Pole pairs:	p = 3
Rotor inertia of one motor:	J_M = 0.000620 kgm^2
Resonance frequency for a single motor connected to an infinite inertia:	f_{INF} = 1.45 kHz
Hollow shaft inertia:	J_{SH} = 0.000220 kgm^2
Hollow shaft stiffness:	K_K = 350 Nm/rad
Viscous friction:	K_V = 0.004 Nms/rad

The motors are supplied from a DSP-based digital servo amplifier [17] capable of performing torque control functions, speed control, and antiresonant compensation. The sampling time is set to T = 100 μs. The motor used as the torque actuator runs in the speed control mode. The other motor is used as a controllable load. Therefore, it is placed in the torque control mode. In this way, the step change of the load torque is achieved by changing the torque reference of the second motor. The block diagram of the servo amplifier is given in Fig. 8.23.

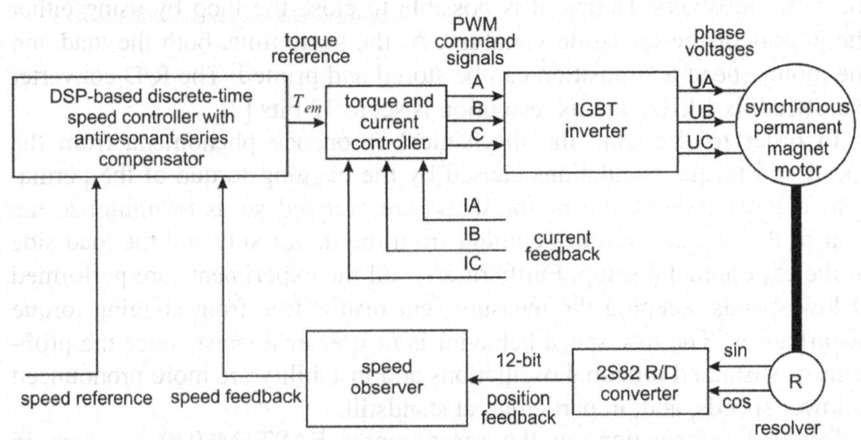

Fig. 8.23. Block diagram of the DSP-based servo amplifier [17].

In Fig. 8.24, the traces of the load speed, motor speed, driving torque, and speed difference $\Delta\omega$ are given, obtained with a PI speed controller without any antiresonant filters. The speed controller gains are reduced three times with respect to their optimized values, suggested in Eq. 4.59. It is verified that any increase in the gains results in unacceptable torsional oscillations. The traces are obtained with the motor-side feedback. The speed reference step of 100 rpm is followed by the load step change of 4 Nm. The speed step (left-hand side of the figure) results in relatively small oscillations, mostly observed in the speed difference $\Delta\omega$. The frequency of torsional oscillations is close to 150 Hz. The load step (right-hand side of the figure) produces sustained oscillation of a larger amplitude. In Fig. 8.25, the same test is repeated with the load-side feedback. This time, the amplitude of torsional oscillations gradually decays.

The experimental traces of the step response obtained with a properly tuned FIR filter and with the motor-side feedback are given in Fig. 8.26. The input step does not incite any torsional oscillations, due to the fact that the filter prevents the mechanical resonator from being supplied from the motor side. The overshoot in the speed and torque, observed on the left-hand side of the figure, is produced by the gains being reduced three times with respect to their optimized settings K_{POPT} and K_{IOPT}. On the other hand, the load step, taking place on the right-hand side of the figure, does produce torsional oscillations of a relatively small amplitude, mostly observed in the speed difference (top trace). The oscillations decay within several hundreds of milliseconds. They cannot be amplified by the loop gains and reenter the system in the form of the driving torque, as the FIR antiresonant filter exhibits an infinite attenuation of the resonance phenomena. In Fig. 8.27, the same test is repeated with the load-side feedback. This time, the speed difference is smaller, and the torsional oscillations are hardly visible. Therefore, there is a potential of increasing the loop gains and bandwidth.

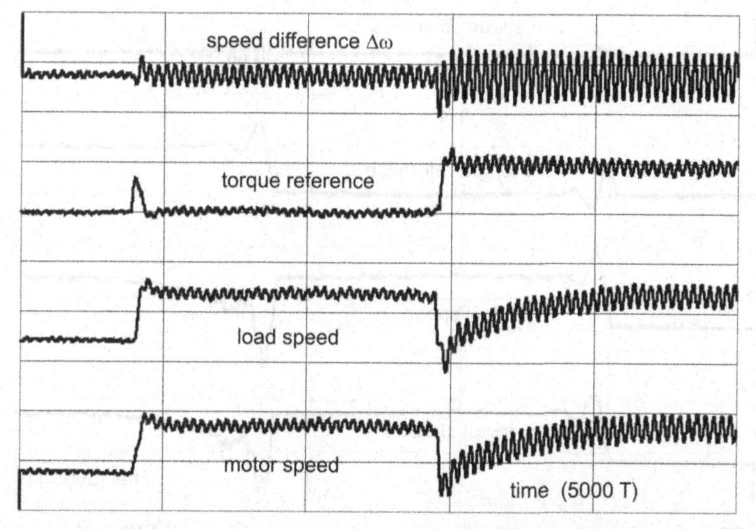

Fig. 8.24. Experimental traces obtained without antiresonant compensation. The loop gain is set to the stability limit. The experimental traces are obtained with motor-side feedback. The speed reference step change is followed by the step change in the load torque. The traces include the speed difference (the uppermost trace, 10 rad/s per div.), torque reference (10 Nm per div.), load-side speed, and motor-side speed (the bottom trace, 10 rad/s per div.).

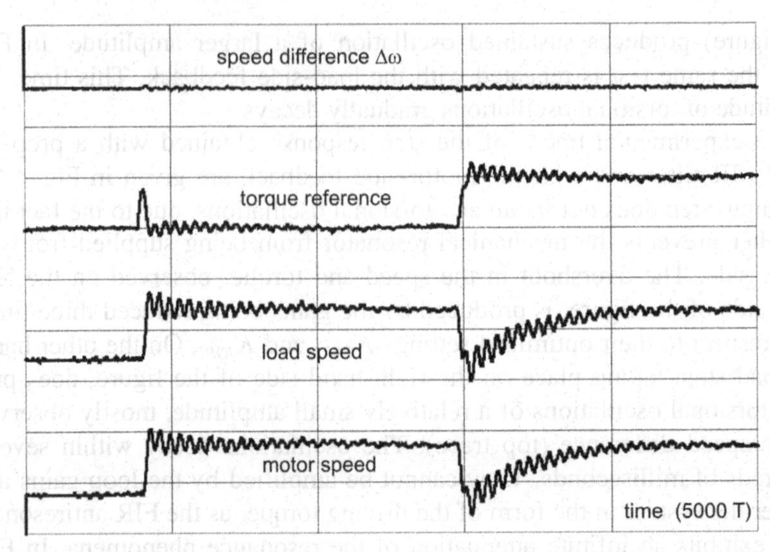

Fig. 8.25. The experiment in Fig. 8.24 is repeated with load-side feedback. Other settings and scaling remain unaltered.

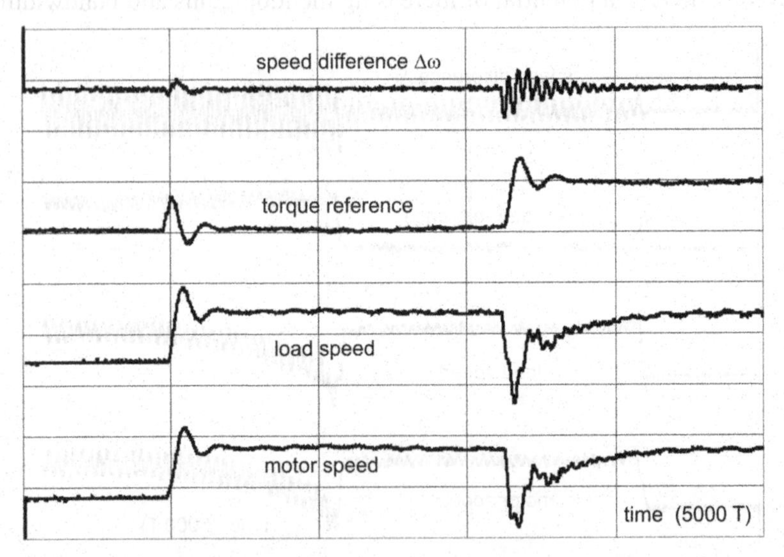

Fig. 8.26. Step response of the system with a properly tuned FIR series compensator and motor-side feedback. The gains, scaling, and settings are the same as in the experiment given in Fig. 8.24.

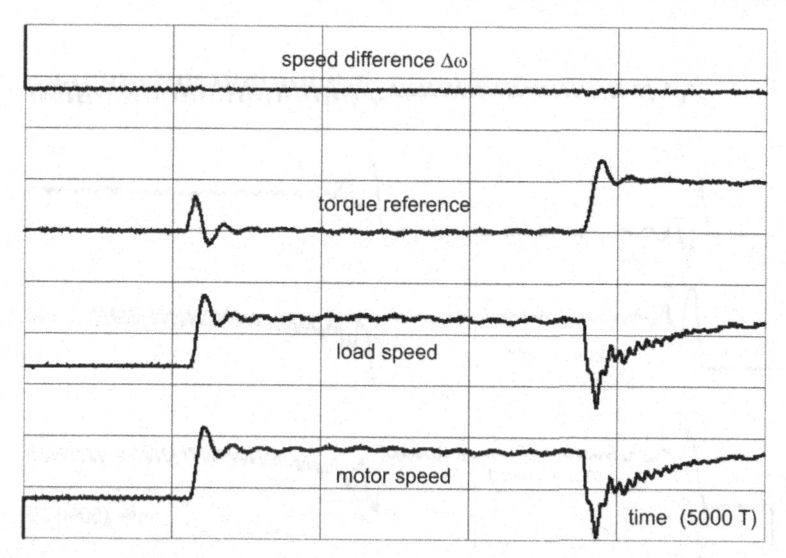

Fig. 8.27. The experiment in Fig. 8.26 is repeated with load-side feedback. Other settings and scaling remain unaltered.

It is worthwhile to test the robustness of both the notch and FIR series compensators with respect to changes in the system parameters. For example, the resonant frequency may change in time, and even fluctuate during the operating cycle, due to changes in the load inertia, machine geometry, and stiffness of the couplings.

In Fig. 8.28, the experimental traces are given for the system with a notch series compensator and the motor-side feedback. The notch central frequency is detuned by 25% with respect to the resonance. Other gains and settings preserve the same values as in the previous experimental runs. Compared with Fig. 8.24, the resonant phenomena are reduced in amplitude. Hence, even a detuned notch filter brings positive effects. On the other hand, sustained oscillations do not decay, and they result in the tracking error, acoustical noise, and accelerated wear of mechanical elements.

Robustness of the system with a detuned FIR antiresonant filter is tested by introducing an error of 25% in setting the delay parameter qT. The speed controller uses the motor-side sensor for the loop closure, and the resulting experimental traces are given in Fig. 8.29. The step in speed produces hardly-visible oscillations in the speed difference $\Delta\omega$. The load step produces torsional oscillations that decay in approximately 1 s.

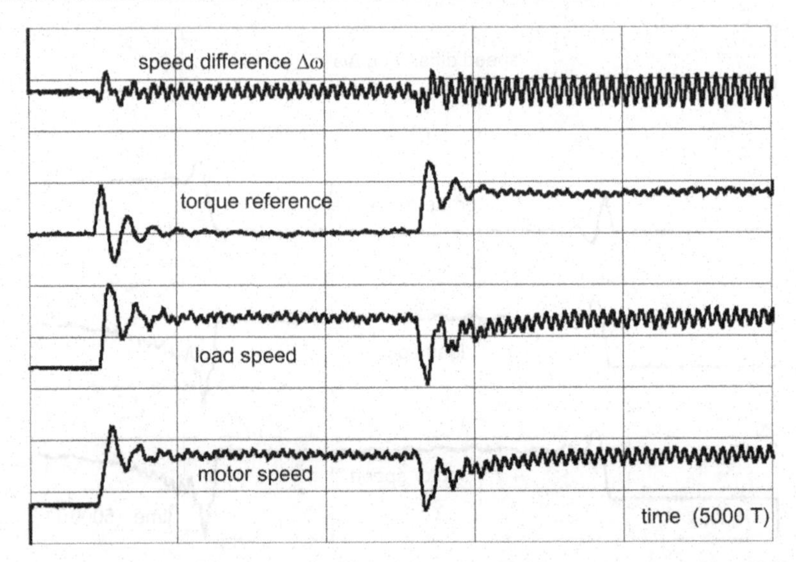

Fig. 8.28. Step response of the system with a notch series compensator and motor-side feedback. The notch filter central frequency is detuned by 25%. The gains, scaling, and settings are the same as in the experiment given in Fig. 8.24.

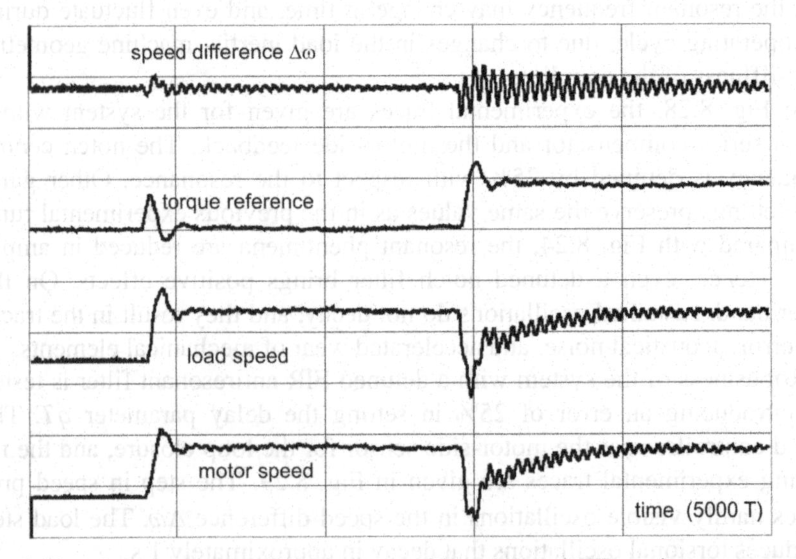

Fig. 8.29. Step response of the system with a detuned FIR series compensator and motor-side feedback. The FIR filter delay qT is detuned by 25%. The gains, scaling, and settings are the same as in the experiment given in Fig. 8.24.

8.10 Sustained torsional oscillations

In practical motion-control systems, the torsional oscillations and mechanical resonance are encountered as sustained oscillations in position, speed, and torque (force), of a smaller or larger amplitude. Due to their frequency ranging from 100 Hz to 1 kHz, these are often heard as an audible noise. In a linear system, sustained oscillations require the closed-loop poles to be positioned on the imaginary axis. Even a small drift of the poles towards the left half-plane causes the oscillation amplitude to decay. In the same way, any departure of the closed-loop poles to the right half-plane results in a progressive increase of the oscillation amplitude. Therefore, sustained oscillations are rarely encountered in systems without nonlinear features.

In a mechanical subsystem subjected to torsional oscillations, the equivalent motion resistance and the internal friction of transmission elements increase with the amplitude of the oscillations. Specifically, the coefficients B (Eq. 1.1) and K_V of the simplified model in Fig. 8.2 tend to increase as the torsional resonance phenomena intensify. According to Eq. 8.5, an increase in mechanical power losses improves the damping of torsional oscillations. In the situation where the oscillation amplitude is small while the damping is insufficient, the amplitude increases, enlarging, at the same time, the mechanical losses and increasing the damping. At a certain point, equilibrium is reached, and the system proceeds with the operation with a constant amplitude of torsional oscillations (*sustained* oscillations). Note at this point that the damping coefficient of the closed-loop poles of a linearized system affects the amplitude of sustained oscillations: the larger the damping, the smaller the amplitude. In most cases, an increase in feedback gains does not result in instability, but it increases the amplitude of sustained oscillations. The reduction in oscillations amplitude shown in Fig. 8.28, compared with that in Fig. 8.24, is an illustration of this phenomenon.

Problems

P8.1

A mechanical resonator consists of two rigid bodies, each one having inertia $J = 0.001$ kgm^2. They are coupled by means of a flexible shaft, with stiffness $K_K = 500$ Nm/rad. The internal (viscous) friction of the shaft is $K_V = 0.01$ Nm/(rad/s). Calculate the resonant and antiresonant natural frequency and relevant damping factors. (Hint: use Eq. 8.5.)

P8.2
The mechanical subsystem in the previous problem is used as the control object of a speed-controlled system. Derive the transfer function $W_P(s) = \omega(s)/T_{em}(s)$ and use Matlab in order to obtain the transient response of the output speed to the torque impulse. Compare the responses obtained with $K_V = 0.01$ Nm/(rad/s) and $K_V = 0.1$ Nm/(rad/s).

P8.3
Consider a speed-controlled system with a PI controller, where the feedback gains K_P and K_I are set to their optimized values, according to the design rule derived in Chapter 4. The sampling time of the system is $T = 300$ μs, the motor and load inertia $J_M = J_L = 0.0007$ kgm^2, the shaft parameters $K_K = 5000$ Nm/rad and $K_V = 0.15$ Nm/(rad/s), and the system parameters K_{FB} and K_M are equal to one. What is the resonant frequency of the torsional resonance modes? Use Simulink to obtain the step response of the output speed. Reduce the closed loop gain in order to obtain a stable, acceptable response of the output speed. Hint: use the model file *P8_3.mdl*, the command file *P8_3cmd.m*, and perform the gain reduction in steps such as $K_{PNEW} = K_P/2$ and $K_{INEW} = K_I/4$.

P8.4
Consider the step response of the output speed obtained in the previous problem with $K_P = K_{POPT}/4$ and $K_I = K_{IOPT}/16$. Estimate the rise time and the closed-loop bandwidth. Calculate the ratio f_{TR}/f_{BW} and verify the *rule of thumb* devised in Section 8.2.

P8.5
In problem P8.3, where the range of applicable gains is sought by reducing K_P and K_I according to the formula $K_{PNEW} = K_P/2$ and $K_{INEW} = K_I/4$, the ratio K_P/K_I is not preserved. Find the explanation for this decision. (Hint: consider the impact of the gains on the natural frequency and damping of the closed loop poles. Refer to the *s*-domain simplified representation of the speed-controlled system, given in Fig. 2.2).

P8.6
In Section 8.3, the transfer function $W_{RR2}(s)$ is obtained in Eq. 8.10, comprising a pair of weakly-damped conjugate complex zeros and one real pole. Confirm the assumption that zeros in Eq. 8.10 do not contribute to oscillations in the driving torque and rotor speed. Suggestion: in order to resolve the problem of the transfer function $W_{RR2}(s)$ being improper, extend the transfer function with the inertial load $1/Js$. Consider the values of $\omega_z = 1$, $2\xi/\omega_z = 0.01$, and $J=1$.

P8.7

Design a continuous-time notch filter with two conjugate complex poles and two conjugate complex zeros, with a notch frequency of $\omega_{NF} = 1$ rad/s, with the damping of zeros $\xi_z = 0.001$ and with the damping of poles $\xi_p =$ 0.5. Calculate the discrete-time equivalent of the filter, assuming that $T = 1$. Verify the amplitude characteristic of the filter by feeding the noise signal to the input and observing the spectrum of the filter output. In order to use the prescribed approach and obtain an estimate of the amplitude characteristics, the spectral energy of the noise must have a uniform distribution.

P8.8

Design the discrete-time notch filter applied as a series antiresonant compensator in a speed-controlled system. The system parameters are $J_M = J_L =$ 0.001 kgm^2, $K_K = 500$ Nm/rad, $K_V = 0.01$ Nm/(rad/s), and $T = 100$ μs. The damping of the notch filter poles is $\xi_P = 0.2$. The notch filter implementation details are given in Section 8.6.3.

P8.9

Consider the *FIR* filter with pulse transfer function $W(z) = 1 + 2z^{-1} + 3z^{-2} + 2z^{-3} + z^{-4}$, supplied with an impulse at $t = 5T$. Verify using Matlab that the impact of the initial impulse vanishes in n sampling periods, where n is the order of the *FIR* filter.

Appendices

A C-code for the PD position controller

```
/*********************************************************************/
/*                                                                   */
/* PD position controller. KP action in direct, KD in feedback path */
/*-------------------------------------------------------------------*/
/*                                                                   */
/* This routine returns the torque reference as a 16-bit signed (int)*/
/* where +/-0x7fff corresponds to +/-Tmax. The feedback signal is    */
/* obtained from the encoder, as a 16-bit "encinput", where 0-0xffff */
/* corresponds to one motor turn. The reference is supplied as a     */
/* 32-bit variable, with the upper 16 bits representing the number   */
/* of turns.                                                         */
/*                                                                   */
/* The signal "yref" corresponds to the internal speed reference     */
/* (y1 in Chapter 6 and Chapter 7). This internal speed reference    */
/* is limited. The speed limit applied is the square root function   */
/* of the remaining path.                                            */
/*                                                                   */
/* The square root function is implemented through piecewise         */
/* linear approximation. To facilitate calculations, the speed error */
/* is separated into the sign and absolute value.                    */
/*-------------------------------------------------------------------*/
/* Input arguments:                                                  */
/*                                                                   */
/*   encinput      16-bit shaft position obtained from an encoder    */
/*                 0-0xffff correspond to one motor turn             */
/*   reference     32-bit position reference, HI part = No. of turns */
/*   kdgain        16-bit derivative gain, internally scaled         */
/*   kpgain        16-bit proportional gain, internally scaled       */
/*                                                                   */
/* Return argument: (output)                                         */
/*                                                                   */
/*   (implicit)    16-bit torque reference, internally scaled        */
/*                 +/- 0x7fff correspond to +/- Tmax                 */
/*                                                                   */
/* Internal static variables:                                        */
/*                                                                   */
/*   position      16-bit torque reference, internally scaled        */
/*                 +/- 0x7fff correspond to +/- Tmax                 */
/*   error         32-bit position error in the output position      */
/*   reftorque     32-bit torque reference - speed controller output */
/*   encold        16-bit past value of the encoder position         */
/*                                                                   */
/* Temporary variables used in calculations:                         */
/*                                                                   */
/*   abserr        32-bit signed position error                      */
/*   yref          32-bit internal speed reference y1                */
/*   encinc        16-bit increment in the encoder position          */
/*   errorsign     16-bit variable keeping the error sign            */
/*   error16       16-bit variable keeping the absolute error        */
/*   temp          32-bit temporary variable                         */
/*                                                                   */
/*********************************************************************/
```

```
/* Piecewise linear approximation of the square root function        */

 enum xpt { X1=598, X2=900, X3=1500, X4=3000,
            X5=7375, X6=18500 };
 enum ypt { Y2=934076, Y3=1145476, Y4=1479076,
            Y5=2091076, Y6=3276701 };
 enum slo { S2=700, S3=556, S4=408, S5=271, S6=225 };
 enum maxmin { WMAX = 5779826, TORQMAX = 0x3fffffff };

 union dugi {
                long lo;
                  struct   { int lower,upper; }in;
                };

int PD_Controller(int encinput, long reference,
                  int kdgain, int kpgain)

{

   static long position,error;
   static long reftorque;
   static int encold;

   long   abserr, yref;
   int    encinc, errorsign;
   int    error16;

   union dugi temp;

   encinc = encinput - encold;
   encold       = encinput;

   position += (long) encinc;

   error      = reference - position;
   abserror   = abs(error);
   if(error<0)  errorsign = -1; else errorsign = 1;

   if(abserror < 0x8000) error16 = (int) abserror;
   else                  error16 = 0x7fff;

   reftorque = (long) (-kdgain * error16);

   if(error16 < X1)              yref = kpgain * error16;
   else   if(error16 < X2)       yref = Y2 + (error16-X1)*S2;
   else   if(error16 < X3)       yref = Y3 + (error16-X2)*S3;
   else   if(error16 < X4)       yref = Y4 + (error16-X3)*S4;
   else   if(error16 < X5)       yref = Y5 + (error16-X4)*S5;
   else   if(error16 < X6)       yref = Y6 + (error16-X5)*S6;
   else   yref = WMAX;
```

```
    if(errorsign == -1) yref= -yref;

    reftorque=reftorque+yref;

    if(reftorque > +TORQMAX) reftorque = +TORQMAX;
    if(reftorque < -TORQMAX) reftorque = -TORQMAX;
    temp.lo = reftorque;

    return(temp.in.upper);

}
```

B ASM-code for the PID position controller

```
;*************************************************************************
;* PID POSITION CONTROLLER with feedforward compensation (W &Acc)       *
;*    NOTCH antiresonant filter of the output - torque reference        *
;*    Platform: TMS320LF240x TI DSP                                     *
;*                                                                      *
;*   This routine implements the PID position controller with           *
;*  the feed forward compensation derived from the 1st and 2nd          *
;*  derivative of the reference profile. The optional notch filter      *
;*  removes the resonant frequencies from the torque reference.         *
;*                                                                      *
;*                                                                      *
;*  INPUTS:                                                             *
;*                                                                      *
;*  Trajectory_HI:  Reference trajectory, number of turns               *
;*  Trajectory_LO:  Reference trajectory, position within one turn      *
;*  RefSpeedFF:     1st derivative (speed feedforward signal)           *
;*  AccelFF:        2nd derivative (torque feedforward)                 *
;*                                                                      *
;*  Abs_Pos_HI:     Position feedback, number of turns                  *
;*  Abs_Pos_LO:     Position feedback, fractional part                  *
;*  ShaftSpeed:     Shaft speed feedback (position derivative)          *
;*                                                                      *
;*  INTERNAL STORED VARIABLES:                                          *
;*                                                                      *
;*  FiltShaftSpeed: Low-pass filtered shaft speed                       *
;*  RefSpeedFilt:   Low-pass filtered reference speed (speed FF)        *
;*  Pos_Err_Old:    Past position error (n-1)                           *
;*  past_werr:      Past value of speed error                           *
;*  int_high:       High part of the 32-bit torque reference            *
;*  int_low:        Low part of the 32-bit torque reference             *
;*                                                                      *
;*  Xnp1 Xnow Xnm1: Notch filter inputs (n+1), (n), (n-1)               *
;*  Ynow, Ynm1:     Notch filter outputs (n), (n-1)                     *
;*                                                                      *
;*  GAINS:                                                              *
;*                                                                      *
;*  kd1, kp1, ki1   PID controller gains                                *
;*  K1_notch.....   Notch filter gains K1-K5                            *
;*  .....K5_Notch                                                       *
;*                                                                      *
;*  TEMPORARY VARIABLES:                                                *
;*                                                                      *
;*  Temp_1, Temp_2, Temp_3, Pos_Err_New                                 *
;*************************************************************************
;*************************************************************************
;*   Find the (limited) position error in 32-bit format:                *
;*                                                                      *
;*   Pos_Err = Trajectory_Hi_lo  - Abs_Pos_Hi:Lo (32-bit, Version)      *
;*   ONLY Pos_Err_New IS USED FURTHER, LIMITED TO +/- 7FFFH RANGE       *
;*                                                                      *
;*************************************************************************
```

```
Find_Position_Error:                    ; Trajectory_HI:Lo keeps the
   spm    0                             ; position reference
   setc   ovm                           ; Abs_Pos_Hi_LO keeps the
   ldp    #Trajectory_Hi                ; 32-bit absolute position,
   zalh   Trajectory_Hi                 ; HI = Turns, LO = position
   adds   Trajectory_Lo                 ; within one motor turn
   ldpk   0
   subh   Abs_Pos_Hi
   subs   Abs_Pos_Lo                    ; Pos_Err_NEw is 16-bit
   sacl   Pos_Err_New                   ; error has to be restricted
                                        ; to +/- 7FFF
   bgez   Positive_Posit_Error
                                        ; must have the
                                        ; ABS value =< 7FFFH
Negative_Posit_Error:
   abs
   sub    #7fffh                        ; If ABS<7FFFH, amplitude=OK
   blz    Correct_PosErrLo_Val
   zac
   sub    #7fffh                        ; If ABS > 7FFFH, limit the
   sacl   Pos_Err_New                   ; amplitude
   b      Correct_PosErrLo_Val

Positive_Posit_Error:
   abs
   sub    #7fffh                        ; If ABS<7FFFH, amplitude=OK
   blz    Correct_PosErrLo_Val
   lac    #7fff                         ; If ABS > 7FFFH, limit
   sacl   Pos_Err_New
Correct_PosErrLo_Val:                   ; Here Pos_Err_New READY

Filter_the_running_speed_and_Trajectory_slope:

    lac    ShaftSpeed,15                ; ShaftSpeed = increment
                                        ; of the shaft position
    add    FiltShaftSpeed,15            ; FiltShaftSpeed =
                                        ; filtered speed
    sach   FiltShaftSpeed               ; low-pass filter,
                                        ; Fbw = 0.11/TSpl
    lac    RefSpeedFF,15                ; filter the reference
                                        ; speed (slope of the
    add    RefSpeedFilt,15              ; position trajectory)
    sach   RefSpeedFilt                 ; first-order filter,
                                        ; Fbw=0.11/Tspl
    zalh   RefSpeedFilt                 ; Find the speed difference
                                        ; between the feedforward
    addh   RefSpeedFilt                 ; speed signal
    subh   FiltShaftSpeed               ; and the shaft speed
    subh   FiltShaftSpeed
    sach   Temp_2
```

```
    zalh   Pos_Err_New                ; (1-z^-1)*Position_Error
    subh   Pos_Err_Old                ; and place result in Temp_3
    sach   Temp_3
    lac    Pos_Err_New                ; OLD  =   z^-1 * NEW
    sacl   Pos_Err_Old

; Find (1 - z^-1) * (Ref_Speed_Feed_Frwrd - Shaft_Speed)

    zalh   Temp_2
    subh   past_werr                  ; past_werr keeps
                                      ; z^-1 * Temp_2
    addh   Temp_2
    subh   past_werr
    sach   Temp_1

; past_werr = z^-1 (Ref_Speed_Feed_Frwrd - Shaft_Speed)
    lac    Temp_2
    sacl   past_werr

;*****************************************************************
;* PID Controller gains:  kp1, kd1, ki1                          *
;* Temp_3 = NONFILTERED Error, Temp_1 = Filtered WError Increment *
;*****************************************************************

Calculate_PID_Control_Actions:

    lt     kd1                        ; Treg = kd1
    mpy    Temp_1                     ; Temp_1=Filtered increment
                                      ; of WError signal
    pac                               ; P -> ACC
    rpt    #10
    apac                              ; Rescale derivative action
    lt     kp1                        ; Treg = kp1
    mpy    Temp_3                     ; PReg =
                                      ; kp1 * Nonfiltered_Error
    rpt    #7
    apac                              ; Rescale proportional
                                      ; control action
    lt     ki1                        ; Treg = ki1
    mpy    Pos_Err_New                ; Preg =
                                      ; Position_Error * ki1
    apac
    apac
;*****************************************************************
;*     At this point, ACCH:ACCL contains the torque increment    *
;*  With Reference == 0, ACCH:ACCL = Torque increment becomes =  *
;*                                                               *
;*  - (kd1 * (1-1/z)^2 * 12   ) * Shaft_Position -               *
;*  - (kp1 * (1-1/z)    * 8   ) * Shaft_Position -               *
;*  - (ki1 * (    1     * 2   ) * Shaft_Position                 *
;*****************************************************************
    ADDH   int_high
```

```
    ADDS  int_low                     ; int_high:int_lo =
                                      ;  Torque Ref.
    SACL  int_low
    SACH  int_high
    addh  Accel_FF                    ; Add Acceleration
    SACH  Temp_2                      ; Feed Forward
;*********************************************
;*  THE TORQUE REFERENCE CONTAINED IN Temp_2  *
;*  EXECUTE ANTIRESONANT NOTCH FILTER         *
;*********************************************

    LAC    k1_notch                   ; Skip notch if K1_Notch = 0

    BZ     Notch_Filter_Done_Or_Skip
;Execute Notch Filter when K1 != 0
    LAC    Temp_2
    SACL   Xnp1
    ZAC                               ; Xnp1 = input (n+1)
    LT     Xnm1                       ; Xnm1 = input (n-1)
    MPY    K5_notch                   ; Xnow = input (n)
    LTD    Xnow                       ; Notch filter coefficients
    MPY    K4_notch                   ; K1 .. K5
    LTD    Xnp1                       ; given in this book
    MPY    K3_notch
    LTA    Ynm1                       ; Ynow = output (n)
    MPY    K2_notch                   ; Ynm1 = output (n-1)
    LTD    Ynow
    MPY    K1_notch
    apac                              ; Coeffs. in format Q12
    rpt    #3                         ; Left shift ACC << 4
    norm   *                          ; and store the result
    SACH   Ynow
    LDPK   0
    SACH   Temp_2
Notch_Filter_Done_Or_Skip:

    clrc   ovm

    lacc   Temp_2                     ; On return, Accumulator
                                      ; (ACCL) contains the PID
    RET                               ; controller output, namely,
                                      ; the torque reference
```

C Time functions and their Laplace and z-transforms

Functions $F(z)$, given in the right column for the relevant $f(t)$, represent the z-transform of the pulse train $f(kT)$ comprising the samples of the time function $f(t)$ at the sampling instants $t = kT$, where T represents the sampling period.

	Time function $f(t)$	Laplace transform $F(s)$	$\mathbb{Z}$ transform $F(z)$
1	$\delta(t)$ (unit impulse)	1	1
2	$\delta(t-kT)$ (delayed unit impulse)	e^{-kTs}	z^{-k}
3	$h(t)$ (unit step)	$\frac{1}{s}$	$\frac{z}{z-1}$
4	$h(t-T)$ (delayed unit step)	$\frac{e^{-sT}}{s}$	$\frac{1}{z-1}$
5	t (unit ramp)	$\frac{1}{s^2}$	$\frac{Tz}{(z-1)^2}$
6	e^{-at} (exponential decay)	$\frac{1}{s+a}$	$\frac{z}{z-e^{-aT}}$
7	$1-e^{-at}$ (exponential growth)	$\frac{a}{s(s+a)}$	$\frac{(1-e^{-aT})z}{(z-1)(z-e^{-aT})}$
8	te^{-at}	$\frac{1}{(s+a)^2}$	$\frac{Tze^{-aT}}{(z-e^{-aT})^2}$
9	$t-\frac{1-e^{-at}}{a}$	$\frac{a}{s^2(s+a)}$	$\frac{Tz}{(z-1)^2}-\frac{(1-e^{-aT})z}{a(z-1)(z-e^{-aT})}$
10	$e^{-at}-e^{-bt}$	$\frac{b-a}{(s+a)(s+b)}$	$\frac{z}{z-e^{-aT}}-\frac{z}{z-e^{-bT}}$
11	$(1-at)e^{-at}$	$\frac{s}{(s+a)^2}$	$\frac{z}{z-e^{-aT}}-\frac{Tze^{-aT}}{(z-e^{-aT})^2}$
12	$\sin \omega t$ (sine wave)	$\frac{\omega}{s^2+\omega^2}$	$\frac{z\sin \omega T}{z^2-2z\cos \omega T+1}$
13	$\cos \omega t$ (cosine wave)	$\frac{s}{s^2+\omega^2}$	$\frac{z(z-\cos \omega T)}{z^2-2z\cos \omega T+1}$
14	$e^{-at}\sin \omega t$ (damped sine)	$\frac{\omega}{(s+a)^2+\omega^2}$	$\frac{ze^{-aT}\sin \omega T}{z^2-2ze^{-aT}\cos \omega T+e^{-2aT}}$

15	$e^{-at}\cos\omega t$	$\dfrac{s+a}{(s+a)^2+\omega^2}$	$\dfrac{z^2 - ze^{-aT}\cos\omega T}{z^2 - 2ze^{-aT}\cos\omega T + e^{-2aT}}$
16	$1-\cos\omega t$	$\dfrac{\omega^2}{s(s^2+\omega^2)}$	$\dfrac{z}{z-1} - \dfrac{z(z-\cos\omega T)}{z^2 - 2z\cos\omega T + 1}$

D Properties of the Laplace transform

Definition	$F(s) = \mathscr{L}\left[f(t)\right] = \int\limits_{-\infty}^{+\infty} f(t)e^{-st}\,dt$
Inversion	$f(t) = \frac{1}{2\pi j}\int\limits_{\gamma-j\omega}^{\gamma+j\omega} F(s)e^{st}\,ds$
Linearity property	$\mathscr{L}\{a\cdot f(t)+b\cdot g(t)\} = a\cdot\mathscr{L}\{f(t)\}+b\cdot\mathscr{L}\{g(t)\}$
s-Domain shifting property	$\mathscr{L}\left\{e^{at} f(t)\right\} = F(s-a)$
Time domain shifting property	$\mathscr{L}\{f(t-T)\cdot u(t-T)\} = e^{-sT}F(s)$
Periodic functions	$\mathscr{L}\{f(t)\} = \frac{1}{1-e^{-sT}}F_1(s)$ $F_1(s)$ is the transform of the function for the first period
Initial value theorem	$\lim\limits_{t\to 0} f(t) = \lim\limits_{s\to\infty} sF(s)$
Final value theorem	$\lim\limits_{t\to\infty} f(t) = \lim\limits_{s\to 0} sF(s)$
First derivative	$\mathscr{L}\left\{\frac{d}{dt}f(t)\right\} = sF(s) - f(0)$
Second derivative	$\mathscr{L}\left\{\frac{d^2}{dt^2}f(t)\right\} = s^2F(s) - s\cdot f(0) - \frac{d}{dt}f(0)$
Integral	$\mathscr{L}\left\{\int\limits_0^t f(t)dt\right\} = \frac{1}{s}F(s)$ Example: $\mathscr{L}\left\{\int\limits_0^t e^{-t}dt\right\} = \frac{1}{s}\mathscr{L}\left\{e^{-t}\right\} = \frac{1}{s(s+1)}$

E Properties of the z-transform

Definition	$F(z)=\mathcal{Z}\left(f(kT)\right)=\sum_{k=0}^{\infty} f(kT)z^{-k}$
Inversion	$f(kT)=\frac{1}{2\pi j}\oint F(z)z^{k-1}dz$
Linearity property	$\mathcal{Z}\{a\cdot f+b\cdot g\}=a\cdot\mathcal{Z}\{f\}+b\cdot\mathcal{Z}\{g\}$
Time shift property	$\mathcal{Z}\{f(t-nT)\}=z^{-n}F(z)$ $\mathcal{Z}\{f(t+nT)\}=z^{n}(F(z)-F_1(z))$ where $F_1(z)=\sum_{j=0}^{n-1} f(jT)z^{-j}$
Initial value theorem	$f(0)=\lim_{z\to\infty}F(z)$
Final value theorem	Provided that the function $(1-z^{-1})F(z)$ does not have any poles on the unit circle or outside the circle, then $\lim_{k\to\infty} f(kT)=\lim_{z\to 1}\left[(1-z^{-1})F(z)\right]$
Convolution	$\mathcal{Z}\{f(t)*g(t)\}=\mathcal{Z}\left\{\sum_{n=0}^{k} f(n)g(k-n)\right\}$ $=\mathcal{Z}\{f(t)\}\mathcal{Z}\{g(t)\}$

F Relevant variables and their units

Variable	Unit	Symbol	Comment
LENGTH	METER	m	- (basic unit)
MASS	KILOGRAM TON	kg t	- 1 t = 10^3 kg
TIME	SECONDS	s	-
ELECTRIC CURRENT	AMPERE	A	-
TEMPERATURE	KELVIN DEGREES CELSIUS	K °C	1 °C = 1 K 0°C corresponds to 273.15 K
ANGLE	RADIAN GRAD	rad °	1 rad = 57.296° = 180°/π 1° = (π/180) rad
ANGULAR VELOCITY	RADIAN PER SECOND REVOL. PER MINUTE	rad/s rpm	1 rad/s = 9.5493 rpm (min^{-1}) 1 min^{-1} = 1 rpm
ANGULAR ACCELERATION, ROTATIONAL ACCELERATION	RADIAN PER SQUARE SECOND	rad/s^2	-
SPEED (RPM), ROTATIONAL FREQUENCY	RECIPROCAL OF SECOND	s^{-1}	360° = 2π rad
VELOCITY	METER PER SECOND METER PER MINUTE	m/s m/min	- 1 m/min
ACCELERATION	METER PER SQUARE SECOND	m/s^2	-
FORCE	NEWTON	N	1 N = 1 kg m/s^2 = 1 J/m
TORQUE	NEWTON METER	Nm	1 Nm = 1 J = 1 Ws
MOMENT OF INERTIA	KILOGRAM TIMES SQUARE METER	kgm^2	1 kgm^2 = 1 Nm s^2
PRESSURE	PASCAL BAR	Pa bar	1 Pa = 1 N/m^2 1 bar = 10^5 Pa
ENERGY, WORK, HEAT	JOULE	J	1 J = 1 Nm = 1 Ws
POWER	WATT	W	1 W = 1 J/s
FREQUENCY	HERTZ	Hz	1 Hz = 1 s^{-1}
ELECTRIC POTENTIAL, VOLTAGE	VOLT	V	1 V = 1 W/A = 1 A · Ω
ELECTRIC CHARGE	COULOMB	C	1 C = 1 As
ELECTRIC	OHM	Ω	1 Ω = 1 V/A

RESISTANCE			
ELECTRIC CONDUCTANCE	SIEMENS	S	$1\ S = 1\ \Omega^{-1}$
MAGNETIC FLUX	WEBER	Wb	$1\ Wb = 1\ Tm^2$
FLUX DENSITY, MAGNETIC INDUCTION	TESLA	T	$1\ T = 1\ Wb/m^2$
MAGNETIC FIELD STRENGTH	AMPERE PER METER	A/m	
ELECTRIC CAPACITANCE	FARAD	F	1 F = 1 As/V
INDUCTANCE	HENRY	H	1 H = 1 Vs/A

References

[1] Aastrom KJ, Wittenmark B (1990) Computer Controlled Systems: Theory and Design, second edition, Prentice Hall International Editions, Englewood Cliffs, NJ

[2] Sheingold DH (ed) and the Engineering Staff of Analog Devices Inc. (1986) Analog-Digital Conversion Handbook, third edition, Prentice Hall Inc, Englewood Cliffs, NJ

[3] Phillips CL, Nagle HT (1984) Digital Control System Analysis and Design, Prentice Hall Inc, Englewood Cliffs, NJ

[4] Naslin, P (1968) Essentials for optimal control, Illifebook Ltd, London

[5] Zah M, Brandenburg G (1987) Das erweiterte Dampfungsoptimum, Ein analytisches Optimierungsverfahren fur regelstrecken mit zahlerpolynom, Automatisierungstechnik 35, Heft 7, pp 275–283

[6] Frohr F, Ortenburger F (1982) Introduction to Electric Control Engineering, Heyden & Son, London

[7] Data Converter Reference Manual (1992) Volume I, Analog Devices Inc.

[8] Shannon C (1948) A mathematical theory of communication, Bell system technical journal, vol. 27, pp 379–623

[9] Marks II RJ (ed) (1991) Introduction to Shannon Sampling and Interpolation Theory, Springer-Verlag, New York

[10] DSP56F807 16-bit Digital Signal Processor, rev. 5 (2001) Motorola Semiconductors, Denver, Colorado

[11] TMS320F2812 Digital Signal Processors Data Manual (2005) Texas Instruments, Dallas, Texas

[12] Doetsch G (1971) Guide to the Applications of the Laplace and Z-Transforms, Van Nostrand Reinhold, New York

[13] Jury, EI (1958) Sampled-Data Control Systems, John Wiley, New York

[14] Jury EI (1964) Theory and Application of the Z-Transform Method, John Wiley, New York

[15] Position encoders, Dr. Johanness Heidenhain GmbH, www.heidenhain.de

[16] vukosavic.etf.bg.ac.yu

[17] DS2000 User's Manual, rev. B, MOOG, 2005. www.moog.com

[18] De Boor C (1978) A Practical Guide to Splines, Springer-Verlag, New York

[19] Sugiura K, Hori Y (1996) Vibration Suppression in 2- and 3-Mass System Based on the Feedback of Imperfect Derivative of the Estimated Torsional Torque, IEEE Trans. Ind. Electronics, vol. IE-43, No. 1, pp 56–64

[20] Brandenburg G, Walfermann W (1985) State Observers for Multi Motor Drives in Processing Machines with Continuous Moving Webs, EPE 1985 Conference Record, pp 3.203–3.208

[21] Profumo F, Madlena M, Griva G (1996) State Variables Controller Design for Vibrations Suppression in Electric Vehicles, PESC 1996 Conference Record, pp 1940–1947

[22] Stein JY (2000) *Digital Signal Processing: A Computer Science Perspective,* John Wiley & Sons Inc, New York

[23] Fastact Servo Motors Data Sheets (2005) MOOG Electric, www.moog.com

[24] Stojic, MR (1984) Design of microprocessor-based system for DC motor speed control, IEEE Trans. on Ind. Electronics, vol. EI-31, No. 3, pp 243–249

Index

G

H

I

J

K

L

M

N

T

U

V

W

Z

Other books in the series:

Power System Oscillations
Graham Rogers
ISBN 978-0-7923-7712-2

State Estimation in Electric Power Systems: A Generalized Approach
A. Monticelli
ISBN 978-0-7923-8519-6

Computational Auction Mechanisms for Restructured Power Industry Operations
Gerald B. Sheblé
ISBN 978-0-7923-8475-5

Analysis of Subsynchronous Resonance in Power Systems
K.R. Padiyar
ISBN 978-0-7923-8319-2

Power Systems Restructuring: Engineering and Economics
Marija Ilic, Francisco Galiana, and Lester Fink, eds.
ISBN 978-0-7923-8163-1

Cryogenic Operation of Silicon Power Devices
Ranbir Singh and B. Jayant Baliga
ISBN 978-0-7923-8157-0

Voltage Stability of Electric Power Systems
Thierry Van Cutsem and Costas Vournas
ISBN 978-0-7923-8139-6

Automatic Learning Techniques in Power Systems
Louis A. Wehenkel
ISBN 978-0-7923-8068-9

Energy Function Analysis for Power System Stability
M. A. Pai
ISBN 978-0-7923-9035-0

Electromagnetic Modelling of Power Electronic Converters
J. A. Ferreira
ISBN 978-0-7923-9034-3

Spot Pricing of Electricity
F. C. Schweppe, M. C. Caramanis, R. D. Tabors, R. E. Bohn
ISBN 978-0-89838-260-0